NEXT GENERATION EVIDENCE
STRATEGIES FOR MORE EQUITABLE SOCIAL IMPACT

Edited by
KELLY FITZSIMMONS
TAMAR BAUER

BROOKINGS INSTITUTION PRESS
Washington, D.C.

Published by Brookings Institution Press
1775 Massachusetts Avenue, NW
Washington, DC 20036
www.brookings.edu/bipress

Co-published by Rowman & Littlefield
An imprint of The Rowman & Littlefield Publishing Group, Inc.
4501 Forbes Boulevard, Suite 200, Lanham, Maryland 20706
www.rowman.com

86-90 Paul Street, London EC2A 4NE

Copyright © 2024 by The Brookings Institution

All rights reserved. No part of this book may be reproduced in any form or by any electronic or mechanical means, including information storage and retrieval systems, without written permission from the publisher, except by a reviewer who may quote passages in a review.

The Brookings Institution is a nonprofit organization devoted to research, education, and publication on important issues of domestic and foreign policy. Its principal purpose is to bring the highest quality independent research and analysis to bear on current and emerging policy problems.

This book is open access under a CC BY-NC-ND 4.0 International Public License

British Library Cataloguing in Publication Information Available

Library of Congress Cataloging-in-Publication Data
Names: Fitzsimmons, Kelly, editor. | Bauer, Tamar, editor.
Title: Next generation evidence : strategies for more equitable social impact / edited by Kelly Fitzsimmons, Tamar Bauer.
Description: Washington, D.C. : Brookings Institution Press, [2023] | Includes bibliographical references and index. | Summary: "Next Generation Evidence features innovative thinking from leaders across policy, philanthropy, research, and practice on how to build a stronger, more equitable data and evidence ecosystem"— Provided by publisher.
Identifiers: LCCN 2023026435 (print) | LCCN 2023026436 (ebook) | ISBN 9780815740520 (cloth) | ISBN 9780815740537 (paperback) | ISBN 9780815740544 (epub)
Subjects: LCSH: Evaluation research (Social action programs) | Social service—Evaluation. | Equality.
Classification: LCC H62 .N596 2023 (print) | LCC H62 (ebook) | DDC 001.4—dc23/eng/20230623
LC record available at https://lccn.loc.gov/2023026435
LC ebook record available at https://lccn.loc.gov/2023026436

NEXT GENERATION EVIDENCE

CONTENTS

Acknowledgments xi

Why This Book and What You'll Learn 1
TAMAR BAUER, KELLY FITZSIMMONS, BETINA JEAN-LOUIS, AND RON HASKINS

SECTION 1

Constructive Dissatisfaction 21
KELLY FITZSIMMONS AND ARCHIE JONES

Engaging the Full Arc of Evidence Building 28
BETINA JEAN-LOUIS

How Researchers Can Make the Evidence They Generate More Actionable 35
REBECCA A. MAYNARD

Year Up: Improving Academic Success and Retention of Professional Training Corps Participants 46
JESSICA BRITT, DAVID FEIN, REBECCA A. MAYNARD, AND GARRETT WARFIELD

Nurse-Family Partnership: Always a Work in Progress 56
MANDY A. ALLISON, GREGORY TUNG, AND DAVID OLDS

Bail Project: Evaluation as Part of Business Strategy *67*
BRAD DUDDING AND TARA WATFORD

Baltimore City Public Schools: Using Data for Equitable CTE Outcomes *74*
BI VUONG

Per Scholas: Navigating COVID with Participant Feedback *80*
PLINIO AYALA

SECTION 2

Reimagining Evidence as Justice *89*
MICHAEL MCAFEE

All Data Is Biased *94*
HEATHER KRAUSE

Five Evaluation Design Principles of Just Philanthropy *103*
CARINA WONG

Can the New Data Economy Give Back to Communities? *111*
CHRIS KINGSLEY

Stop Extracting: Our Data, Our Evidence, Our Decisions *119*
ROBERT NEWMAN WITH DYLAN EDWARDS, JORDAN MORRISEY, AND KIRIBAKKA TENDO

De-Risking Data: Equitable Practices in Data Ethics and Access *126*
AMY O'HARA AND STEPHANIE STRAUS

Lakota Perspective on Indigenous Data Sovereignty *137*
DALLAS M. NELSON, DUSTY LEE NELSON, AND TATEWIN MEANS

Building Evidence and Advancing Equity: A Call to Action for Local Government *144*
CARRIE S. CIHAK

How Funders Can Center Evaluation Norms on Equity *153*
TRACY E. COSTIGAN AND RAYMOND MCGHEE JR.

Leadership Is Cause; Everything Else Is Effect *161*
LOLA ADEDOKUN

The Case for Cash *170*
NISHA G. PATEL

ParentCorps: A Collaborative Evidence-Building Strategy Guided by Family Voice *180*
LAURIE MILLER BROTMAN, SHANIKA GUNARATNA, ERIN LASHUA-SHRIFTMAN, AND SPRING DAWSON-MCCLURE

SECTION 3

The Power of Community Voice *189*
DANIEL J. CARDINALI

Data Justice and the Risks of Data Sharing *194*
MARIKA PFEFFERKORN

Philanthropy's Rightful Role in Evaluation: Fostering Learning and Empowerment *201*
JOHN BROTHERS

Earning Community Trust in Data-Driven Interventions at The Duke Endowment *209*
RHETT MABRY

The Duke Endowment Summer Literacy Initiative: Questions Matter and Methods Should Match *218*
HELEN I. CHEN

Criminal Justice Lab: Embedding Upstream Behavioral Health Solutions into the Criminal Justice Process *225*
KATY BRODSKY FALCO AND CHRISTOPHER LOWENKAMP

Pace Center for Girls: Advancing Equity through Participant-Centered Research and Evaluation *231*
MARY MARX AND KATIE SMITH MILWAY

SECTION 4

The Unfinished Business of Evidence Building: Directions for the Next Generation *243*
BRIAN SCHOLL

Facing Evidence Fears: From Compliance to Learning *265*
CHRISTOPHER SPERA

How to Better Measure Poverty: The Comprehensive Income Dataset *270*
KEVIN CORINTH AND BRUCE D. MEYER

New Federal Strategies to Strengthen Data Analytics Capacity of States, Localities, and Providers *278*
KATHY STACK AND GARY GLICKMAN

The Power of Nonprofit Sector R&D: Creativity during COVID *287*
NEAL MYRICK

The Value of Pre-Analysis *295*
DAVID YOKUM AND JAKE BOWERS

A Mountain of Pebbles: Effectively Using RCTs in the Public and Nonprofit Sectors *306*
JAMES MANZI

CEO: Credible Messengers in Re-Entry Services *311*
AHMED WHITT

Noggin: Learning Impact Evidence in a Multimedia Children's Platform *317*
KEVIN MIKLASZ, MAKEDA MAYS GREEN, AND MICHAEL H. LEVINE

First Place for Youth: Aligning Strategy, Data, and Culture to Drive Impact *327*
ERIKA VAN BUREN, MATT LEVY, AND JANE SCHROEDER

Gemma Services: Generating Actionable Evidence for Practitioners *335*
PETER YORK

SECTION 5

Systems Must Change: Dismantling, Disrupting, and Reimagining Evidence *349*
MICHAEL D. SMITH

The Next Generation of Evidence-Based Policy *356*
JENNIFER L. BROOKS, JASON SAUL, AND HEATHER KING

Evidence and Impact in a Post-Covid World *364*
VERONICA OLAZABAL AND JANE REISMAN

The Private Impact Economy: The Global Movement to Traceable Societal Impact *374*
BRIAN KOMAR AND ANDREW MEANS

Beyond the Evidence Act *381*
DIANA EPSTEIN

Building More Haystacks, Finding More Needles *391*
RYAN MARTIN

Transforming Government *399*
MICHELE JOLIN AND ZACHARY MARKOVITS

Democratizing Evidence *408*
VIVIAN TSENG

Stanford RegLab-Santa Clara County: Academic-Public Health
Collaboration for Rapid Evidence Building *413*
SARA H. CODY AND DANIEL E. HO

Camden Coalition: Healthcare and Public Health Data
Integration during COVID *422*
AARON TRUCHIL, CHRISTINE MCBRIDE, AUDREY HENDRICKS, NATASHA DRAVID,
AND KATHLEEN NOONAN

UnitedHealthcare Community & State: Using Housing to
Improve Health—A Social Impact Investing Strategy *429*
ANDY MCMAHON AND NICOLE TRUHE

About the Contributors *437*

Index *467*

ACKNOWLEDGMENTS

We are deeply grateful to the many partners who contributed their time, insights, passion, talents, and humanity to this project.

We dedicate this book to practitioners, those who invest their time and talents in the social and education sectors to deliver, improve, and innovate on the services they provide for the students, program participants, and communities they serve. They are the too-often overlooked and undersupported leaders working hard on the front lines. We want to honor and acknowledge their work. It is in many ways their collective body of work and their efforts that have motivated this book.

For the eighty-four authors who were able to join us in telling their stories, we are inspired by both the practical and visionary insights they shared for achieving more meaningful and equitable outcomes for communities. We were struck by the candor and transparency with which some authors explained the evolution in their thinking. We are grateful for their passion and dedication, and for the time they took to contribute these essays and use cases. We are honored to have them as partners in field building efforts to improve the evidence ecosystem.

A special thank you to Ron Haskins, so widely recognized for his immeasurable contributions to advancing the field of evidence-based policy in this country. We thank him here for being our partner on this effort, for not closing the door when we so insistently banged on it with our message, and for joining us to lift the voices of practitioners to articulate loudly and clearly in *Next Generation Evidence*. He understood from early on that there is no evidence-based policy without evidence-based practitioners. We are proud that our unexpected partnership is helping make that message more coherent and actionable. We appreciate support from Brookings Institution's Center on Children and Families in hosting and shaping the 2019 convenings, in particular Morgan Welch, and from Brookings Institution Press in producing this book.

We appreciate guidance from the Next Generation Evidence Advisory Committee for their suggestions on field trends and potential authors: Amy O'Hara, Carrie Cihak, Lola Adedokun, Marika Pfefferkorn, Nisha Patel, Paul Shoemaker, Ron Haskins, and Vivian Tseng. We are deeply grateful to our Project Evident board of directors for ongoing leadership in supporting our work and vision: Archie Jones, Dan Cardinali, Idara Umoh, Mark D'Agostino, Paul Shoemaker, Rhonda Evans, and Sandy Petty.

Project Evident also wishes to thank all our philanthropic partners for their support over the years, particularly Arnold Ventures, the Bainum Family Foundation, the Bank of America Charitable Foundation, the Barr Foundation, the Bill & Melinda Gates Foundation, Carnegie Corporation of New York, the Chan Zuckerberg Initiative, the Doris Duke Charitable Foundation, the Edna McConnell Clark Foundation, MacKenzie Scott & Dan Jewett, the Overdeck Family Foundation, the Robert Wood Johnson Foundation, and the William T. Grant Foundation.

We appreciate the diligent efforts of the Project Evident team, especially founding team members Matt Hillard, Naima Pittman, Gabriel Rhoads, and Bi Vuong, and those team members who shepherded this volume through the publication process, especially Dustin Sposato for his editorial insights, and for keeping track of the many balls in the air throughout. Thanks are due also to Joel Horwich, editor extraordinaire, Betina Jean-Louis, and Zach Baum for reviewing the many manuscripts and providing feedback to authors early in the process. And for her expert editorial eye, we thank Katie Smith Milway for her great work in organizing the many themes in this volume and sharpening the message throughout. We are grateful for the contributions of so many at Project Evident in doing whatever it took to complete this project, often rushing just to wait but doing so in the Project Evident way, with good cheer and humility.

We would like to acknowledge that, individually, we are on Wampanoag, Massa-adchu-es-et, Wappinger, and Munsee Lenape Land.

And, finally, we thank our families for their patience, support, and good humor. From Kelly: Denis, Ailish, Finn, Jack, Ruth, Sean, Scot, and Michael, I am so grateful for life with you. From Tamar: Ronnie, Rosa, Gili, and Maya, you know what you mean to me!

—Kelly and Tamar

NEXT GENERATION EVIDENCE

INTRODUCTION

WHY THIS BOOK AND WHAT YOU'LL LEARN

TAMAR BAUER, KELLY FITZSIMMONS, BETINA JEAN-LOUIS, AND RON HASKINS

*N*ext *Generation Evidence* is about five ideas.
  The first set of essays and use cases illuminate the often **overlooked role of public agencies, school districts, and nonprofit organizations** that deliver social and educational services—what we refer to throughout this book as *practitioners*—in systematically gathering, analyzing, and acting on data and evidence to improve and innovate their own programs, what we refer to throughout this book as *continuous evidence building*.

  The second section describes the **importance of embedding equity** (i.e., equitable processes and equitable outcomes) throughout the work of gathering, analyzing, and acting on evidence, and the third section addresses the **need to elevate community voice**—listening to and involving the ideas and feedback of people served, and sharing results early and often across evidence building to inform the process with respect and relevance.

  In the fourth section, our contributors tackle the power and promise of **embracing a continuous research and development-like approach** to the use of data and evidence across the social and education sectors. They highlight the importance of undertaking more frequent and diverse

5 Principles	Next Generation Evidence
1.	Centers on Practitioners and the Communities They Serve
2.	Connects Equity with Data and Evidence
3.	Elevates Community Voice
4.	Embraces a Continuous R&D-Like Approach
5.	Reimagines Evidence to Broaden Its Definition and Use

activities for learning, testing, and improving outcomes to generate more actionable results, with internal and external validation. And, in the fifth section, the authors speak to **reimagining evidence** to broaden its definition and use.

Together, these five approaches for accelerating social impact are bundled into one concept that Project Evident calls the *Next Generation of Evidence.* We are living through a moment that requires thinking and acting more boldly to address this country's deep economic and racial disparities, which are exacerbated by still-raw events of recent years. There is an urgency to get smarter and be more inclusive about how and where data and evidence can help, and where existing approaches perpetuate inequities. These *Next Generation* ideas can strengthen the use of data and evidence to accelerate improvements in social and educational outcomes for the 100 million Americans living today without economic security.[1]

Creating more actionable evidence is critical. By this, we mean evidence that is useful to practitioners and meaningful for program participants. We mean evidence that draws on the voices and experiences of those closest to the problem being addressed to help answer questions related to their needs. And we mean applying such evidence in more robust and consistent ways to improve economic and well-being outcomes for program participants.

We offer this volume of essays by practitioners, policymakers, activists, researchers, and philanthropists as follow-up to a first-of-its-kind convening in 2019 around making continuous evidence building by practitioners a social sector norm. Cohosted by Brookings Institution's Center on Children and Families and Project Evident in Washington, D.C., many ideas in this book surfaced during those discussions.[2] We see this book as a "prequel" to Ron Haskins and Greg Margolis's *Show Me the Evidence,*[3] which focused on the work of the Obama administration in creating tiered evidence initiatives that were useful in bringing up the question: "Where is the evidence pointing us in the social sector?"[4]

But these programs were more focused on one-off, third-party evaluations and less, if at all, focused on building cultures of learning and continuous evidence building among practitioners and their capacity and infrastructure to consistently act on that evidence. Without investing more resources and attention to the latter—the building and use of evidence—we fall short of realizing the promise of the next stage of evidence: to reimagine and rebuild a more equitable society. As coauthor Haskins observed during the 2019 conference and elsewhere, "Social science has been much more effective at showing what does not work than what does work. Thus, program developers, social scientists, and policymakers need to up their game and develop effective solutions to growing problems."[5]

This book is about how to help the field "up its game," via a series of insights and cases addressing the following themes:

1. WHY AND HOW EVIDENCE-BUILDING NEEDS TO CENTER ON PRACTITIONERS AND THE COMMUNITIES THEY SERVE

First and foremost, practitioners "should be active leaders in evidence building, not at the mercy of research and evaluation shops but in partnership with them, their funders, and their program participants, aligning their goals and interests," write Project Evident founder Kelly Fitzsimmons (a coauthor of this introduction) and founding board chair Archie Jones. "No one cares more than nonprofit leaders that their theories of change work as intended," the guiding principle for creating Project Evident (chapter 1.1).

This is because the innovation and testing needed to build effective social and education policies will require investments in the full cycle of evidence building. The cycle begins with early-stage people-and-technology investments (which have been under-resourced to date) to lead and facilitate evidence building. And it ranges to later-stage investments in frequently high-cost, third-party empirical evaluations, including well-designed and implemented randomized controlled trials (RCTs), which require practitioner insights to set useful parameters. A critical element will be investing in this work on a ***continuous*** basis for ongoing learning. Evidence collection and analysis is useful only when followed with evidence take-up; when practitioners play a greater role in all forms of evidence building, greater relevance accrues.

Examples for this book are drawn from practitioners who are relatively advanced in their evidence journeys. Some may speak to mature third-party evaluations and others may speak to smaller tests or internal analyses. Regardless of the level of the study, they all count as important case examples. We have not drawn examples from practitioners just beginning their evidence work, who tend to be undercapitalized and often left out of the data and evidence discussion. We need to do more as a field to fund and build the capacity of these practitioners so they, too, can participate and contribute use cases of their own in the future.

Essayists in this section include Fitzsimmons and Jones, former Harlem Children's Zone evaluator Betina Jean-Louis, who is a coauthor of this introduction, and University of Pennsylvania professor emeritus Rebecca Maynard, a leader in the design and conduct of RCTs. They all share their journeys to practitioner-centered evidence building.

Use cases include practitioners employing multiyear, randomized studies, like the story of workforce development nonprofit Year Up.[6] A small RCT evaluation showed blockbuster results in improving academic performance for Year Up's Professional Training Corps participants. The project included three partners—Year Up, Abt Associates, and the University of Pennsylvania—who used standard research methods yet novel approaches centered on practitioner needs and realities. For example, the needs of staff informed the research questions, "the usefulness and use of the final products deviate from a typical evaluation, and all parties relied on feedback loops to provide strategic tweaking of plans and timely use of findings."

This section also includes cases of earlier-stage approaches, such as a pre-quasi-experimental evaluation for nonprofit Nurse-Family Partnership (NFP). Here, University of Colorado's David Olds, founder of NFP, Mandy Allison, and Gregory Tung integrate "scientific evidence into practice design" and ground "research in the reality of the practice world" to innovate the nurse home-visiting approach for pregnant women who have had previous live births. They describe the formative development and pilot testing of the innovation, laying out their approach to model innovation (see the first figure in chapter 1.5). The Bail Project, a criminal justice organization, roots continuous learning in its theory of change and collects ongoing evidence of the program's efficacy.

In the case of Baltimore Public Schools, its City School's Office of College and Career Readiness drove the data gathering and analysis to develop a four-year strategy for career and technical training, interrogating existing data sets to set priorities. For technical trainer Per Scholas, which pivoted to remote learning at the onset of the pandemic, Plinio Ayala observes: "For researchers, especially those focused on mounting gold standard evaluations like the ones that Per Scholas has hosted . . . before, we would suggest that our [COVID] project shows that evidence-building can come in many forms. In this case, a rapidly-constructed and fielded implementation analysis focused on participant and practitioner voices fostered a profound new shift in direction" (chapter 1.8).

We need more examples like these, including practitioners across all stages of the evidence-building continuum. "Everyone has a greater potential to win," writes essayist Jean-Louis, when "practitioners become more active in the evidence space and are strategic in making the best use of tools along the arc of evidence that has historically been the province of classic researchers" (chapter 1.2).

2. CONNECTING EQUITY WITH DATA AND EVIDENCE

At Project Evident, we believe that (1) evidence can be a promising and powerful driver of equity; (2) equitable evidence practices will result in better data and evidence building and use, and, ultimately, stronger outcomes; and (3) equity must be considered both in the way evidence and data are built and used and in the types of outcomes social and educational interventions seek to address. Many of the book's authors share ideas and examples that align with or support these beliefs.

Michael McAfee from PolicyLink calls for creating "a new vision of evidence—evidence as justice, evidence as truth." He says: "If evidence is not leading us inexorably toward justice, we are not maximizing the use of evidence." To create this new paradigm, McAfee continues: "We must first ask ourselves some vital questions: What does it take to reverse 400 years of systemic oppression? What does it take to undertake a truly equitable redesign of a country built upon genocide, stolen land and slave labor? If we don't ask ourselves these questions before we set out to gather evidence, we will miss the destination. Evidence today is a microscope. We need it to also be a telescope" (chapter 2.1).

Heather Krause of We All Count observes that "the worst equity problem we're dealing with in data at the moment is that we're making prejudiced choices but don't understand how." With concrete examples shattering the myth that "data offers an objective, bias-free way to make decisions," Krause offers a roadmap for using data for racial equity by being transparent and intentional about the choices that are made at every single step of a data project. Carina Wong, a social impact advisor formerly at the Bill and Melinda Gates Foundation, offers six design principles to improve a strategy's equity orientation in her effort to advance more just philanthropy, writing: "Equity will continue to be elusive if we dance around the edges of racism and power dynamics and fail to address these issues in our strategies, organizations, and systems" (chapter 2.3).

While noting that the social sector typically uses "data to define, limit and control programs and organizations rather than to interrogate, explore and empower them," Chris Kingsley at Annie E. Casey Foundation highlights initiatives in Los Angeles, New York City, and Cuyahoga County, Ohio, that take seriously the needs of agency and nonprofit practitioners and their clients "that use data as a flashlight and not as a hammer." Similarly, AMP Health's Robert Newman, Dylan Edward, Jordan Morrisey, and Kiribakka Tendo propose alternatives to the enduring tendencies in field evaluations in sub-Saharan Africa to extract data. Georgetown University Massive Data Institute's Amy O'Hara and Stephanie Straus describe work that also addresses participant agency and inclusion: the Civil Justice Data Commons[7] seeks to increase equitable access to data by applying the best practices of data governance to civil courts. This discussion is part of a broader essay emphasizing the need to build *social license*, which "exists when the public trusts that data will be used responsibly and for societal benefit" (chapter 2.6).

Equity and inclusion are often considered in conjunction with data ownership. Tatewin Means and coauthors Dallas Nelson and Dusty Lee Nelson, with South Dakota's Thunder Valley Community Development Corporation, in their essay on Lakota data sovereignty, describe a "new and emerging idea to all of the Indigenous communities around the world and specifically in the United States." They quote Liz La quen náay Kat Saas Medicine Crow: "Information, data, and research about our peoples—collected about us, with us, or by us—belong to us and must be cared for by us" (chapter 2.7). This essay is about inclusion and agency, emphasizing the importance of communities having ownership over their own data, a concept embraced in Project Evident's Actionable Evidence Framework.[8] In this vein, when describing her work to advance racial equity in King County, Washington, Carrie Cihak pushes back on the myth "that local governments need to set aside data and evidence to work with community." Rather, she calls on metropolitan areas to do the hard work of challenging our data and evidence practices to be "more driven by, inclusive of, and responsive to communities." She posits that King County and other local governments "cannot become anti-racist organizations that contribute to building a pro-equity future without co-creating and innovating with community, and that includes how we use data and evidence."

Across the federal government, there is a renewed focus on racial equity in evidence building as part of an unparalleled commitment to "an ambitious whole-of-government equity agenda" with the "Executive Order on Advancing Racial Equity and Support for Underserved Communities Through the Federal Government," 1/20/21.[9] This order builds on the existing requirements of the Foundations for Evidence-Based Policy-Making Act of 2018 (the "Evidence Act").[10] Together, the Evidence Act and the Executive Order create a new directive to strengthen use of data and evidence by explicitly considering racial equity.

From a funder perspective, Tracy Costigan and Raymond McGhee of the Robert Wood Johnson Foundation share learnings over time on centering evaluation norms on equity, underscoring that "centering equity does not mean abandoning rigor." Lola Adedokun, formerly with the Doris Duke Charitable Foundation and now with the Aspen Institute, calls for supporting the next generation of leaders to advance equity. "Just as we recognize physician scientists as practitioner scholars—with academies in place to recognize and preserve their leadership (for example, National Academies of Sciences, Engineering, and Medicine), the same standards and expectations should be set for practitioners leading the way in building evidence in the social service sector."

Among use cases, Nisha Patel from Washington University in St. Louis flips the lens from eradicating child poverty to achieving guaranteed minimum income levels, citing cases and evidence where practitioners' cash distributions make a difference. Notably, the monthly federal child tax credit payments during the pandemic reduced monthly child poverty by nearly 30 percent. Meanwhile, COVID-19's disproportional toll on low-income communities of color highlighted deep inequities. Consider ParentCorps, a nonprofit that engages parents as partners to strengthen early childhood education. ParentCorps, like many organizations, went into crisis management mode in March 2020 when COVID-19 engulfed New York City and forced school closures. Almost overnight, staff transitioned ParentCorps's programs and evidence building to virtual activities, considering each point of contact with families as an opportunity to assess need and inform rapid adaptation.

We applaud these directions to strengthen data and evidence with equity, and Project Evident offers its support in helping bring this body of work forward. With insights from and in partnership with field leaders across the nation, including BCT Partners, Equitable Evaluation Initiative, leadership from Seattle, Washington's King County, We All Count, Erika Van Buren, Amy O'Hara, Peak Grantmaking, Spectrum Health, Hopelab, and others, we are refining Project Evident's *Data Equity and Evidence Guide* to offer much-needed support in building staff capacity to strengthen evidence building by integrating equity.[11,12]

3. ELEVATING COMMUNITY VOICE

One important way to advance equity, which merits its own set of essays and use cases, is to involve community members across every step of evidence building, from defining questions to gathering, interpreting, and applying data, to sharing results, an approach that will contribute to a stronger evidence process.

Dan Cardinali, formerly of Independent Sector, in chapter 3.1 calls on evidence builders to "agree upon and accept ways in which people in communities, especially those that are structurally marginalized, define what individual and collective human and environmental flourishing looks like for themselves, their loved ones, and their neighborhoods," and then build evidence in service of those goals. In this way, the institutions designated to serve communities earn the trust of people in them. Building trust, says Cardinali, "is one of the most pressing adaptive challenges of our day."

Marika Pfefferkorn's subsequent chapter (3.2) tells a story of data justice in the Twin Cities as "the opposite of what many governmental bodies, nonprofit agencies, private companies and technical assistance providers put forth as 'community engagement.'" It is the story of a school district and police district that pivoted to better, transparent consultation with community constituents in response to initial community outrage that an agreement to share data would lead to racial profiling. The case demonstrates that if community partners are involved when technological solutions are brought into the mix, fair and just data practices can result. Says Pfefferkorn of Midwest Center for School Transformation: "Data fixes generated by systems built on injustice will most likely replicate those injustices.

Communities disproportionately injured by bad data practices need to be at the center of discussions and designing any use of technology that purports to address those injuries."

Trust is key, and respecting communities' agency builds such trust. John Brothers of T. Rowe Price Foundation calls on philanthropy to "find evaluative approaches that help communities use their own data for their own self-determination while at the same time building the capacity of our under-resourced community-based organizations to measure and grow their impact." Rhett Mabry of The Duke Endowment writes about unintended consequences when philanthropy fails to listen to community, describing a disappointing evaluation of a program to increase kinship placements for foster youth: "In our zeal to test this new, promising approach . . . we failed to adequately engage caseworkers to capture their input, before subjecting the approach to rigorous trial." In contrast, the Duke Endowment and partners deeply engaged practitioners and community members when making decisions about how to scale the Rural Summer Literacy Initiative, an unusual multiyear collaboration designed to help United Methodist congregations improve early childhood literacy in North Carolina's rural communities. For example, one early childhood education site intentionally adjusted some of their teaching practices to meet the tactile learning styles preferred by their Native American students (chapter 3.5).

New York University's Criminal Justice Lab also focuses on trust in a case about developing a health diversion tool to address the large intersection between public safety and public health (54 percent of arrestees made five or more visits to the emergency room during the study timeframe). The lab was careful to use language that would encourage people to answer honestly—placing all the questions in a framework of health rather than criminal behavior—in a tool that law enforcement would be comfortable using. The tool garnered promising results from initial testing in Indiana and Illinois.

And the case of Pace Center for Girls, which tracks participatory research back to its roots with W.E.B. Dubois, demonstrates how involving Pace girls and their communities in identifying and pursuing research questions, and seeking feedback as a regular part of evidence building, has increased both the relevance of Pace research and speed to findings. At the same time, it has seeded a culture of deep listening throughout the organization that has boosted girls' self-advocacy and efficacy.

4. EMBRACING AN R&D-LIKE, CONTINUOUS USE OF DATA AND EVIDENCE

This book will suggest how to be intentional in continuously using data and evidence to innovate, improve program implementation, and assess impact, thinking of it as an R&D function that informs and is informed by strategy. With increasing demand for more relevant, rapid-cycle evidence; the confluence of data science and evaluation; and calls for more frequent testing and learning, it is time to think of evidence as a core process—like an R&D function. In CEO Brian Scholl's wide-ranging chapter on the unfinished business of evidence building, he articulates the promise and pitfalls of evidence building, extolling the need for researchers to "work backward" from practical outcomes to design worthwhile studies.

Working backward from practical and intended outcomes means starting with a **logic model or theory of change** and interrogating assumptions with data to frame test questions. Implementing this testing approach requires **a practice of continuous, disciplined data collection**, guided by the theory of change or logic model. Quality data collection gathers facts and feedback across populations in a given program, including vulnerable and underrepresented groups, to understand respective barriers to participation and success. Supporting strong data collection and use calls for **reliable information architecture** that makes it easy to develop and test hypotheses, develop solutions, and "play back" insights to frontline staff and communities providing the data.

Being intentional about data use and the questions we want to test or evaluate is critical for better and more equitable decision making, for innovation, improvement, and the development of new solutions and to assess impact—all of which should inform and be informed by strategy. Building on elements of continuous learning and classic R&D, this practice includes activities such as developing and testing hypotheses more rapidly, understanding differential impacts on the population served, and grounding test questions in the theory of change or logic model (i.e., How do we know this activity will lead to this desired outcome?).

A range of R&D-like evidence-building activities are illustrated in essays and practitioner use cases throughout this book, including those of criminal justice nonprofit Center for Employment Opportunities, a later-stage organization, that illustrate the value of establishing and staffing internal R&D capacity. Children's media innovators at Noggin (chapter 4.9) believe the future of education progress lies in giving children access to "digital teachers and role models that [they] truly love." Noggin uses multiple strategies to quickly iterate on content and ensure it continually improves. The case of First Place for Youth highlights their continuing journey to generate knowledge and impact that catalyzes programmatic and system-level impact on one "north star" outcome: life sustaining, living wage employment for youth aging out of foster care.

Meanwhile, the case of Gemma Services, a youth-oriented psychiatric care program, demonstrates how insights were gained to improve student outcomes through their work with BCT Partners. Building on algorithms commonly used by organizations like Netflix or Amazon, Gemma designed recommendation engines sensitive to inherent bias in order to help practitioners make better decisions related to student's needs. This approach produced more precise and contextualized information for practitioners.

Among essayists, Brian Scholl also emphasizes that evidence comes in many forms and the key is "to find the highest quality research appropriate to the question, circumstances and problem, but not shy away from tackling questions that add value even if the research methods aren't the cleanest" (chapter 4.1). Chris Spera, formerly of Abt Associates and now of Arbor Research Collaborative for Health, calls for a shift from evaluating program information for compliance purposes to engaging in a "tug-of-war" between using evidence for learning and program improvement versus accountability (chapter 4.2). University of Chicago's Kevin Corinth and Bruce Meyer offer a research tool that can advance both learning and accountability goals, discussing how the new Comprehensive Income Dataset can better measure poverty by overcoming the limitations of any single data source that measures income or well-being (chapter 4.3).

Meanwhile, coauthors Gary Glickman and Kathy Stack of Tobin Center for Economic Policy at Yale University speak to the need to fund data access, integration, and use across local, state, and federal governments to assess real problems and progress (chapter 4.4). This work is underway with the new federal appetite for a more systemic approach, particularly in education research,[13] as demonstrated by the 2018 evidence act and the Biden administration's related executive orders and guidance,[14,15] and as noted in section 5, with major new federal investments available to fuel more of this work. These combined federal initiatives will help deepen and implement strategic directions, including a focus on connecting strategy with evidence.[16] We also see a need for greater investment in infrastructure to enable more systemic use of data and evidence across all sectors to increase learning and knowledge. For example, governments should include dollars for continuous learning, research and development, and evidence building as a core part of grants across education and social sector programs.

Remaining essayists zero in on research approaches, some refined in the crucible of COVID-19 response. Neal Myrick, reflecting on Tableau Foundation's grantmaking, observes: "When encouraged and funded well, we saw that research and development (R&D) and the real-time use of data could help nonprofits solve our world's most complex challenges—even during the toughest of times . . . This model for supporting and driving continual learning and change isn't just suitable for a pandemic response—it's a best practice for future social impact work" (chapter 4.5). David Yokum and Jake Bowers, from Policy Lab at Brown University, discuss the power of a pre-analysis plan to ensure the right questions get asked (chapter 4.6). Industry leader Jim Manzi, of Applied Predictive Technologies and Foundry AI, offers insights to guide better use of RCTs in the social sector, highlighting the dangers of drawing conclusions from a single RCT or trying to generalize proof of benefits that are specific to a context. At the same time, he asserts that RCTs may be underused where they can helpfully demonstrate impact and potential for replication (chapter 4.7).

Manzi's observation underscores that we also face a crisis of replication, including challenges with scale and sustainability.[17,18,19] Replication issues may be exacerbated by evaluations focused on research priorities with less input from practitioners and stakeholders. In a 2005 synthesis of the research literature on implementation science, the National Implementation Research Network observed: "All the paper in file cabinets plus all the manuals on the shelves do not equal real world transformation of human service systems through innovative practice."[20] We see an opportunity to better leverage practitioner-level insights throughout this work, from developing, scaling, and sustaining evidence-based interventions to newer approaches that emphasize scaling *impact* rather than programs.[21]

5. REIMAGINING EVIDENCE: EXPANDING ITS DEFINITION AND USE

Our final cluster of essays and use cases speak to expanding both the definition of evidence and approaches to creating it, as well as increasing **use** of evidence by engaging with end users to find out how to make evidence more actionable. Our authors' ideas aim to help organizations accelerate to insight, collaborate on evidence creation, and increase transparency, what coauthors Jennifer Brooks and Impact Genome Project at University of Chicago's Jason Saul and Heather King call "breaking open the black box" (chapter 5.2).

In chapter 5.1, AmeriCorps's Michael Smith reflects on his experience championing the role of evidence in the past decade, saying: "We must demand that governments, businesses, nonprofits and philanthropies do more to shift the massive amount of dollars to solutions that have measurable evidence of impact." But he wants to see those dollars support a much broader, more inclusive definition of evidence. "We have to also expand our understanding of what constitutes evidence, grow our tent so more diverse voices and perspectives are included and evolve our concept of what classifies as an evidence-based solution from solely programs that meet immediate needs to policy reform that dismantles, disrupts and reimagines the broken systems that have failed far too many."

To build evidence that is more relevant, timely, and cost-effective, we must broaden its definition to include not only statistical but also practical significance, and include input from multiple stakeholders. We must reimagine evidence to consider context, confidence level, size of impact, speed to insight, and cost of implementation. "We have seen [discourse around] evidence play out in many ways recently—from climate change debates, to disinformation/misinformation around COVID-19, to the U.S.'s story on racial justice," say coauthors Veronica Olazabal, BHP Foundation, and the American Evaluation Association's social impact advisor Jane Reisman (chapter 5.3). "Evidence in these broader debates shows that evidence-based decision making is about more than generating proof through credible research efforts . . . it's about diverse perspectives, mindsets, uptake, use and management." Meanwhile, Brian Komar speaks to building evidence for environmental, social impact, and governance (ESG) efforts, identifying four steps for improving the quantity, quality, and interoperability of the information we use as evidence of impact (chapter 5.4).

Companion essays in this section speak to reimagining of evidence building by a variety of actors in multiple fields. The Office of Management and Budget's Diana Epstein observes that the alignment of evidence with strategy "is an opening to bring the evidence-builders and the strategic planners together from the outset. This has typically not been done in Federal agencies, but the Evidence Act offers a new framework within which evidence-building priorities are aligned with strategy and envisioned together from the start" (chapter 5.5).

Ryan Martin, with the National Governor's Association, speaks to the need for more small-sample studies to find dependent variables—"needles in haystacks"—in the spirit of fostering "a climate in Congress and elsewhere where failure is acceptable, evidence building is prioritized and those running programs adapt based on what has been learned" (chapter 5.6). Meanwhile, Results for America's Michele Jolin and Zachary Markovits describe how evidence is fueling a quiet revolution in cities across the United States that have embraced data-driven transformation, noting that the new infusion of trillions of dollars from the Federal American Rescue Plan,[22] Infrastructure Investment and Jobs Act,[23] and Inflation Reduction Act[24] can be used to build the necessary "test, learn and improve infrastructure" we need to address the most intractable problems of tomorrow. Cincinnati and Tulsa are two examples of cities able to use their earlier investments in data infrastructure to respond quickly when the pandemic struck (chapter 5.7).

And Vivian Tseng, formerly of the William T. Grant Foundation and now at the Foundation for Child Development, calls for incorporating the basic principles of democracy into evidence initiatives to give communities meant to benefit from government policies and programs "access to the evidence, a say in identifying which problems require more evidence, and . . . a seat at the table in interpreting the evidence and determining what it means for government action and spending" (chapter 5.8).

The Stanford RegLab case relays the benefits of collaboration in evidence building. It shows how the Santa Clara County Public Health Department teamed across sectors, with academics at Stanford University, to develop the people, health, and information processes for rapid evidence building to respond to the COVID-19 pandemic (chapter 5.9). A second case in this vein shows how collaboration between the Camden County Health Department and nonprofit Camden Coalition, a multidisciplinary nonprofit working to improve care for people with complex health and social needs, advanced the region's pandemic response. The Camden Coalition put their Health Information Exchange (HIE), which connects siloed data across health systems, in the service of the county's COVID-19 response. The positive results have spurred conversation about HIE's broader use to support non-COVID-19 programming and to create a more robust ecosystem of regional care (chapter 5.10).

CONCLUSION

Each of us brings a different perspective to our shared focus on how to achieve better, more meaningful, and more equitable outcomes for communities. Together, we feel a sense of urgency and optimism that, as a nation, we can and will do better. To get there, we will need to turbocharge investments in practitioner-centric evidence building and use, and focus on continuous learning and R&D practices. Many of these practices have taken root in the commercial sector, yet they have not become standard and supported practice in the social and education sectors. It is time for funders to help establish a new norm. They should support practitioner-generated evidence, both for their own use in learning and funding decisions and to help close the data and evidence divide.

In the words of Scholl in section 4: "Evidence, for the most part, is an exercise in innovation: how to make processes work better, how to develop better products or combinations of services." But too often, we lose our way in the process. "When we in the evidence community talk about building evidence, so often our conversation goes to the math and the statistics of it all: experiments, treatment effects, causal estimates, randomization protocols, and so on. Those are so important in so many ways, but also so unimportant in so many other ways."

"In my mind," Scholl continues, "it is the organizations, the institutions and the people that really matter . . . [not] as some kind of easy lip service—the people really do matter. The wrong people at the top (leadership) can dead-end any efforts to generate evidence. Wrong people generating evidence get to all the wrong questions and all the wrong answers using all the wrong methods. Wronged people at the bottom (beneficiaries or constituents) bear the consequences of getting policies and programs wrong. . . . Evidence is critical to getting our work to work."

We agree with Scholl that "marginalizing evidence generation can create distortions that hurt people and society, and can undermine trust." It eats at the core of a functioning democracy. So, we applaud funders, policymakers, researchers, evaluators, and technical assistance providers who are embracing new partnerships with practitioners to create actionable evidence, evidence that is equitable and useful to those closest to problems being addressed. We encourage more to join in this work. And, as AMP Health's Robert Newman notes, "Practitioners, for their part, must recognize their own role: *Our data. Our evidence. Our decisions*" (chapter 2.5).

We hope this book, full of insights, experience, and expertise, will give voice to practitioners and move readers to help advance the *Next Generation Evidence* for greater social impact.

NOTES

1. For Love of Country: A Path for the Federal Government to Advance Racial Equity, *PolicyLink*, July 2021.

2. More details on the convening, including the full agenda, can be found at https://www.projectevident.org/next-gen-evidence-convening.

3. *Show Me the Evidence: Obama's Fight for Rigor and Results in Social Policy*, Haskins and Margolis, Brookings Institution Press (2014).

4. By examining the evidentiary foundations for social and educational programs, and tying more dollars to programs with the most rigorous evidence that they succeeded, these tiered evidence efforts redirected public dollars to programs with greater evidence of impact. See, for example, The Maternal, Infant, and Early Childhood Home Visiting (MIECHV) program (US DHHS), Investing in Innovation Fund (i3) (US Ed), two of the thirteen Obama tiered evidence initiatives in Show Me the Evidence.

5. Ron Haskins, "Evidence-Based Policy: The Movement, the Goals, the Issues, the Promise," in Evidence-Based Social Policy: The Promise and Challenges of a Movement, Special Editor Ron Haskins, *The Annals of the American Academy of Political and Social Science* 676 (July 2018): 36.

6. Britt et al., "Improving Academic Success and Retention of Participants in Year Up's Professional Training Corps.," *Project Evident* (July 2021).

7. Amy O'Hara and Stephanie Straus, "The Civil Justice Data Commons," *Georgetown University Law School.*

8. "Actionable Evidence Framework," Project Evident, 2021.

9. Executive Order 13985, "Advancing Racial Equity and Support for Underserved Communities Through the Federal Government," DCPD-202100054, January 20, 2021.

10. U.S.C 115-435, "Foundations for Evidence Based Policy Act of 2018," United States House of Representatives, January 14, 2019.

11. Project Evident, "Comments to the Office of Management and Budget on Methods and Leading Practices for Advancing Equity and Support for Underserved Communities Through Government," United States Office of Management and Budget, July 6, 2021.

12. The updated DEEG will more fully integrate equity into guidance for all aspects of the data pathway, from theory of change creation through data collection and use to data sharing and communication.

13. Adam Gamoran and Kenne Dibner, "The Future of Education Research at IES," *National Academies Press*, 2022.

14. Office of the President of the United States, "Memorandum on Restoring Trust in Government Through Scientific Integrity and Evidence-Based Policymaking," DCPD-202100096, January 27, 2021.

15. Executive Order 14007, "Executive Order on the President's Council of Advisors on Science and Technology," DCPD-202100094, January 27, 2021.

16. Kelly Fitzsimmons, "Supporting Effective Policymaking through the Development of Strategic Evidence Plans," *Project Evident*, September 3, 2020.

17. James Kim, "Making Every Study Count: Learning from Replication Failure to Improve Intervention Research," *Educational Researcher* 48, no. 9 (2019): 599–607 (on replication in education, one of three articles in a special issue of *Educational Researcher*).

18. Stephanie Bell et al., "Challenges in Replicating Interventions," *J Adolesc Health* 40, no. 6 (June 2007): 514–520 (on an NIH initiative looking at replication issues for HIV prevention).

19. Kumpfer et al., "Strategies to Avoid Replication Failure with Evidence-Based Prevention Interventions: Case Examples from the Strengthening Families Program," *Evaluation & the Health Professions* 43, no. 2 (2018): 75–89 (on strategies to avoid replication failure in parenting interventions).

20. Fixen et al., "Implementation Research: A Synthesis of the Literature." University of South Florida, Louis de la Parte Florida Mental Health Institute, The National Implementation Research Network, 2005.

21. McLean et al., "Scaling What Works Doesn't Work, We Need to Scale Impact Instead," *London School of Economics*, September 7, 2020.

22. Rachel Dietert and Ben Reynolds, "How States Are Putting American Rescue Plan Dollars to Work," The Council of State Governments, August 24, 2022.

23. U.S.C. 117-58, "Infrastructure Investment and Jobs Act," United States House of Representatives, November 15, 2021.

24. U.S.C. 117-169, "Inflation Reduction Act," United States House of Representatives, August 16, 2022.

SECTION 1

CENTERING ON PRACTITIONERS AND THE COMMUNITIES THEY SERVE

For too long, practitioners have been the caboose of the evidence train, when they should have been the engine.

—KELLY FITZSIMMONS AND ARCHIE JONES,
"CONSTRUCTIVE DISSATISFACTION"

Where we stand determines what we see. And, for too long, social and education impact evaluators and researchers have stood outside programs and organizations, studying elements of their practice and its relationship to results, instead of inside or alongside, listening to the practitioners and those they serve to understand their definitions of success.

This first set of essays reshapes our perspective on evidence building. The authors, including Project Evident founder Kelly Fitzsimmons and founding board chair Archie Jones, former Harlem Children's Zone evaluator Betina Jean-Louis and University of Pennsylvania economist Rebecca Maynard, a leader in the design and conduct of randomized controlled trials, share their journeys to practitioner-centered evidence building. They lift up the role of practitioners—public agencies, school districts, and nonprofit organizations—that deliver social and educational services. And they describe ways and means to anchor studies in practitioner experience, loop back to practitioners early and often with findings, and help practitioners build measurement capacity within their organizations to collect and analyze data in an ongoing manner to inform daily decision making. This is in striking contrast to relying primarily on the one-and-done approaches of periodic external evaluations, which too often deliver insights months or years after decisions need to be made.

Following their essays, you will find cases of practitioners in partnership with researchers, funders, and data scientists that have taken these methods to heart—workforce development nonprofit Year Up; maternal-child health nonprofit Nurse-Family Partnership; criminal justice nonprofit The Bail Project; Baltimore Public School District; and virtual technology training nonprofit Per Scholas—highlighting their approach to continuous evidence building and the results they have obtained.

Questions raised and addressed in this section include:

1. How is practitioner-centric evidence building different from traditional evaluation?
2. How can practitioner-centric evidence building lead to better, more meaningful, and equitable outcomes for communities?
3. How does practitioner-centric evidence building change the role and perspectives of communities, funders, researchers, and policymakers?

CONSTRUCTIVE DISSATISFACTION

KELLY FITZSIMMONS AND ARCHIE JONES

KELLY'S STORY

I (coauthor Kelly) began my career in the social sector over twenty-five years ago at a large multi-service organization in Boston that offered an array of education, workforce, health, and early education programs to single-parent families, adults, and students experiencing poverty. During my time there, we developed a program that directly engaged parents in building early literacy skills to prepare children for kindergarten. The program was highly successful and popular with mothers and their children. The state of Massachusetts expressed interest in expanding it, but told us we needed a third-party evaluation as a condition of funding. We could not afford to foot the bill on our own, so I approached one of the few foundations that supported evaluation work at the time. The foundation declined my request, and the program officer went on to tell me that it would be "a complete waste of money to fund the evaluation because the target population was too risky, too transient, and anyway—those moms don't really care about their kids."

While the callousness of this particular program officer was unusual, versions of this story, unfortunately, are not. Rather than being fueled by productive partnership focused on learning and improvement and dedicated to achieving stronger and more equitable outcomes, evaluation in the social sector was largely top down, driven by the objectivity (and bias) of the expert, set to a thumbs up or thumbs down meter, and not predicated on trust or learning. In the words of Ballmer Group co-founder Connie Ballmer, "I didn't realize that evaluation was so punitive."

As I migrated from direct services into philanthropy, I learned how much donor and political intent shaped the social sector's approach to building evidence, for better or worse. During the early years of venture philanthropy, there was a call for greater emphasis on performance data but little exploration of whether that performance data translated to real programmatic outcomes for program participants. In fact, many considered evaluation the polar opposite of innovation—an overly academic exercise that was sluggish and a waste of money. Then the fixation with scale set in (a fixation I wholeheartedly and naively adopted), and much of the social sector began spreading interventions to new communities without asking sufficient questions regarding how an intervention and its outcomes in one community might translate to another.

As the notion of evidence crept into philanthropy's investment theses and frameworks, and the notion of evidence-based policy caught on through the George W. Bush and Barack Obama administrations, an overreliance on a few evaluation methods—most notably randomized controlled trials—emerged. This had the unintended consequence of creating a fixation with "getting on the list" of preferred service providers by practitioners, and creating an incentive for evaluators to skew studies toward what met the framework requirements (and, therefore, what might be more lucrative) rather than generating the right evidence for the right people at the right time.

For a period of time, I served as a coach to CEOs and executive directors who were working on their strategies for impact. One of the CEOs I coached received findings from a multiyear randomized control study of their work that found positive results in two of three intended outcomes. Instead of celebrating wins and focusing on what could be learned from the results, the CEO was terribly anxious about how their funder would react to two out of three. They worried their funder's reaction would be thumbs up or thumbs down versus positively embracing what worked and learning from what did not.

Practitioner-grantees of philanthropy and government, whose success is their grantmaker's success, have too often and for too long feared challenging donor guidance on evaluation design or choice of evaluator because donors held the purse strings and, thus, the power. Practitioners often lacked the knowledge and proof points to advocate for measurement approaches best adapted to their context. Nor did funders offer them resources to build their own evidence plans. Instead, funders standardized an approach and plugged grantees into it. Hence, practitioners most responsible for delivering quality programs became least empowered to shape their evidence agendas.

Of course, there were exceptional cases. Many funders pointed to maternal-child health nonprofit Nurse-Family Partnership's (NFP) own approach as a gold standard for evaluating social programs. As I sat at a turning point in my career, I realized a glaring irony. NFP was lauded as the exemplar, yet NFP's approach to building evidence—continuously, with smaller studies leading to increasing confidence, more data collection, more improvement, and bigger studies at multiyear intervals with blockbuster impacts—was at odds with what our evaluation and funding industry supported as a rule.

For far too long, practitioners have been the caboose of the evidence train when they should have been the engine. They should be active leaders in evidence building, not at the mercy of research and evaluation shops but in partnership with them and with their funders, aligning their goals and interests. No one cares more than practitioners that their theories of change work as intended. But that requires change. So, as someone who grew to embrace the power and possibility of better evidence building over my career, I became what my friend and Project Evident alum Dr. Charles Carter would call "constructively dissatisfied," and reached out to trusted colleagues to explore new thinking. One of those colleagues was the coauthor of this essay, Archie Jones, who came from the private equity world, working with a team of excited dreamers on the prospect of bringing innovation to the social sector.

ARCHIE'S STORY

That's right! I (coauthor Archie) was a venture investor on a mission to make it easier to scale high-impact nonprofits. The idea was simple: increase the amount and types of capital available to promising nonprofits so they could scale faster and more broadly to accelerate the pace of innovation and impact. I found myself asking a question similar to Kelly's: How do you know which innovations are ready to scale, and how do you measure how efficiently and effectively they scale after you have invested? How can you avoid defaulting to the usual standards of measuring inputs and outputs?

Inspired by social entrepreneurs whom my venture philanthropy team funded, and frustrated with the lack of strong data and evidence and tools for making informed decisions, I also believed that better evidence would lead to better strategy, and better strategy would lead to higher impact. I wanted to found an organization for this purpose, hoping to provide a set of tools, frameworks, and processes for the sector to make better use of data in efforts to achieve impact. Like most innovators, I pursued collecting better, cheaper (but not cheap), and more quickly accessible data and evidence. My definitions:

Better: More strategic about what data to collect and use (leading indicators on outcomes)

Cheaper: Using tools and technology to lower the cost of implementing a data strategy

Faster: Testing early and often to support the continuous refinement of an innovation

My team and I also envisioned data and evidence as being tools investors and funders could use to achieve more equitable outcomes. Data and information gaps either increase perceived risk or mask inherent risks, with either scenario stifling investment. This means that underserved communities actually are communities where we have underinvested—not because there aren't superior returns to be gained and impact to be had but because we overestimate the risk—or, more importantly, do not fully understand the risk-reward relationship. Data and evidence are crucial to better risk analysis and stronger, more equitable returns.

I knew that social entrepreneurs struggling to bring their ideas to life need capital—not just financial capital but, more importantly, information capital, which allows them to iterate and refine their hypotheses and test high-impact and sustainable solutions. Helping social entrepreneurs design an evidence strategy that will ensure a steady flow of information capital gives them a tremendous advantage. It not only helps them report to others on progress but, more importantly, to create a stronger, internal culture of learning.

THE STORY OF PROJECT EVIDENT

And so, with initial support from the Edna McConnell Clark Foundation and joined by five other funders, we—Archie and Kelly—founded Project Evident in 2017 to transform the evidence ecosystem by elevating practitioners and the communities they serve. After extensive research and interviews with nonprofits, funders, and policymakers, we decided on a platform of shared services (consulting, tools, and technical assistance) staffed with leaders from philanthropy, technology, policy, analytics, and nonprofit management and working with existing service providers. Though our platform would help practitioners prepare for rigorous third-party RCTs when appropriate, our primary focus was on exploring a wider range of evidence that focused on practitioner learning needs and continuous improvement. We blended this platform of services with field-building efforts to contribute to a healthier ecosystem that would enable a practitioner-centric, R&D approach for the social and education sectors and change the incumbent approach to evaluation.

Our core offering at Project Evident is the strategic evidence plan (SEP), which we developed alongside social sector organizations with the recognition that the field needed a new, strategic approach to continuous evidence building that went beyond the one-study-at-a-time mindset. SEPs are designed to advance actionable, practical knowledge needed to build and scale solutions, and to listen to the voices of practitioners and community members—making the process of building evidence more equitable. As we have grown and recognized needs in the field, we have added a number of other direct service and technical assistance offerings to support nonprofit organizations, funders, intermediaries, government agencies, and education agencies.

We also recognize that practitioner-centric and actionable evidence building is not required just at the organizational level. The evidence ecosystem currently lacks the incentives and enabling conditions that would support effective data infrastructure development and evidence-building activities. And social investors lack incentives to explore opportunities in unfamiliar areas with unfamiliar people—they need actionable evidence to analyze risk. We also must address the broader policies, structures, and orthodoxies in the ecosystem to support more actionable, equitable, and continuous evidence building and to assure that all forms of capital can flow more freely to traditional and emerging markets of social return.

This shift requires that we educate stakeholders, including practitioners, funders, policymakers, researchers, and technical assistance providers to promote investment in research and development that will inform a more productive market for outcomes. We use proof points from our direct services work to help the broader field understand, in a user-friendly way, the key elements of continuous evidence building and what they look like in practice. Our ecosystem efforts to scale knowledge and practice span a range of activities, including the development of this book.

THE NEXT GENERATION OF EVIDENCE

We share this moment with other field leaders, who are fundamentally rethinking impact and what that means across multiple sectors. Our vision and work at Project Evident contribute to a next generation ecosystem that supports more equitable and actionable evidence building. We hope, on this journey, to collaborate, humbly, with other leaders on:

- Understanding the structural drivers of racial inequity, and using data to help us spotlight gaps and move swiftly toward more equitable outcomes.
- Helping policymakers further embrace evidence and make long-term commitments to data gathering and use beyond evaluation, including building their staff capacity and changing regulations to make evidence building an allowable federal cost across all government agencies.
- Using actionable evidence to broaden selection criteria for grantees and reassess funding approaches like cash transfers and making flexible, unrestricted funding a philanthropic norm.
- Broadening impact investment practice to consistently value nonfinancial returns like increasing equity and ESG.
- Aiding researchers, evaluators, and technical assistance providers as they reassess their roles and examine their own practices as organizations like the Equitable Evaluation Initiative, We all Count, Equal Measure, and many others who have contributed to this book advocate for fresh approaches.

- Learning alongside practitioners who are leading the way, serious about the impact they are delivering and eager to know where their theories of change are working, where they are not, and how to improve.

We look to field leaders as potential partners in this endeavor to help practitioners achieve sustainable funding, better infrastructure, and conditions that enable program success. And we invite those on a similar journey to channel constructive dissatisfaction into a commitment to evidence building to accelerate positive and sustainable change as we navigate a world of uncertainty.

ENGAGING THE FULL ARC OF EVIDENCE BUILDING

BETINA JEAN-LOUIS

INTRODUCTION

In an ideal world, passion for a cause, dedication to equity, and certitude about the rightness of one's approach would be fully predictive of a given intervention's impact. In the real world, my experience as a research and evaluation professional shows otherwise. Our strongest weapon *to ensure that do-gooders are **actually** doing good* and that efforts are not just well-intentioned but effective is ***evidence.*** Evidence improves the likelihood that those who do social impact work, fund the work, and support policies that expand the work will do so in ways that advance social and racial equity and improve the lives of under-resourced children and families.

Those of us who seek to do good must ensure that precious resources, which include practitioner and funder investments but, more importantly, the time, hope, and trust of the individuals served by social impact efforts, are, indeed, well utilized. Government, the philanthropic sector, researchers and academics, and practitioners and educators all have roles to play in improving and increasing the use of evidence. Understandably, others have figured more prominently than practitioners, particularly researchers and academics, who are well entrenched in the evidence space.

The historical discrepancies in leadership and representation relate to many factors, including:

- Practitioner prioritization of investments in direct service provision rather than research and evaluation.
- Gaps in resources and expertise that have exacerbated inequities over the years.
- Historical power imbalances that have privileged the research questions and concerns of others over those of community-based organizations (Northridge and others 2005).

Nevertheless, over time, practitioners and educators have stepped into the data sphere in greater numbers. They have increased their data capacities, presence in gatherings where they were previously underrepresented (such as conferences, webinars, and other opportunities to discuss data collection and use), and advocacy efforts relating to data. I have seen an increase in the existence and uptake of data-focused technical assistance offerings—such as Project Evident's Talent Accelerator and office hours, and the mini-course on evaluation I recently co-taught to Promise Neighborhoods grantees with the Urban Institute—speaking to the hunger of service providers to increase their knowledge and to improve understanding of their own data.

The greater embrace of evidence is fantastic, and the field's willingness to do more than just tell practitioners and educators to "go forth and collect evidence" is key to even deeper engagement. As practitioners endeavor to build and flex data muscles, more funding, practical support, strategies, and capacity building focused on how best to engage in the right evidence building with data would greatly improve effectiveness. Fortunately, we can build on the foundational work government, philanthropy, researchers and evaluators, and nonprofit leaders already are doing to foster engagement of direct service providers in the evidence space.

ENGAGEMENT IN EVIDENCE: NONPRACTITIONERS

In *Show Me the Evidence: Obama's Fight for Rigor and Results in Social Policy*, R. Haskins and G. Margolis (2014) discuss the public sector's push for evidence, describing how several government agencies pushed grantees to utilize evidence-based practices and to collect data that would expand the evidence base. Federal agencies and policymakers have sought to operationalize these directives by developing guidelines that assist organizations in categorizing the strength of evidence, helping providers: (1) assess claims of scientific backing for interventions they are considering replicating; and (2) weigh the costs of gathering and analyzing evidence against the potential benefits.[1]

The philanthropic sector also has increased their focus on evidence, and many funders have reframed their guidelines to incentivize grant applicants and recipients to include considerations of prior evidence in their programmatic choices and—even if not as thoroughly supported—to encourage evidence building. Some foundations have stepped forward to help organizations build their evaluation infrastructure and engage in research,[2] and this has been helpful in preparing practitioners to play a bigger role in data conversations.

ENGAGEMENT IN EVIDENCE: PRACTITIONERS

Practitioners are essential to the evidence space, and their engagement can make a world of difference. My experiences as a consultant and as the lead evaluator at the Harlem Children's Zone have taught and retaught this lesson: any attempts to use data to improve social impact cannot work without the buy-in and partnership of direct service leaders and frontline staff. As practitioners and educators become more active in the evidence space and are strategic in making the best use of tools along the arc of evidence that has historically been the province of data scientists, everyone has a greater potential to win.

I have engaged organizations in employing the full arc of evidence, making use of **performance measurement**, **formative and summative evaluation**, **quasi-experimental studies**, **random-assignment research**, and **cost-benefit analysis** to guide and describe the work (McCarthy and Jean-Louis 2016). Each point along the arc has its benefits but its limitations, too, relating to costs, rigor, efficiency, and timing.[3] It is important to be tactical in determining what works best for each project or organization at its particular moment of development. That may well be in flux. Investments in exploration of a particular program may legitimately involve performance measurement, followed by evaluation, followed by random assignment, and then performance measurement approaches again as targeted changes are made. Evidence creators and users must make decisions about the types and rigor of the proof desired. An organization or community-based initiative with multiple components may have several programs that *are at different places on the developmental pipeline and evidence continuum* at a given time. As organizations continue to serve participants and to innovate, continuous evidence building is required, even for interventions with the benefit of having been proven by rigorous testing.

As the founding evaluation director at the Harlem Children's Zone® (HCZ) and during my eighteen years as their lead researcher and evaluator, my role was to build a data ecosystem that would allow HCZ to: (1) track individual-level engagement and outcomes for children and adults at scale; (2) make informed decisions about resource utilization; and (3) develop the legitimacy to play critical roles in philanthropic, youth development, education, and policy debates centering on poor children and families. To do all that successfully, the organization needed evidence, and lots of it. HCZ was able to influence so many arenas impacting the lives of poor children and families (for example, community and place-based, health, child welfare, and education) because the organization endeavored to adhere to these principles: Do good work; have good data; and stay ready.

That approach was made possible by knowledgeable evaluation staff, a commitment to data work from the non-evaluators in the organization, and flexible funding for the research and evaluation structure from several supporters. Continuing data collection, analysis, and discussion leads to continuous improvement in programs and the readiness to answer the call when fellow practitioners and educators, funders, policymakers, and others are actively seeking out solutions. HCZ's Healthy Harlem initiative, an anti-obesity program with prevention and targeted intervention components, provides an example of the investment made in evidence building.

Healthy Harlem data work included:

- Mutual selection of key performance indicators by program and evaluation staff and engagement in performance measurement activities led by administrators.
- Evaluation activities that included dissemination and review of student surveys; collection and review of BMI data; and focus groups and interviews with youth and staff.
- Engaging consultants who helped identify proven practices and strategies that other initiatives had used and consider how best to integrate their lessons learned.

- A random assignment study—led by Mathematica Policy Research—that resulted in Tier 1-level proof of the effectiveness of the program's Get Fit component in reducing mean body mass index and decreasing the percentages of middle and high school youth classified as overweight and obese (Mabli and others 2020).

While HCZ has had many successes in the evidence space, the organization also learned some hard lessons along the way. Early on, HCZ entered a good faith partnership with local schools to have HCZ's school-based employees work together with school and district staff to complete district-mandated hearing, vision, and body mass index screening. HCZ helped engage medical professionals for the screenings, devoted a great deal of person hours to coordinating student travel to school-based testing spaces, documented screening results on paper, and happily shared all screening notes with district staff for planned data entry.

HCZ staff members anticipated receiving organized lists of students who might need subsequent medical intervention, ready to aid students and families in obtaining any needed supports (for example, eyeglasses, additional testing), only to be floored by the discovery that the very information they had collected *could not be shared* because of confidentiality constraints. This stymied plans to have staff members lean in after the screenings to ensure that follow-up could occur (for example, that HCZ staff could check in with the families of children who needed glasses and remind students to wear their glasses). This incident made the need to specify the rules of engagement for data-related collaboration—in writing—very apparent, a lesson that was extremely helpful as evidence-building work continued with other collaborators over the years.

CONCLUSION

Back in 2002, when I started working at the Harlem Children's Zone and building their evidence infrastructure, few community-based organizations had chosen to provide that same level of support to evaluation and research. Today, more direct service providers have established internal evaluation capacities and are working with external researchers and evaluators in more engaged and informed ways. While the road to deeper engagement in building and using evidence is not always smooth, many direct service providers understand that ambitious plans for greater impact (in scale and influence) demand dedicated and continuing attention to gathering, analyzing, and using evidence. Some key practitioners have joined the vanguard that is pushing for the use of evidence relating to social impact initiatives, recognizing the need to use data for continuous improvement, evaluation, and research.

Increasingly, direct service providers are willing to take the reins and engage in research and evaluation, but they need continuing support, capacity building, funds, and guidance to do so. Many practitioners know that evidence building is not in opposition to accomplishing the primary work—which has been and always will be providing high-quality direct services to children and families—but in support of it. The picture is nuanced. While programmatic endeavors continue to hold special weight, the importance and urgency of the work needed to address the country's problems require a rigorous and robust evidence gathering and review infrastructure. Scale demands evidence, both for fundraising and to allow a complete understanding of increasingly complex pathways of activities, inputs, outputs, and results. Do-gooders always will prioritize the work, but they also will need to prioritize inquiry and be strategic in their evidence building. Taken together, that is what is needed to change the world.

NOTES

1. The Department of Education's four tiers of evidence provide one example of such guidelines, ranging from Tier 1, the most stringent level of support for program effectiveness—requiring support by at least one random assignment experiment that yields at least one statistically significant positive finding—to Tier 4, which requires providing a rationale for why outcomes are likely to improve based on high quality prior research and continuing evaluation.

2. See, for example, *The Edna McConnell Clark Foundation's Youth Development Fund: Results and Lessons from the First 10 Years* (Ryan and Taylor 2013), which includes a discussion about the need to help grantees build capacity to evaluate their programs.

3. Centers for Disease Control, "Types of Evidence," addresses some of these issues www.cdc.gov/std/Program/pupestd/Types%20of%20Evaluation.pdf, as do the two figures in Olds and others (2013).

REFERENCES

Haskins, R. and G. Margolis. *Show Me the Evidence: Obama's Fight for Rigor and Results in Social Policy*. Washington, DC: Brookings Institution Press. 2014.

Mabli, J., M. Bleeker, M. K. Fox, B. Jean-Louis, and M. Fox. "Randomized Controlled Trial of Healthy Harlem's Get Fit Program: An After-School Intervention for Childhood Overweight and Obesity in the Harlem Children's Zone." *Childhood Obesity* 16, no. 7 (2020): 479–87.

McCarthy, K., and B. Jean-Louis. *Friends of Evidence Case Study: Harlem Children's Zone*. 2016. Washington, DC: Center for the Study of Social Policy.

Olds, D., and others. Improving the Nurse-Family Partnership in Community Practice." *Pediatrics* 132, suppl. 2 (2013): S110–S117.

Northridge, M. E., K. Shoemaker, B. Jean-Louis, B. Ortiz, and others. "What Matters to Communities? Using Community-Based Participatory Research to Ask and Answer Questions Regarding the Environment and Health." *Environmental Health Perspectives*, January 2005. https://p2infohouse.org/ref/52/51647.pdf.

Ryan, W. P., and B. E. Taylor. *The Edna McConnell Clark Foundation's Youth Development Fund: Results and Lessons from the First 10 Years*. Report prepared for the Edna McConnell Clark Foundation. September 2013. www.emcf.org/fileadmin/media/PDFs/EMCF_ResultsandLessonsReport_2001-2012.pdf.

HOW RESEARCHERS CAN MAKE THE EVIDENCE THEY GENERATE MORE ACTIONABLE

REBECCA A. MAYNARD

Most professional evaluators seem highly motivated by the prospect that the results of their studies will improve outcomes for individuals and for society. However, recent work on the use of evidence shows a surprisingly large gap between these aspirations and the instrumental use of the evidence produced (Honig and Nitya 2012; Farley-Ripple and Jones 2015; Penuel and others 2017; Tseng and Coburn 2019). Concurrently, the public policy and philanthropic communities have increased substantially their emphasis on evidence-based policy and practice. In turn, this has fueled the imperative and seeded opportunities for rethinking approaches to evidence production to yield better and better used results (Haskins and Margolis 2015; Fedorowicz and Aron 2021; Zhang and others 2017). Practitioners and evaluators are having to adapt their usual practices to these new standards for production and use of evidence.

In this paper, I share snippets of my personal journey from a classically trained micro-economist to a "roll up your sleeves" evaluation partner of practitioners who are intent on improving the efficiency and effectiveness of their programs. I still use, primarily, the tools acquired in graduate school (updated as the evaluation field has matured). However, over time, I have found myself "flipping the script" to shift greater priority toward the evidence needs of the program partners, while still ensuring I maintain high standards for evidence and meet essential requirements of funders. As a result, I have been able to sidestep many of the commonly perceived barriers to engaging in more actionable evidence building enterprises (Brooks and others 2019; Donaldson, Christie, and Mark 2015).

My graduate training was a fairly traditional mix of economic theory and its application in public finance policy, coupled with an atypical experience working on the Rural Negative Income Tax Experiments (commonly referred to as the NIT). This assistantship afforded me the opportunity to work hand in glove with faculty members, peers, and field staff who were breaking new ground. None of us had experience running large-scale social experiments or designing and carrying out complex study designs—for example, those entailing large amounts of primary data collection and the design and implementation of statistical programs to address the nuances of study designs that included multiple sites, children nested within families, longitudinal surveys, and merging data from disparate sources (Levine and others 2005; Maynard 1977). Federal funders, program administrators, field data collectors, senior faculty research directors, and graduate students all were breaking new ground in our jobs. This was a quick lesson in the value of stakeholder engagement and teamwork.

Working on the NIT provided the foundation for a somewhat unusual career path. I became skilled at designing evaluations that yielded credible impact estimates but rarely provided much detail about on-the-ground experiences of those in the study sample to contextualize the findings or guide improvements in the design or implementation of the policies. Over the course of my career, I have encountered a few defining experiences that caused me to rethink how I approach my work as an evaluator. Here are three examples.

Example 1. While at Mathematica, I worked on a study looking at the impact of supported work programs for ex-addicts, ex-offenders, school dropouts, and welfare recipients. In advance of our revealing the actual impact findings, field staff at MDRC who oversaw the ten supported work programs in our study ranked the programs in order of "expected" impacts—rankings that seemed reasonable to those of us on the study team. We all were shocked when the study findings revealed that these rankings were the opposite of reality. This was a reminder that easily observable markers of program performance may not be reliable indicators of program impacts (Gueron and Rolston 2013; Maynard 2015).

Example 2. I was working on evaluations of programs intended to lower the incidence of repeat pregnancies and improve economic prospects for teen mothers and their children. At the time, conventional wisdom among the social welfare and academic communities was that these young mothers needed supportive, nurturing environments (Quint and Riccio 1985). At the same time, policymakers were arguing that it was important to institute financial incentives and basic supports that would encourage young single mothers on welfare to pursue education and training that would improve their long-run economic prospects.

Unsurprisingly, we encountered strong community pushback on plans for a federally funded randomized controlled trial to determine the consequences of eliminating the current exemption from work requirements for first-time teenage parents applying for federal welfare benefits. In response, federal program staff, our research team, and the local welfare office staff co-developed a logic model to guide the study design and implementation. This model included engaging with the teenage mothers and their caseworkers and welfare office directors throughout the course of the study—hearing their perspectives and capturing their first-hand experiences in ways that proved to be critically important for making meaning of the study findings.

By the end of the study period, a majority of the young mothers subjected to the new requirements for continuous engagement in school, work, and/or training as a condition of receiving the full welfare benefit reported feeling the policy was not only fair but helpful. The reasons they gave all related to the fact that the requirement was accompanied by supportive case management and other services to help them raise their child and improve their own lives (Maynard 1995; Maynard and Rangarajan 1994). Many also saw the requirement as an "escape hatch" from controlling partners or family members who preferred the young mothers remain dependent.

By broadening our evaluation agenda to include a rich (though not outrageously costly) qualitative study component, we were able to piece together a coherent explanation for why a policy that initially seemed draconian became preferred by both welfare workers and program participants. Most notably, we conducted periodic rounds of case conferences with welfare office staff (that is, meetings where case workers provide status updates on selected cases and invited reactions/input from peers and supervisors); focus groups with teenage mothers; and periodic meetings with case managers and office directors (for example, to discuss policies, practices, challenges, successes) (Polit 1992).

Example 3. A third defining experience was the design and implementation of evaluations of four of the Title 10 Abstinence Only Education Evaluation programs (Devaney and others 2002). This pulled on every lesson of my then twenty-something years of program evaluation experience. The Title V Abstinence Only Until Marriage legislation (Section 510 (b) of Title V of the Social Security Act, P.L., 104–193, provided major funding for health and sex education programs that promoted abstinence until marriage as the only way to avoid sexually transmitted diseases and unwanted pregnancies, and it prohibited any teaching related to contraceptive effectiveness or access. This policy had exceedingly strong support from the Christian Coalition and other conservative groups like Focus on the Family, and equally strong opposition from the health and sex education advocates and organizations like Planned Parenthood and the Alan Guttmacher Institute (Darroch and others 2000).

To successfully design and launch an evaluation, it was critical to fully understand the logic behind the views of those who endorsed the policy, those who opposed the policy, and those in between, and to ensure the perspectives of all three groups were well represented in the study design. Moreover, the study design needed to be recognized by all sides as nonpartisan; the goal of the study was to learn about and understand the consequences of the policy relative to the status quo and to be able to communicate the findings to all sides in a manner that was respectful and useful to them. This meant we needed to design a study that was centered on the common goal of supporting the sexual and social emotional health of youth.

To achieve this goal, we needed to invest heavily in stakeholder outreach at three levels. First, we needed to engage with the various constituents—for, against, and neutral toward abstinence-only education—and incorporate their beliefs and fears into a program logic model. Second, we needed to recruit and engage communities that had received Title 10 funds to partner with us in experimentally testing the abstinence-only programs against their usual health and sex education practices—a task that required building trust and a shared commitment to the evidence-building agenda. Third, we needed to make sure we maintained sufficient communication with sex education providers, students, and school administrators to ensure we would understand and be able to communicate to others the mechanisms through which the Title 10 programs altered (or did not alter) outcomes for students. This required a lot of listening, learning, and documenting to arm the study team with knowledge to support a cogent, evidence-based interpretation of the findings—whatever they were. Community members needed assurances that our goal was not to "prove" anything but, rather, to learn what difference the choice of sex education strategy made for students and to advance understanding of health and sex educators, parents, and community members about what it was about one approach versus the other that made the difference (assuming there was one). In the final analysis, there is neither evidence that abstinence-only education improves outcomes for youth nor evidence that it is harmful relative to comprehensive sex education (Ott and others 2007; Trenholm and others 2007).

Through these types of experiences, I have developed a vigilance around designing studies that will be useful to the practitioners and policymakers regardless of whether the tested concept yields the expected result. First and foremost, the goal is to learn something that will help them improve their program, policy, or product. This was an explicit motivation behind the development and improvement research I worked on in partnership with colleagues at Abt Associates, doctoral students at the University of Pennsylvania, and Year Up staff (Britt and others 2021; Fein and others 2020). This work produced concrete illustrations of some of the most important lessons of my career—lessons for researchers and researchers in training, lessons for practitioner partners, and the inevitability of limitations in any evidence-building enterprise.

The following are some of the major take-aways from the one slice of evidence building we featured in the Year Up case study.

1. Engaging in a collaborative process to prioritize investments in evidence building was extremely beneficial. It not only grounded the research team in the intricacies of the program—its goals, culture, participants, staff, partners, opportunities, and constraints—it paved the way for collaboratively identifying and prioritizing "rooms for improvement" in Year Up's Professional Training Corps program model that would be the focus of quick-turnaround evaluations to inform short-run improvement efforts and lay the groundwork for longer-term plans for evidence building.

2. Designing evaluations to inform program improvement regardless of the study findings can improve both the efficiency of evidence generation and the likelihood that the resulting findings will be used. This begins with exploring what questions practitioner partners would want answered under a range of plausible study findings—for example, if a strategy for boosting program retention proved not to be effective or even harmful.

3. It was important to balance staff and participant burden with its potential return. For us, this meant prioritizing essential data needs with "nice to know" data dreams. Having strong, trusting practitioner partners, we were able to fill many data gaps through opportunistic encounters and very targeted "mini data collection efforts." We also worked with program staff to plan implementation strategies that were minimally disruptive to operations and respectful of the informants, including providing tokens of appreciation such as modest gift cards or food.

3. We held fast on study design features critical for to generating credible evidence to inform program practices features, while being flexible in their implementation. For example, we insisted on testing program improvement strategies using a randomized controlled trial. Yet, we were very flexible on issues like sample size, assignment ratio, and timing of randomization, we provided options for "pairing" and "separating" participants, and we accommodated staff recommendations to exclude certain participants from randomization.

4. We not only encouraged programs to iterate on their improvement strategies between testing cycles, but we actively facilitated information sharing among program staff within and across sites between cycles regarding their professional judgments about what was and was not working. We also encouraged staff to modify their improvement strategies based on the shared experience. We did not provide interim impact findings, which (by design) were intended to inform summative conclusions about the progression of improvements over time.

5. All study findings were shared first with Year Up staff working directly with the study team in "draft" briefing documents—one package targeted on senior management and a second package

targeted at site staff. Each packet consisted of three parts: 1) a very summary "pre-read," which was a high-level overview of the study, findings, and recommendations; 2) a slide-deck to support an on-line briefing; and 3) a more detailed slide-deck sent out immediately following the briefing containing more detailed information about the study design and findings.
6. We collected artifacts created and/or used by program staff to support their program improvements and, with encouragement of the Year Up national staff and site program directors, created an indexed compendium of tools to support Year Up's use of the study findings to improve outcomes throughout its network of local programs.

The research team has prepared detailed technical reports on the study background, methods, and findings (Fein and others 2020), presented on its work at numerous professional conferences, and published academic and policy articles on the study methods and findings (Britt and others 2021).

Variations of these evaluation strategies are reflected in many other recent and ongoing evidence generation efforts by social scientists trained in various disciplines and research traditions—a sample of which is reflected in the twelve case studies commissioned as part of the Actionable Evidence Initiative supported by the Bill and Melinda Gates Foundation. For example, Julie Martin and Elisabeth Stock (2021) conducted a rapid cycle evaluation of an interactive digital learning strategy for improving academic mastery and social emotional learning skills of high poverty middle school youth; D. Bradley and S. Burkhauser (2021) report a partnership among the Mid-West Regional Educational Laboratory (REL Midwest), the Minnesota Department of Education, and four state-approved alternative education programs for students judged to be at-risk of not graduating from high school in a Networked Improvement Community aimed at identifying evidence-supported strategies for improving instructional practices through small-scale, quick-turnaround testing of strategies for collecting and using data to make real-time shifts in strategies that improved student outcomes; and P. York (2021) reports on a partnership among two programs serving youth aging out of foster care—First Place for Youth and Gemma Services—to use evaluation and data science to build predictive and prescriptive models that use real-time data to guide case management and support of youth in ways that improve their outcomes.

Increasingly, many in the evaluation community have been moving toward more nimble applications of their research tools than is common. Historically, evaluators have gravitated toward using their tools in service of independent evaluations aimed at general knowledge development or applying them in service of "thumbs up or down" accountability. Increasingly, evaluators are recognizing the power in applying their skills and resources to support practitioners to improve their decision making—thus, increasing their emphasis on actionable evidence. Through effective curation and dissemination of the products of such evidence building, we likely also will accelerate the pace and impact basic education and social science knowledge. This does not require a shift in methods; rather, it requires asking the right questions, gathering and using credible data, matching the questions to the methods, and reporting in a timely and accessible format. The odds of producing actionable evidence and having it used goes up significantly when practitioners are invested partners in the effort (Brooks and others 2019).

It has been amazing to watch even strong, long-established organizations learn and improve as they work to craft a strategic evidence plan. Even more rewarding has been the excitement among program partners once they have prioritized their needs for evidence and arrived at creative options for generating and using it.

REFERENCES

Bradley, D., and Burkhauser, S. "Lessons from a Minnesota Networked Improvement Community." Boston, MA: Project Evident. September 21, 2021. https://static1.squarespace.com/static/58d9ba1f20099e0a03a3891d/t/61426c66ac7327450a8141a7/1631743085075/MN+SAAP+NIC+Actionable+Evidence+Case+Study+Sept21.pdf.

Britt, J., Fein, D., Maynard, R., and Warfield, G. "Improving Academic Success and Retention of Participants in Year Up's Professional Training Corps." Boston, MA: Project Evident. July 7, 2021. www.projectevident.org/updates/2021/7/26/improving-academic-success-and-retention-of-participants-in-year-ups-professional-training-corps.

Brooks, J. L., Boulay, B. A., and Maynard, R.A. "Empowering Practitioners to Drive the Evidence Train: Building the Next Generation of Evidence." Boston, MA: Project Evident. March 27, 2019. https://static1.squarespace.com/static/58d9ba1f20099e0a03a3891d/t/5cf81b7b4e259a0001c7a4a5/1559763835719/Empowering+Practitioners+to+Drive+the+Evidence+Train_+Building+the+Next+Generation+of+Evidence.pdf.

Darroch, J. E., Landry, D. J., and Singh, S. "Changing Emphases in Sexuality Education in U.S. Public Secondary Schools, 1988–1999." *Family Planning Perspectives* 32, no. 5 (2020): 204–11, 265. PMID: 11030257.

Devaney, B., Johnson, A., Maynard, R., and Trenholm, C. *The Evaluation of Abstinence Education Programs Funded Under Title V Section 510: Interim Report.* Princeton, NJ: Mathematica Policy Research, Inc. 2002. https://eric.ed.gov/?id=ED471712.

Donaldson, S. I., Christie, C. A., and Mark, M. M. (Eds.). *Credible and Actionable Evidence: The Foundation for Rigorous and Influential Evaluations.* Los Angeles, CA: Sage Publications. 2015.

Farley-Ripple, E., & Jones, A. R. (2015). Educational Contracting and the Translation of Research into Practice: The Case of Data Coach Vendors in Delaware. *International Journal of Education Policy and Leadership* 10, no. 2, n2. https://files.eric.ed.gov/fulltext/EJ1138590.pdf.

Fedorowicz, M., and Aron, L. Y. *Improving Evidence-Based Policymaking: A Review.* Washington, DC: Urban Institute. 2021. www.urban.org/sites/default/files/publication/104159/improving-evidence-based-policymaking-a-review_0.pdf.

Fein, D., Maynard, R., Baelen, R., and others. *To Improve and to Prove: A Development and Innovation Study of Year Up's Professional Training Corps.* Rockville, MD: Abt Associates Inc. 2020.

Gueron, J. M., and Rolston, H. *Fighting for Reliable Evidence.* New York: Russell Sage Foundation. 2013. https://files.eric.ed.gov/fulltext/ED610239.pdf.

Haskins, R., and Margolis, G. *Show Me the Evidence.* Washington, D.C.: Brookings Institution Press, 2015.

Honig, Meredith I., and Venkateswaran, N.. "School-Central Office Relationships in Evidence Use: Understanding Evidence Use as a Systems Problem." *American Journal of Education* 118, no. 2 (2012): 199–222. www.journals.uchicago.edu/doi/full/10.1086/663282?casa_token=-u6UPJnBiToAAAAA:McLtNdE4zBDz4unl7v1KL25UgBoJNmuIeBHehp13-PGuRCm0r0-wYT_OUzpeXRnc-ylCpok9TqA.

Martin, J., and Stock, E. *Examining Family Playlists' Impact on Student Social Emotional Learning and Science Mastery through Short-Cycle RCTs.* Boston: Project Evident. 2021. https://projectevident.org/wp-content/uploads/2022/08/PowerMyLearningActionableEvidenceCaseStudySept21.pdf.

Maynard, R. "Teenage Childbearing and Welfare Reform: Lessons from a Decade of Demonstration and Evaluation Research." *Children and Youth Services Review* 17, no. 1–2 (1995): 309–32. https://citeseerx.ist.psu.edu/viewdoc/download?doi=10.1.1.574.7021&rep=rep1&type=pdf.

Maynard, R. A. "Review of *Fighting for Reliable Evidence* by J. M. Gueron and H. Rolston." *Journal of Policy Analysis and Management* 34, no. 1 (Winter 2015): 243–47. www.jstor.org/stable/43866094?casa_token=Qhghtt23KlgAA AAA:lRAs6djZQOf3K6cRxuCPGi8c-8BM3ATmKrhU5enZgnutQu9bq bTi3CPepp77WyZPHdHrdF53X8YfIOjBm23zJ7VnCSc5oxtuEvT K7YHRqVV4_SzfP7g.

Maynard, R., and Rangarajan, A. "Contraceptive Use and Repeat Pregnancies among Welfare-Dependent Teenage Mothers." *Family Planning Perspectives* (1994): 198–205. https://aspe.hhs.gov/sites/default/files/private/pdf /74136/report.pdf.

Maynard, Rebecca A. "The Effects of the Rural Income Maintenance Experiment on the School Performance of Children." *American Economic Review* 67, no. 1 (1977): 370–75. https://aspe.hhs.gov/sites/default/files/private /pdf/74136/report.pdf.

Ott, M. A., and Santelli, J. S. "Abstinence and Abstinence-Only Education." *Current Opinion in Obstetrics & Gynecology* 19, no. 5 (2007): 446–52. https://doi.org/10.1097/GCO.0b013e3282efdc0b.

Penuel, W. R., Briggs, D. C., Davidson, K. L., Herlihy, C., and others. "How School and District Leaders Access, Perceive, and Use Research. *AERA Open* 3, no. 2 (2017). 2332858417705370. https://journals.sagepub.com/doi /pdf/10.1177/2332858417705370.

Polit, D. "Barriers to Self-Sufficiency and Avenues to Success among Teenage Mothers." Princeton, NJ: Mathematica Policy Research, Inc. 1992. TPD# 2584. https://aspe.hhs.gov/reports/barriers-self-sufficiency-avenues-success -among-teenage-mothers.

Quint, J. C., and Riccio, J. A. "The Challenge of Serving Pregnant and Parenting Teens. Lessons from Project Redirection." 1985. https://files .eric.ed.gov/fulltext/ED275777.pdf. Updated version 1988. https:// files.eric.ed.gov/fulltext/ED299444.pdf.

Levine, R. A., Watts, H., Hollister, R. G., Williams, W. and others. "A Retrospective on the Negative Income Tax Experiments: Looking Back at the Most Innovative Field Studies in Social Policy. In *The Ethics and Economics of The Basic Income Guarantee*. Ashgate, 2005. https://works.swarthmore .edu/fac-economics/347/.

Trenholm, C., Devaney, B., Fortson, K., and others. *Impacts of Four Title V, Section 510 Abstinence Education Programs, Final Report*. Princeton, NJ: Mathematica Policy Research. 2007. www.mathematica.org/publications /impacts-of-four-title-v-section-510-abstinence-education-programs.

Tseng, V., and Coburn, C. "Using Evidence in the US." In *What Works Now*, edited by A. Boaz and H. Davies, 351–68. Policy Press: 2019. https://

wtgrantfoundation.org/library/uploads/2019/09/Tseng-and-Coburn-_What-Works-Now.pdf.

York, P. "Generating On-Demand, Actionable Evidence for Front-Line Practitioners at First Place for Youth and Gemma Services. Boston, MA: Project Evidence. 2021. www.projectevident.org/updates/2021/7/26/generating-on-demand-actionable-evidence-for-front-line-practitioners-at-first-place-for-youth-and-gemma-services.

Zhang, X., Griffith, J., Pershing, J., Sun, J., and others. "Strengthening Organizational Capacity and Practices for High-Performing Nonprofit Organizations: Evidence from the National Assessment of the Social Innovation Fund—A Public-Private Partnership. *Public Administration Quarterly* (2017): 424–61. www.jstor.org/stable/pdf/26383392.pdf?casa_token =yLdRkiQ4D_AAAAAA:7-B5AdFYdjNLX2HyCeIKgeZNT3dDaWT KtaPTj3oVPfq4lqUZJRg-ZG2BQ_xkHs5iRo7cQA2jFGNmFyYgkC-pt 6TIGRuYjU2ZTntrBMBdY17lvAcufWQ.

YEAR UP

IMPROVING ACADEMIC SUCCESS AND RETENTION OF PROFESSIONAL TRAINING CORPS PARTICIPANTS

JESSICA BRITT, DAVID FEIN, REBECCA A. MAYNARD, AND GARRETT WARFIELD

Year Up and researchers at both Abt Associates and the University of Pennsylvania collaborated to come up with evidence-informed solutions to staff concerns about academic challenges faced by participants in the Year Up's Professional Training Corps (PTC) program. The collaboration led to the launch of a "mini-study" using random assignment and extant data to explore signals of and impediments to participants' success in their college courses and to devise and rigorously test strategies for more quickly identifying and addressing those impediments.

With support from the study team, PTC staff identified the improvement strategies for testing and then implemented those strategies in three sites with a random subset of participants in two successive enrollment cohorts. Other participants received the usual coaching strategies and supports. Staff working with the improvement strategies group were encouraged to alter their strategies between cohorts 1 and 2 based on their experiences with the first cohort.

Year Up is a national nonprofit organization with a mission to close the opportunity divide by ensuring that young adults gain the skills, experiences, and support that empowers them to reach their potential through careers and higher education. Year Up's core model serves young adults age eighteen to twenty-four with a high school degree or equivalent who are, otherwise, disconnected from higher education and quality job opportunities. Participants enroll in a year-long program. They spend the first six months in classroom training in basic and technical skills and the second six months in internships with corporate partners. The goal is for these internships to culminate in full-time jobs related to their technical training.

The PTC is a second-generation version of Year Up's core program and operates in partnership with career-focused colleges with the goal of achieving impacts comparable to the core program but at a substantially lower cost. The Year Up team for the improvement study was led by Dr. Garrett Warfield, chief research officer, and Jess Britt, senior director of research and evaluation. **Abt Associates** is a research and evaluation organization that specializes in applying systems analysis and social science techniques to social and economic problems. Their research, monitoring, and evaluation practice is known for interdisciplinary approaches. The Abt study team included Dr. David Fein, principal associate for social and economic policy; Azim Shivji, senior analyst; and Phomdaen Souvanna, research associate. The research team from the **University of Pennsylvania** included Dr. Rebecca Maynard, professor of education and social policy, and Rebecca Baelen, a PhD student.

The PTC evaluation grew out of a shared experience with the Pathways for Advancing Careers and Education (PACE) evaluation, a large-scale, long-term impact evaluation commissioned by the Administration for Children and Families. Abt's PACE evaluation team, led by Dr. Fein, actively solicited Year Up's participation in the PACE evaluation as one of eight fully developed, seemingly high performing career and technical programs targeting low-income adults.

A PRACTITIONER-CENTERED APPROACH

The PTC evaluation arose from concerns that poor academic performance was impeding program retention and completion. In addition to Year Up's interest in generating good outcomes for participants generally, the PTC financial model calls for over 90 percent of program revenues to come from employer-sponsored internships for participants during the second six months of their program participation. Year Up staff believed that more timely identification of and support for students struggling with their academic courses could make a meaningful difference. Thus, Year Up was especially keen on an evaluation that could inform strategies to strengthen its participant coaching so as to ensure timely identification of academic challenges and provision of support to address them.

Although the technical methods used in the PTC evaluation were not novel, the way the evaluation partners approached the work differed from a typical evaluation. The process was practitioner-centric. The interests and needs of Year Up staff determined the research questions—namely, bolstering academic performance and boosting program retention—and all parties relied on feedback loops to provide strategic tweaking of plans and timely use of findings.

Codeveloping an Evidence Agenda with Practitioners and Students. The partners winnowed a long list of concerns to three focal issues for study. The winnowing process entailed a series of stakeholder engagements, careful review of readily accessible historical program data, and multiple brainstorming sessions with Year Up's National leadership team.

The evaluation team conducted interviews and focus groups with a diverse array of PTC stakeholders, including Year Up national and local staff and college and employer partners. These conversations had several purposes. First, they established a connection with key players. Second, they provided contextual information useful in planning, implementing, and interpreting findings from the study. Third, they elicited valuable information on key stakeholders' priorities for program improvement. Finally, they were a source of information to help interpret study findings as they emerged.

Selecting the Research Methods. The team used random assignment and extant data to test field-generated, low/no-cost strategies for improving participants' academic success. This strategy was an efficient way to generate highly credible evidence on the effectiveness of the improvement strategies program staff designed.

Four principles guided these choices. First, since a major challenge for the PTC was its high operational costs, the strategies tested needed to be inexpensive to implement. Second, program and evaluation staff needed to agree that, if effective, the strategies could be implemented successfully in all PTC locations. Third, the evaluators aspired to be able to judge the success of the program in near real-time to support continuous improvement. Fourth, it needed to be possible to test the strategy in a manner that would produce highly credible evidence of effectiveness.

Designing Improvement Strategies for Testing. The evaluators led the research design process, working closely with their Year Up partners and local site directors in planning the improvement strategies to be tested, selecting sites, enrolling and randomizing the sample, and interpreting and disseminating findings. They supported the PTC site staff on detailed planning and implementation of the improvement strategies to be tested and assumed responsibility for monitoring their implementation. They also collaborated on the design and coordination of data collected from PTC staff and participants to support the evaluation.

Based on a brief qualitative assessment, the team hypothesized that academic performance was a major contributor to high program attrition and, thus, strengthening, and monitoring support to struggling students could be an effective response. The team then selected three local PTC programs that differed in the nature and degree of challenges faced to develop and test low-cost, high-promise strategies for improving academic success and program retention during the first six months of the program (i.e., the classroom training phase).

The improvement strategies tested were designed by the PTC program staff in the three study sites with guidance from Year Up National and support from the evaluation team. In addition to working directly with program staff to design the strategies, the team applied insights gleaned from focus groups, interviews, and reviews of the literature.

Transparency and Respect in Participant, Stakeholder, and Site Staff Engagement. The study benefited from the fact that all Year Up staff serve as coaches to participants. As a result, Year Up staff working on the evaluation had first-hand knowledge of local operational practices, which enabled those using the enhanced coaching practices to more easily determine which strategies tried in cycle 1 were not helpful and which were. Their input was critical to their work with the study team to adjust the enhanced coaching strategies for testing in cycle 2 and for making meaning of the final study findings. For example, the study team worked closely with local program staff to assemble a binder of tools coaches found useful for training and supporting staff and other coaches to more effectively identify and support students who were struggling during this academic phase of the program.

A Stable Research Collaboration with Qualified and Complementary Partners. The stability and complementarity of roles within the research partnership helped it generate high-value output. Senior members of the team from each of the three organizations brought extensive program evaluation expertise.

Year Up leads guided work on identifying the focal topics for research, coordinating with Year Up national and the local PTC sites on implementation issues, and planning dissemination and follow-up on the research findings. Abt's staff led a number of technical and administrative tasks—such as coordinating various studies that were co-occurring, IRB and data management planning, and data processing and analysis, while the UPenn researchers assumed primary responsibility for the design and implementation of the improvement study as well as a cost analysis of one PTC program.

Project Cost and Efficiencies. The PTC evaluation is part of a larger evaluation initiative. Funding for a suite of evaluations conducted under two grants available to this partnership totaled a little over $2 million. The study team estimated that the improvement study focused on academic monitoring and supports cost about $400,000 in external funds—an amount that would have been larger had it not been for the high-functioning partnership that provided opportunities to economize by coordinating efforts across various partnership activities.

RESULTS

Participants in the improvement strategies group were substantially more likely than their counterparts in the usual strategies group to successfully complete the learning and development phase of the program and to enroll in one or more college courses during the internship phase of the program. On average, participants who received the improved coaching strategies had nearly 10 percentage points higher rates of retention through the end of the Learning and Development (L&D) phase of the program than their counterparts who received the usual coaching (see figure 10.4-1). Even more notable is the fact that retention gains were much larger for the second cycle of testing (14 percentage points) than the first (4 percentage points). Similarly, a significantly higher percentage of those in the improvement strategies group than their counterparts in the usual strategies group continued to enroll in college courses after entering internships (67 versus 54 percent overall, and 84 versus 64 percent for the second cohort).

Coaches working with participants assigned to the improvement strategies group reported substantial changes in their approach to coaching. They were nearly four times more likely than coaches working with those engaging the usual strategies group to spend most of their coaching time on academic issues (43 versus 11 percent) and more than three times more likely to refer par-

FIGURE 1.4.1 Retention of Participants in Improvement Strategies Group versus the Usual Strategies Group through the End of the Learning and Development Phase by Testing Cycle. *Source*: Fein and others (2020). Data on retention from Year Up's management information system.

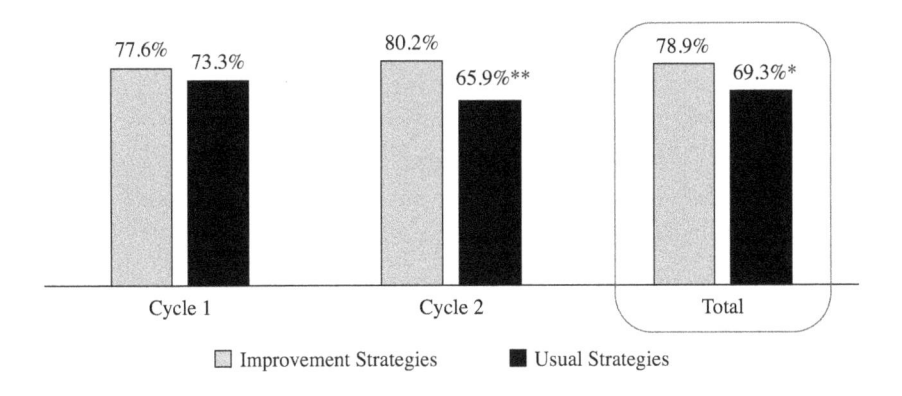

	Cycle 1	Cycle 2	Total
Improvement Strategies	77.6%	80.2%	78.9%
Usual Strategies	73.3%	65.9%**	69.3%*

ticipants to tutoring (46 versus 14 percent) (Maynard and others 2018). They were only slightly less likely than coaches working with the usual strategies group to report spending their coaching time on generic Year Up topics commonly addressed during group coaching. Notably, participants assigned to the improvement strategies group rated the quality and level of support received from PTC staff higher than did their counterparts whose coaches followed the program's usual strategies.

Year Up shared the main study findings with both its national and local site staff at critical junctures during evaluation. The primary method of sharing study findings and recommendations was timely, end-of-study online briefings of about an hour in duration. These typically were preceded by a brief pre-read summarizing the study, findings, and recommendations and a post-read document providing more detail—both formatted in PowerPoint slides to honor the strong preference of Year Up's practitioners for PowerPoint presentations over technical white papers.

The team also shared emerging findings with Year Up staff on an as-needed basis to support strategic decisions. For example, the team briefed the national staff on the findings of field efforts aimed at prioritizing evaluation topics and broad evaluation plans, and provided high-level general feedback to national and local staff in conjunction with other phone or in-person encounters as requested.

The main evaluation report covering the full suite of studies on the PTC (Fein et al. 2020) included four recommendations to Year Up and other training providers. The central recommendation was to modify the program's approach to coaching to include a deliberate focus on academic goals, achievements, and challenges. A second recommendation was to offer formal staff training on academic coaching strategies. The emphasis would be on improving early identification of academic challenges and devising timely strategies to help participants address them. A third recommendation was for program staff to be on the alert for additional ways of identifying academic challenges that would complement asking participants directly. The final recommendation was for Year Up national staff to consider other applications of the evaluation-based improvement process used in this study.[1]

Year Up's Response to the Study Findings. Year Up national staff and local staff involved in the study reported liking the approach used in this evaluation, citing its collegial and relatively low-burden nature. More importantly, Year Up national and participating PTC programs are using evidence from the evaluation to improve practice. Staff at study sites reported they still use the coaching practices developed and tested in the study, as well as the system they created for documenting and sharing participant information. Using these coaching practices has improved academic oversight, facilitated early detection of academic challenges, and increased retention in the study sites. Year Up also rolled out features of the improved coaching strategies, including a binder of tools assembled as part of the evaluation effort, to all its programs nationwide (Baelen, Britt, and others 2020), and staff have continued to iteratively adapt these shared materials to local contexts.[2]

The COVID-19 pandemic necessitated major shifts in coaching strategies to accommodate online delivery. The study team has not been able to assess the degree to which the enhanced focus on academic issues has been embodied in the online coaching formats or to isolate potential confounding influences of the myriad operational and contextual shifts arising due to the pandemic. Thus, the applicability of the study findings to the current environment is unclear.

REFLECTIONS

This study differed from the typical program evaluation in several important ways. First, it focused squarely on issues of immediate concern to practitioners—in this case, Year Up leadership and staff. Second, the study's success owes in large measure to a well-functioning partnership among program management, student-facing staff, and the evaluation team. Third, the evaluation team—comprised of Year Up staff and external researchers—was able to apply its experience and tools in a manner that produced highly credible evidence with minimum burden on program participants or staff. Fourth, the team provided timely feedback from the study in formats useful to frontline program staff.

This work was made possible in large part due to flexibility on the part of funders. One funder, the Social Innovation Fund, allowed the team to restructure the research agenda to delay implementation of a traditional impact evaluation of the PTC program in the Philadelphia site so they could "braid" the evaluation they funded through an IES Development and Improvement grant. The flexibility in funding and evaluation design allowed the team to include three programs in the "rapid-cycle" improvement study prior to launching the summative evaluation.

The resulting improvement study proved useful to the program and offers an example of practitioner-research partnerships that yield credible and actionable evidence. This type of evidence generation was much more valuable than if the partnership had prioritized a traditional impact evaluation of a program model still working to address known performance shortfalls. It also was more valuable than purely descriptive and anecdotal evidence.

Site staff drove decisions about the improvement strategies for academic monitoring and support that would be tested. Within broad guidelines, Year Up site staff were empowered to design strategy changes that meshed with their local contexts. They also were encouraged to modify their strategies for the second cycle of testing based on experiences in the first cycle, reinforcing the notion that they had been invited to participate in a program improvement effort. The evaluation team used light-touch monitoring of the enhanced coaching and supports during the study period but strategically timed monitoring to encourage continuous reflection by program staff while also yielding adequate contextual information to support the study.

The external evaluation team drew heavily on its Year Up partners for guidance in designing and communicating with local staff. This guidance included counseling in the program language and in protocols for meeting preparation, conduct, and follow-up (for example pre-reads; tailored protocols; timely and conventional formats for follow-up). Products of the evaluation included not only conference calls and post-reads to present study findings but also a compendium of tools that were assembled, tailored, or otherwise created by program staff working with participants in the improvement strategies group. This compendium has since been adapted for use throughout Year Up as part of its adoption system-wide of lessons from the study.

Many factors contributed to the success of this partnership. Three were especially critical. First, all parties shared a commitment to using the available study resources to help Year Up improve its ability to close the opportunity divide for young adults from disadvantaged backgrounds. Second, all parties were willing and able to adjust their project roles and responsibilities as needed to keep the effort on track. Third, all parties had tremendous respect for and trust in one another and for the youth whose welfare was at stake.

NOTES

1. The evaluation work discussed in this paper was supported by grants from the Social Innovation Fund and the Institute of Education Sciences (IES Grant Number R305A150214). This article does not necessarily reflect the views of the funders. The primary study reports include two for the Social Innovation fund: one on early implementation of the Philadelphia Program (Fein and Maynard 2015) and one expanded on in this improvement study and including a cost analysis of the PTC (Maynard and others 2018). Two final products to the Institute of Education Sciences included a summative report (Fein and others 2020) and a compilation of tools and guidance from documents developed or adapted for use in the improvement study discussed here (Baelen and others 2020). Arnold Ventures has supported longer-term follow-up of the study sample, which will be reported on in the future.

2. Britt and others (2021) provides a fuller discussion of methods, findings, and resulting actions of the improvement study.

REFERENCES

Baelen, R., J. Britt, R. Maynard, P. Souvanna, and G. Warfield. "To Improve and To Prove: Tools to Improve Academic Monitoring and Support for Young Adults." Rockville, MD: Abt Associates Inc. 2020. https://drive.google.com/file/d/1dPmmPC5wNo3i39mdTbmm4sKkM3x_9qfo/view.

Britt, J., D. Fein, R. Maynard, and G. Y. Warfield. "Improving Academic Success and Retention of Participants in Year Up's Professional Training

Corps." Boston, MA: Project Evident. 2021. https://static1.squarespace.com/static/58d9ba1f20099e0a03a3891d/t/61000d46065c2d3b39c0cdbc/1627393354762/Year+Up+Actionable+Evidence+Case+Study+July21.pdf.

Fein, D., R. Maynard, R. Baelen, A. Shivji, and P. Souvanna. "To Improve and to Prove: A Development and Innovation Study of Year Up's Professional Training Corps." Rockville, MD: Abt Associates Inc. (2020). https://www.abtassociates.com/files/insights/reports/2021/tip-report-1-21-21.pdf.

Fein, D. J., and R. A. Maynard. "Implementing Year Up's Professional Training Corps in Philadelphia: Experience in the First Year and a Half." Bethesda, MD: Abt Associates Inc. (2015). https://drive.google.com/file/d/1s3r-YvJwq7jhwUwvBd1BXXuvk0NA2hwF/view.

Maynard, R., R. Baelen, P. Souvanna, D. Fein, and A. Shivji. "Final Evaluation Report for Year Up's Professional Training Corps Program in Philadelphia." Rockville, MD: Abt Associates. (2018). https://americorps.gov/sites/default/files/document/2018_10_25_Greenlight_Year_Up_PTC_PHL_SIF_Final_Report_FINAL_508_ORE.pdf.

NURSE-FAMILY PARTNERSHIP

ALWAYS A WORK IN PROGRESS

MANDY A. ALLISON, GREGORY TUNG, AND DAVID OLDS

Nurse-Family Partnership (NFP) is a national nurse home visiting program that serves first-time, low-income mothers and their children until the child reaches age two. Three randomized clinical trials conducted in different communities and contexts have shown that NFP is effective at improving maternal and child health and life-course outcomes. In spite of NFP's grounding in randomized clinical trials with decades of follow-up, we consider NFP to always be a work in progress. This means the program will continuously require additional formative development and testing of clinical innovations and expansions to reach and better serve those most likely to benefit. The practice and research worlds are sometimes seen as not compatible or being somehow misaligned. For our practice-based activities to have the greatest impact, we must effectively incorporate scientific evidence into practice design and decision making. For research to be relevant, it must be grounded in the reality of the practice world and produce findings that are relevant and actionable. Instead of being at odds with one another, practice and research efforts will be stronger if the two disciplines are effectively integrated. In this chapter, we illustrate our approach to integrating practice and research efforts by outlining the way we have approached the development and testing of a version of NFP for pregnant women who have had previous live births, or multiparous mothers.

Our work began with funding from the Maternal Infant and Early Childhood Home Visiting (MIECHV) program in collaboration with Tribal Nations partners who encouraged flexibility surrounding program eligibility that respected cultural traditions and norms.[1] Given continued requests from community partners for NFP to serve multiparous mothers and their children, we have embarked on a series of research projects to adapt and refine NFP and test the effectiveness of NFP for this population. Serving multiparous mothers in NFP on a large scale will require a significant investment of resources from community partners implementing NFP, philanthropic organizations, and state and federal governments. Therefore, strong evidence for effectiveness is needed to determine whether these resources for delivery of NFP should be directed toward multiparous mothers, particularly when there is a risk of diverting resources from first-time mothers for whom NFP is known to be effective.

INNOVATION DEVELOPMENT AT NFP

In 2013, Dr. David Olds and colleagues published an article describing the process used to continue to study and improve NFP in community practice. Figure 1.5.1 shows this model for innovation development in NFP. In this piece, we describe the formative development and pilot testing of the innovation—NFP for mothers with previous live births. Our next steps are to conduct rigorous testing of the version of NFP for mothers with previous live births with a quasi-experimental design study and a randomized clinical trial.

To provide a real-world perspective to program innovations, NFP has organized an Innovations Advisory Committee (IAC) comprised of over 130 NFP home visitors, supervisors, and staff from around the United States who volunteer their time to address specific topics, such as substance use disorder, cultural awareness, and maternal morbidity and mortality. At the beginning of the formative study of NFP for mothers with previous live births, an IAC was formed to advise the research team regarding all aspects of study design and implementation.

FIGURE 1.5.1 **NFP Model for Innovation Development**

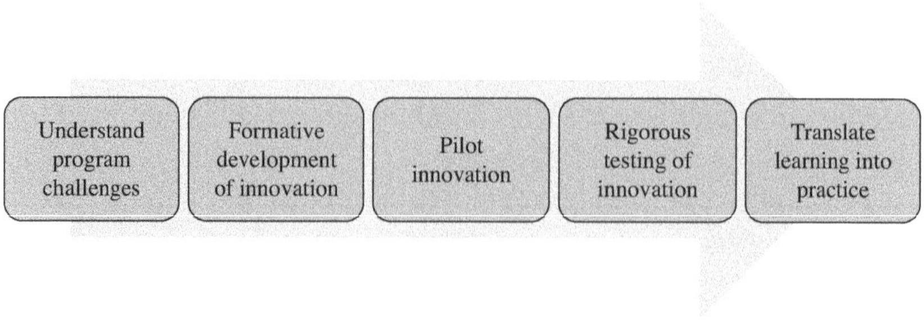

Understand program challenges → Formative development of innovation → Pilot innovation → Rigorous testing of innovation → Translate learning into practice

OBJECTIVES OF FORMATIVE STUDY OF NFP FOR MOTHERS WITH PREVIOUS LIVE BIRTHS

From 2017 through January 2021, we conducted a formative study with the following objectives developed in collaboration with NFP implementing partners: 1) determine the feasibility and learn ways of improving implementation of NFP for multiparous mothers experiencing risk factors for poor birth, parenting, and child development outcomes;[2] 2) evaluate existing criteria, referral sources, and process for defining and recruiting the target population of multiparous mothers; 3) assess and enhance collaboration and coordination of care between NFP and community stakeholders; 4) learn from NFP teams' experiences serving multiparous mothers and multiparous mothers' experiences in NFP to identify and strengthen program elements critical to serving this population; and 5) identify and integrate successful practices for serving multiparous mothers to inform the creation of program elements and educational materials.

STUDY SITE SELECTION

All community partners currently implementing NFP were invited to participate in the formative study. Interested sites completed a brief application, and sites were evaluated for readiness to participate in the study using specific criteria.[3] A total of thirty-one sites in fifteen U.S. states applied to participate and met the criteria.

DATA COLLECTION

We conducted a series of interviews from selected sites to understand the experiences of nurse home visitors, nurse supervisors, and other staff members in serving multiparous mothers. We also conducted interviews of key organizational partners, such as primary care providers and social services. The purpose of these interviews was to understand the challenges, barriers, and opportunities faced by NFP in the implementation of the program with multiparous mothers and to understand how collaboration with organizational partners might address these challenges. We interviewed multiparous mothers referred to NFP who had declined to enroll or had enrolled in the program and dropped out. Our goal in conducting these mother/client interviews was to learn: 1) how clients who participate in NFP experience the program and what important factors shape their perspectives, what they value in the program, what they do not value in the program, and how these factors influence their retention; and 2) why some multiparous mothers chose not to participate in NFP after being referred, paying attention to variation in non-participation by institutional partners such as primary care doctors and human service agencies. The findings from these interviews were used to develop recommendations to improve enrollment, engagement, and retention of multiparous mothers.

We gathered quantitative data from multiple sources, including data collected as part of routine program implementation and housed by the NFP National Service Office's data system, referral spreadsheets developed specifically for the formative study, and surveys of nurse home visitors and supervisors. Our goal was to understand the characteristics of multiparous mothers who were referred to NFP; the characteristics of those who chose to enroll; NFP sites' collaboration with healthcare providers; and the differences in clients with previous live births versus clients who were first-time parents.

In addition, we gained a wealth of information from monthly conference calls with the participating NFP sites. These calls were facilitated by a nurse consultant with extensive NFP experience. On these calls, the consultant provided updates about the formative study and nurses shared experiences in serving multiparous clients and recommendations for additional resources and adaptations. The consultant also reviewed each site's referral and enrollment numbers and discussed strategies for collaborating with referral partners.

RESEARCH AND PRACTICE INTEGRATION

Updates and findings from all data collection efforts were reviewed by the research team at weekly meetings to validate findings, adjust research approaches to best meet the needs of study sites, and inform ongoing program adaptations to more effectively serve multiparous clients. The research team, including the study nurse consultant, created brief research summaries. The study nurse consultant shared research updates and the research summaries with the IAC subcommittee at their monthly meetings and returned the IAC subcommittee's feedback to the research team. The research team also shared the research summaries and conducted webinars with the study sites to inform them of research findings on an ongoing basis. Finally, the study nurse consultant facilitated ongoing communication between the research team and participating NFP sites.

CHALLENGES AND RESPONSES

We encountered some challenges that are likely to be encountered by others conducting research in real-world settings. First, we experienced tension between the time required to conduct rigorous data collection and analysis and the need for rapid feedback for continuous quality improvement. For example, we wanted to use rigorous qualitative methods that reduce the likelihood of bias and ensure an accurate synthesis of people's perspectives and experiences to understand how to adapt NFP for women with previous live births. However, we also wanted to be able to use what we were learning from sites participating in the formative study to suggest ideas for improvement or changes they could make in real time. Our response to this tension was to build in intentional elements to enhance the translational nature of the project. We prioritized analysis that was especially relevant to ongoing program improvement efforts and, when needed, generated preliminary analysis to inform time-sensitive program improvement efforts.

Second, NFP sites participating in the formative study experienced competing demands, including the research team's needs and changing requirements dictated by funders, supporting community agencies, and state and local governments. Our response to this was to use both formal qualitative and quantitative methods and informal data gathering through monthly meetings facilitated by the nurse consultant to understand sites' competing demands and how this may have affected their ability to serve multiparous mothers and participate in this research. We communicated to the NFP sites that whatever they were experiencing was part of our learning.

Third, at the beginning of the study, the research team did not clearly communicate the objectives of the formative study and what this type of study could and could not tell us; therefore, some NFP sites had unrealistic expectations regarding the study results. For example, some sites expected to learn whether NFP was effective for multiparous mothers based on our findings from the formative study. We had to clarify that determining effectiveness or impact requires a comparison group and doing so would be addressed at a later phase in the research process. We learned about the importance of setting clear expectations at the beginning of the study and have developed processes for ensuring we do this with our research going forward.

Finally, many project partners have expressed a strong desire to move forward with expanding eligibility of NFP for multiparous mothers before we have the results of the next phase of our research that will measure the effectiveness of NFP for this new population. This desire is based on positive personal experiences serving multiparous mothers and their families, belief that NFP will be effective for this population, and a sense of urgency to reach and serve more families facing adversity. Our response has been to acknowledge partners' strong desire to do good and to communicate that we do not want to displace service to first-time mothers, among whom we know NFP does good based on the original trials and other studies of NFP in community practice.

FORMATIVE STUDY FINDINGS

The thirty-one NFP sites participating in the formative study enrolled 1,571 pregnant women with a previous live birth.[4] These enrolled women represent 37 percent of the multiparous mothers referred to NFP as part of the formative study. This "conversion rate" from referral to enrollment for multiparous mothers is similar to the conversion rate of 35 percent for first-time mothers. Sites that routinely employ a "warm handoff" (that is, a referral made when the potential client hears about the program from a trusted source and has the opportunity to ask questions) had conversion rates that were higher. Multiparous mothers experienced more nurse-assessed risks and were referred to needed services more frequently compared to first-time mothers. Despite experiencing more risks and having more needs than first-time mothers, rates of program retention were higher for multiparous mothers, with an 80 percent retention through pregnancy, 63 percent through child age six months, 55 percent through child age twelve months, and 50 percent through child age eighteen months compared to 77 percent, 58 percent, 48 percent, and 37 percent, respectively, for first-time mothers.

While the formative study did not include a comparison group of similar women who did not receive NFP, we examined outcomes based on data collected by the NFP nurse as part of routine program delivery. The proportion of multiparous mothers enrolled in NFP who reported smoking decreased from 17.2 percent at intake to 14.0 percent at thirty-six weeks Estimation of Gestational Age (EGA). Among primiparous mothers enrolled at the same sites during the same time period, the proportion of those who reported smoking was 7.9 percent at intake and 6.7 percent at thirty-six weeks EGA. The proportion of multiparous mothers who delivered preterm was 14.2 percent compared to 12.4 percent of first-time mothers. Among multiparous mothers, 0.10 percent reported that the index child had been admitted to the hospital for injury and 0.10 percent for ingestion compared to 0.15 percent for injury and 0.07 percent for ingestion among children of first-time mothers. These data suggest that outcomes for multiparous women are similar to those for first-time mothers served by NFP. However, the lack of an equivalent comparison group of families who did not receive NFP prevents us from determining if NFP is truly effective for this population.

NFP nurses and community providers described the need for stronger collaborative relationships with community partners to better serve multiparous clients. Nurses and mothers also recommended additional nurse resources and training to effectively meet multiparous families' needs. Flexibility with visit schedule, length, location, and content was particularly important for engaging and retaining multiparous clients. At its core, NFP is intended to activate a mother's instinct to nurture her child. NFP nurses report that some first-time mothers express this verbally as a desire to "do right by this child." For some multiparous mothers, they expressed a desire to "get it right this time." We interpret this as an acknowledgment that experiences with their previous children might not have gone well but they desire a better experience with their current pregnancy. This consistency in how the NFP program activates a mother's desire to nurture her children suggests it may also produce effects with multiparous mothers.

APPLICATION OF RESULTS AND NEXT STEPS

Based on our findings from the formative study, the NFP nursing and education teams have developed additional education for nurses,[5] materials for nurses to use with their multiparous clients,[6] and an online learning community to support nursing teams serving multiparous mothers and their families. In addition to developing additional support for nursing teams serving multiparous clients, we are continuing to conduct research to determine the effectiveness of NFP for this expanded population. We have partnered with three sites that participated in the formative study to conduct a quasi-experimental design study to measure the impact of NFP for multiparous mothers. Using data from health plans' billing and electronic medical records, we will compare pregnancy, maternal, and child health outcomes for women who received NFP with a similar group of women, matched on sociodemographic and health characteristics, who did not receive NFP. We also are working with collaborators in Florida to explore the use of state-level data, including perinatal and infant risk screens, birth certificates, and centralized intake and referral system data to measure the reach and impact of NFP for multiparous mothers. Finally, we are pursuing funding to support a randomized clinical trial of NFP for multiparous mothers.

While a randomized clinical trial remains the gold standard for determining effectiveness of an intervention and findings from randomized clinical trials are the foundation for the current NFP program implemented in the United States and other countries, conducting these trials requires extensive resources, including money and time. The quasi-experimental study has increased risk of noncomparable intervention and control groups but has the advantage of requiring less time and money to complete compared to randomized clinical trials. We will use the findings from the quasi-experimental study to guide our next steps while awaiting the results of the trial. We have developed a timeline and a plan for how we will incorporate our findings from the studies into plans for NFP program implementation.

REFLECTIONS

In this project, we aligned program adaptation with research by embedding research and the dissemination of findings to inform ongoing program adaptation and improvements into the project. This goes beyond typical

Alignment with NextGen Evidence Principles

Principle	In This Case...
Centers on Practitioners *Grounded in practitioner needs, challenges, learning questions, and decisions. Examples: allows practitioners to make evidence-informed decisions in a timely manner; reflects the context in which practitioners operate; rigor is aligned to practitioner needs.*	The motivation for this translational research project came from needs communicated by practitioners. We also structured the project so that data gathering, analysis, results, and implications were all conducted within a real-word context to maximize the relevance of our results to practitioners. We included translational components in the research project such as practitioner/expert guidance and validation of the research process and results as well as dissemination products specifically intended to meet the needs of practitioners to inform program improvement efforts.
Embraces an R&D Approach *Builds practitioner capacity to continuously build evidence, advance a learning agenda, and translate evidence into action.*	Our entire project was grounded in one of the core principles of NFP, which is being evidence-based. We embrace the individual expertise and perspective of practitioners and researchers. We structured this project to integrate those perspectives and expertise into a project that adhered to high standards of research, practice, and the translation of evidence into action.
Elevates the Voices of Communities *Addresses the needs and challenges of communities, especially groups that face systemic disadvantages, and incorporates input from community stakeholders throughout evidence building and evaluation processes.*	We designed the qualitative portion of the project to capture diverse perspectives from community partners. This included qualitative interviews with partners from various sectors, including health care, child welfare, social services, mental health, substance treatment, and housing services. We also included qualitative interviews with multiparous mothers who participated in NFP and referred mothers who declined to enroll. Through these interviews, we incorporated the experience, perspectives, and expertise of diverse stakeholders into the project. While we learned a great deal from this effort, our project would have benefited from additional stakeholder and patient perspectives in determining the implications of our findings and the specific program improvement elements moving forward.

quality improvement efforts by enhancing the systematic collection of data and the rigor of data analysis to inform ongoing program development and improvements. More accurate data should lead to more effective decision making and, ultimately, greater program impact. The use of scientific evidence in program design and improvement is a foundational principle of NFP. The commitment to this foundational principle and a shared commitment to improve the lives of mothers and their children brought alignment between the objectives and activities of the practitioners and researchers, and we were collectively able to accomplish more as a result. Practitioners who want to be the most effective should partner with researchers to generate the best data and findings to inform their decision making. Researchers who want their work to be relevant and have an impact should partner with practitioners to have their results inform decision making. Funders and policymakers who want their priorities to be achieved should facilitate partnerships between practitioners and researchers.

NOTES

1. NFP replication in Australia with aboriginal and Torres Strait Islander people also has embraced NFP's serving multiparous mothers.

2. Risk factors include previous pre-term births, previous low birth weight infant, homelessness, mental illness, substance use, previous or current involvement with child welfare, less than high school education or GED, history of IPV, medical complications, developmental disability, and being nineteen years old or younger with the current pregnancy.

3. Criteria for participating in the formative study were: 1) secure funding for the three-year length of the study; 2) commitment from all staff and no conflicting projects; 3) commitment to collaborate with primary care and child protective services; 4) ability to enroll multiparous mothers without displacing first-time mothers; 5) demonstrated proficiency with using the NFP's client Strengths and Risks (STAR) assessment framework; 6) minimum of three full time nurse home visitors or part-time equivalents; 7) agree to meet with research consultants monthly; and 8) agree to participate in qualitative data gathering such as focus groups and interviews.

4. Enrolled women had the following characteristics: mean age of 27.8 years; 76 percent unmarried; 28 percent with less than high school completed; 25 percent Hispanic/Latinx, 39 percent Black/African-American and 49 percent white; 50 percent with depression and 27 percent with anxiety based on validated screeners. These characteristics were similar to those of enrolled

first-time mothers at the same sites except mothers with previous live births were older and more likely to be married and have a positive anxiety screen.

5. The topics for additional nurse education include case management, addressing the needs of other children in the home, engaging community partners, and addressing concerns for child maltreatment.

6. The additional materials developed for nurses to use with clients include topics such as birth planning for subsequent children, introducing a new baby to the family, parenting styles, and past experiences with breastfeeding.

BAIL PROJECT

EVALUATION AS PART OF BUSINESS STRATEGY

BRAD DUDDING AND TARA WATFORD

Most Americans recognize that the use of cash bail to detain people pretrial must change.[1] That is because there is no place in the criminal legal system where money more clearly buys justice than bail. A person accused of a crime where bail is set pays the entire bail amount set by a judicial officer or a deposit in exchange for their liberty. Tying freedom to financial ability upends the presumption of innocence, tears lives apart, and perpetuates racial and economic disparities. But while there is growing consensus that reform is needed in pretrial systems, a shared vision for what this looks like is less clear.

Founded by Robin Steinberg in 2017, The Bail Project (TBP) is using data and on-the-ground experience to create a tangible model for what a world without cash bail can look like. In this world, there is a default presumption of release and strong procedural protections to protect a person's rights and liberty if the government seeks pretrial detention. In this world, people have access to community-based resources to help them get back to court and meet other essential health, housing, and employment needs. Finally, crimes of poverty are decriminalized, and people are not subject to burdensome pretrial conditions and surveillance like electronic monitoring.

To get to this place, TBP seeks nothing less than broad adoption of its needs-based and community-based model as the foundation for reimagining pretrial justice.

In 2019, I was hired as the chief impact officer, along with chief data officer, Tara Watford, with a mandate to codify TBP's evidence-building practices to show our model is a more effective and just alternative to cash bail. We were fortunate to be joining a nonprofit with a pedigree for evidenced-based leadership. TBP is the offspring of two other start-ups founded by Steinberg that demonstrated results for reducing incarceration for marginalized individuals in the criminal legal system. Starting in 1997, The Bronx Defenders pioneered an innovative model of public defense that approached the legal representation of low-income people through a holistic lens, identifying the underlying causes of a person's criminal justice involvement and deploying interdisciplinary teams of attorneys, social workers, and advocates to address them. In 2018, a RAND study found that, over the course of ten years, this holistic model prevented over 1 million days of incarceration.[2]

But while holistic defense proved effective on many fronts, Steinberg knew that cash bail remained the decider of many people's cases, and that is where The Bronx Freedom Fund came in. Using philanthropic dollars, the Freedom Fund, created in 2007, was able to post bail for people who could not afford it, leveling the playing field and preventing pretrial incarceration. After ten years, the results spoke for themselves: not only did the vast majority of individuals return to court without having any money on the line, but over 50 percent of the cases were dismissed when people could defend themselves from a position of liberty. The resulting stories and data were critical in pushing the case for bail reform into the mainstream, and Steinberg utilized those lessons and strategies to launch The Bail Project.

After four years, this simple model, which we call Community Release with Support, is operating in twenty-seven metro areas in sixteen states across the United States. The model is defined by four essential components: 1) an individualized needs assessment conducted by TBP client advocates that documents what the client, as well as their support network, voluntarily identifies as their needs to return to court; 2) automated and personal court reminders for clients as well as free transportation assistance to and from court; 3) connections to community resources for clients to help them address self-identified needs surfaced during the intake process; and 4) community capacity building to facilitate collaboration between CBOs, and, in some cases, seed funding for local organizations that can continue the work after TBP exits a jurisdiction. These efforts allow individuals to regain their freedom and resolve their court cases with improved outcomes while reducing jail populations and mitigating the harms of wealth-based detention on low-income people and communities of color.

Since its inception, TBP staff have helped over 25,000 people return to their jobs, homes, and communities, preventing more than one million days in jail and saving almost $2 million in pretrial detention costs. We have supported clients' attendance at more than 85,000 court dates, with a court appearance rate of 92 percent, even though they have no financial obligation to us. A staggering 32 percent of TBP clients have all of their charges dismissed, and of the clients who reach a final disposition, 92 percent are not required to spend any additional time incarcerated. In short, TBP is proving that cash bail is unnecessary and unjust. Steinberg's demonstrated leadership to inextricably connect mission to results cannot be overemphasized as a driver for these outcomes. It is a vital ingredient in the recipe for creating social change and the reason Tara and I were so excited to join TBP's cause to disrupt the money bail system and challenge a system that criminalizes race and poverty.

Our evidence-building objectives were clear: 1) create a user-centered platform to reliably collect client and jurisdictional data; 2) nurture data practices with staff to optimize bailouts and improve service quality to clients, and 3) generate rigorous proof points that demonstrate our model is an effective alternative to cash bail and motivates change in pretrial systems. Tara and I were not starting this work with a blank slate. Both of us had experience at our previous organizations pursuing similar goals, and we also had the benefit of an existing business plan created by senior TBP staff and the Bridgespan Group in 2018. The business plan exhibited the DNA of all Robin Steinberg's start-ups: a deep commitment to acting on a learning agenda and building a dataset to demonstrate why the criminal legal system should change.

The business plan laid out aggressive milestones for program expansion: dramatically increasing clients served and building a rigorous evidence base over a five-year period.

> Sidebar: Based on our theory of change, TBP's learning agenda is to create strong proof points to challenge prevailing counter-narratives and show that:
>
> - Individuals can be released pretrial and will **return to court** without: a) putting up any of their own money for bail, or b) restrictive release conditions such as electronic monitoring.
> - Our model results in fairer **case outcomes** for individuals than the current system of unaffordable cash bail.
> - Our model will not, in aggregate, pose an increased threat to **public safety**, and through positive impact on clients' outcomes, may actually *improve* public safety over the long term.
> - Our model will **reduce the bias and racial disparities** that are part and parcel of the criminal legal system and disproportionately impact low-income communities and people of color.
> - Our model is more **cost effective** for jurisdictions than unaffordable cash bail or other alternatives.

The evidence strategy centered around two core sets of activities: first, build a strong internal capacity for data collection, measurement, and research; and second, embark on a multi site external impact evaluation. While the business plan provided a template for defining this work, it also posed a challenge because Tara and I were not participants in its development. We also quickly learned that TBP had secured funding to study the feasibility for conducting a multi site evaluation and had selected an evaluation firm to partner on the research. Still early in our tenure at TBP, Tara and I grew concerned about the organizational capacity to coordinate both sets of activities simultaneously. It would be difficult to expand sites rapidly and implement TBP's program with fidelity while conducting an impact evaluation. Presented with these circumstances, we chose not to slow down the planning process but to become more involved in the feasibility work and participate in the design of the proposed impact evaluation.

The feasibility study was completed in collaboration with the evaluator in late 2019. It achieved its purposes of identifying several TBP sites eligible for an impact evaluation and selecting a methodology for determining impact. The criteria we considered included the scale required to generate an

RCT sample, the strength of local stakeholder relationships, the accessibility of court outcomes data, the ability to generate subgroups, the status of policy context that could interrupt operations, and the capacity of the site to optimize bailouts with fidelity. Given the developing maturity of TBP sites, Tara and I insisted the design of the evaluation include a robust formative stage to test program fidelity with an implementation study and provide ample time to optimize data automation and a process for generating a randomized sample. Applying a "toll gate" approach, we would proceed with the impact stage of the evaluation only if a site could reach scale without risk to client reach and program quality, and could demonstrate an efficient and equitable client randomization process.

With our feasibility study in hand, we presented our evaluation plan to a funder interested in the impact of cash bail on case and life outcomes of people incarcerated before trial. Different perspectives quickly surfaced about the purpose of the evaluation. Our desire was to test our theory of change: If we bail out clients *who meet our eligibility criteria* and provide them with court reminders, transportation, and connections to voluntary support services, they are more likely to meet their court obligations and resolve their court cases more favorably while saving county governments millions of dollars in jail and court costs.

Conversely, the perspective of the funder and evaluator was to evaluate outcomes from an intervention that effectively eliminates the impact of cash bail. This type of study would require bailing out a significant portion of detainees awaiting trial in the county jail; however, as originally designed, TBP's model does not allow for this level of system penetration. Additionally, it is challenging for an experiment implemented at this scale to adequately support the needs of individuals released from jail and their successful return to court given existing social service infrastructure at the local level.

While TBP strives to provide bail assistance to as many people as possible, we are acutely aware that the existing network of community supports and other social resources, which we do not manage or control, is not yet designed to fully address all clients' needs. Thus, TBP client advocates are diligent about applying decision criteria (that is, client needs, contacts, court history, case history, bail amounts) about who we can bail out and actively support during the court case.

Despite limitations on who TBP can serve, we believed our theory of change was the most practical and systematic implementation of an intervention that models the future of a just and humane pretrial justice system and could meaningfully contribute to growing research on the ineffectiveness of cash bail. Ultimately, we could not reconcile the key research questions with the funder's expectations. Expanding TBP's target population could potentially compromise the effectiveness of our model for clients and add unsustainable risk and operational stress to our organization. After a good deal of dialogue with the funder and evaluators, we respectfully decided to part ways—on good terms, I might add—and reconsider our evaluation strategy.

Months after this decision, the pandemic took hold, fundamentally changing the context for how every organization pursued its mission. TBP temporarily paused its operations so it could adapt to a remote environment and assess the impact of COVID-19 on the criminal legal system. Jail decarceration temporarily became an emerging national trend due to the compassionate release of detainees and the slowdown of police activity. Jail populations now included more people with serious charges and higher needs. TBP bailouts declined in the early stage of the pandemic, as did the potential to generate robust research samples.

Pursuing evaluative work during this uncertain period seemed risky and operationally challenging. Instead, TBP saw an opportunity to double down on improving data quality and driving up model fidelity. We introduced a new user-centered version of our database, codified and trained staff on quality standards, and rolled out program monitoring tools that encouraged staff learning from collected data. We considered how we could leverage our existing SMS platform to collect and respond to clients' perceptions about our model's effectiveness. Finally, TBP realized its goal of ending cash bail in Illinois when the legislature passed the Pretrial Fairness Act in 2021 eliminating bail setting in January 2023. TBP is now partnering with a local organization to implement a community release with support program and will rigorously evaluate the model's fidelity and effectiveness for clients over a one-year period. In these topsy-turvy times TBP is clearly following the adage: inside every crisis there is also opportunity.

As Tara and I reflect back on almost two years at TBP, an external evaluation strategy is a difficult undertaking. As defined by our business plan, our initial timetable to execute on internal and external research goals was too ambitious. Second, as is mentioned throughout this book, it is vitally important for practitioners to drive their evaluation strategy and remain an equal partner in the workflow and data sensemaking. From the start, TBP designed its own evaluation strategy as part of its business planning and hired a data team, including me and Tara, who could work alongside evaluators and produce our own internal research. Third, never stray too far from your theory of change, no matter how tempting the research question. All nonprofits have limits on the people they can serve, and these limitations are necessary to run an effective and efficient program without creating harm for participants. Finally, senior leadership focused on results and supportive of research and staff learning is a critical ingredient for an organization that wants to meaningfully contribute to systemic change.

Emerging from Robin Steinberg's long-standing commitment to generating evidence, TBP's leadership team embodies a mindset that continually links our mission to results. And these results will, ultimately, lead to the dismantling of the cash bail system in America and create a world where people's needs are addressed by community led institutions rather than carceral systems.

NOTES

1. Pretrial Justice Institute and Charles Koch Institute. Lake Research Partners conducted a nationwide survey of 1,400 registered voters, including oversamples of 200 African American and 200 Latino registered voters (MoE + 3.1%), May 2–17, 2018.

2. James A. Anderson, Maya Buenaventura, and Paul Heaton, "The Effects of Holistic Defense on Criminal Justice Outcomes," *Harvard Law Review* 132, no. 3 (January 2019): 819.

BALTIMORE CITY PUBLIC SCHOOLS

USING DATA FOR EQUITABLE CTE OUTCOMES

BI VUONG

In early 2020, Baltimore City Public Schools (City Schools) launched an initiative to deepen the review of its Career and Technical Education (CTE) programming over a four-year period to improve employment and earnings for students after they graduate. As City Schools' chief executive officer Sonja Brookins Santelises explained, "The challenges presented by the pandemic have further accelerated what we already knew to be true: the world our students enter upon graduation requires a different level of preparation than what we have traditionally provided to them. . . . Career readiness—and CTE specifically—is yet another forum for developing knowledge and skills in a meaningful, personal context."[1]

City School's CTE programming review included efforts to improve the alignment of its CTE offerings to regional labor market demands that pay a family-sustaining wage; strengthen the quality of programs to align with employer needs, work-based learning opportunities, and supports for transition after high school; improve the geographic distribution of programs and resources across the city to ensure equity in student access; and maximize the use of existing resources.

City Schools partnered with Project Evident to delve more deeply into their student data to help identify challenges and opportunities as part of their strategic planning process. With those recommendations in hand, City Schools and Project Evident are continuing their work together to develop and support the implementation of a four-year career readiness strategy.

City School's Office of College and Career Readiness—which oversees implementation of the CTE programs in the district—was committed to taking a data-driven approach in developing a four-year CTE strategic plan. However, owing to the timing of the work, City Schools experienced both financial and technical capacity constraints. As a result, several organizations came to the table to ensure the project's success. These partner organizations included Baltimore City Public School's Office of College and Career Readiness, which was responsible for sharing data, coordinating site visits, engaging the community, and engaging in decision making and trade-off discussions; Project Evident, which served as an external consultant to City Schools and was responsible for conducting data analysis, reviewing internal processes and documents, providing recommendations for implementation, and supporting with community engagement and program implementation; and Baltimore Fund for Educational Excellence with the support of Annie E. Casey Foundation, which provided the philanthropic support that allowed City Schools to undertake this work.

City Schools and Project Evident engaged in this project in the spirit of true partnership, as both partners were focused on improving the opportunities and outcomes of the students and families. Both parties recognized and appreciated the fact that there would be moments where they would have to work with imperfect information; they were aligned in the belief that parent, school, and community feedback was a critical source of information; and they both recognized the importance of being responsive to City Schools' timeline.

To meet City Schools' goals within that timeline, Project Evident set up an infrastructure that allowed City Schools to securely upload relevant data and background documents, and maintain a shared work plan. Both partners also agreed to a set of partnership management principles around points of contact, communication cadence and preferences, and a shared commitment to meeting key deadlines.

This approach to the engagement ensured that the work was relevant and timely. Just as importantly, however, was the evidence used to help City Schools make different trade-off decisions. City Schools sought to extract insights from existing data sets to determine what CTE programs to offer going forward and, prioritizing equitable access, determine how the programming should be distributed geographically across the city. To ensure that City Schools had the most relevant information to support its decision making, over a four-month period between January and April 2020, City Schools partnered with Project Evident to answer a series of critical questions, including: How do current CTE programs align with in-demand, living wage jobs in the region? How can our pathways be realigned to reflect trends in labor market demands and student interests? How can our programs be situated across the city to allow for equitable access for City Schools' students?

To address these questions, Project Evident provided analytical and decision making support to City Schools. First, Project Evident analyzed data on employment and wage projections for occupations in industries targeted by the CTE programs to assess labor market demand and earning potential, and analyzed recent student- and school-level data to assess the demand, distribution, and capacity of program offerings, as well as student access, engagement, and outcomes. Project Evident complemented these analyses by having conversations with students, teachers, principals, district leaders, and CTE staff to capture their perspectives on existing policies and processes and best practices. Finally, Project Evident mapped all programs and schools against City Schools' Community Condition Index—a measure of a community's access to resources—and students' ability to access the program based on travel time on public transportation.

Project Evident took all this data and used it to develop recommendations for the next three school years to either sunset, scale back, maintain, or expand CTE offerings with a lens toward equity. Project Evident also led workshops with school leadership and staff to support decision making and to discuss trade-off and implementation considerations.

In the second part of the engagement, City Schools sought to assess the feasibility of these recommendations by collecting, generating, and reviewing additional evidence. Over an eight-month period, with support from Project Evident, City Schools engaged in over 100 meetings to gather feedback from students, parents, school staff, alumni, community members, and elected officials. This feedback ranged from the specific programs a school needs to better support its students to the values the community felt were most important for a career readiness program. They then created seven different program placement scenarios based on a combination of facility conditions and size, student access, human resource allocations, labor market demand, community feedback, and financial feasibility. And

City Schools developed a series of progress monitoring and accountability metrics to ensure the strategy is being implemented and City Schools is achieving its intended outcomes.

As a result of this additional work, City Schools developed a four-year career readiness strategic plan, "Career Readiness: A New Pathway Forward."[2] Supplementing the strategy is a four-year financial model to ensure resources were available to support implementation.

At the time of writing, City Schools and Project Evident are deep into implementation planning. A high-level implementation plan, which includes actions that need to be taken by human resources, facilities, finance, academics, and student supports, has been developed, and a series of cross-district planning and progress monitoring meetings is being planned to ensure the plan is executed as intended.

Finally, to ensure that City Schools has the capacity to continue to monitor its progress against its four-year plan, by the conclusion of the engagement, City Schools' staff will have been trained on how to use an action planning template and will have integrated this into its daily work. Project Evident also will transfer its analytic code and final data files for City Schools to use and replicate.

TRANSPARENCY AND STAKEHOLDER INPUT

The partnership experienced a few challenges, but one in particular could have set the progress back. As scenario planning and feasibility testing were being conducted, a key stakeholder was left out of the process regarding facilities planning. However, as partners had communicated at the outset to all stakeholders that nothing would be finalized until all voices were heard, the team was able to resolve this by quickly integrating previously excluded personnel into conversations and adjusting the plan based on their input.

This approach to transparency in decision making helped the district team navigate other political challenges and build trust, as they were able to show along the way how various stakeholder feedback had been incorporated into their final strategy. Various stakeholder groups, including staff whose positions may change or be eliminated, expressed gratitude for being able to be a part of the discussion. The political challenges are not over for the team, but they have built a lot of goodwill through this process.

Another challenge of this work was the data quality and infrastructure. It took a significant amount of time to understand the nuances of how school data were recorded to be able to tease out data quality issues. However, additional team members from the research and evaluation side had first-hand knowledge and understanding of school systems data as well as CTE-specific data and indicators. This helped not only in making sense of the data; it also informed recommendations for how to best move forward. Rather than be derailed by data quality issues, the team prioritized building a stronger infrastructure that would allow City Schools to better track outcomes, monitor implementation, and pivot when needed.

RESULTS

As a result of this partnership, City Schools has created a four-year career readiness plan that aims to achieve its four goals. To strengthen rigor and relevance, the plan calls for increasing the number of in-demand living wage seats from approximately 3,800 to 7,950 by the end of the four-year plan, and growing the number of programs that put students on a path to in-demand occupations. It also entails strengthening partnerships with workforce organizations and creating greater alignment with college readiness efforts, including career-specific certifications and dual-enrollment opportunities. And to increase equity, the plan calls for relocating or adding programs across the city to ensure a more equitable distribution.

More immediately, the partnership has resulted in the reallocation of approximately $1 million in SY2020–2021 to bolster the equipment, materials, and curricular resources available to students and teachers in high-demand, high-wage pathways. During this same year, SY2020–2021, City Schools reduced redundancy in its workforce to create more programmatic efficiencies while simultaneously allocating additional teachers to schools where there is demonstrated student demand. For SY2021–2022, City Schools will be redesigning five curricular pathways to better align with industry demands and expanding career readiness opportunities to additional students. An outline of its implementation plan for SY2021–2022 can be found in "Career Readiness: A New Pathway Forward."

REFLECTIONS

The most unique and powerful parts of this engagement were the scenario planning and implementation support that was aligned to the district's timeline. Often, research stops at the foundational analysis and recommendation stage without moving into actionable steps and support, which makes it difficult for practitioners to understand the critical path to move forward. Additionally, by conducting research on the district's timeline rather than on the researcher's timeline made it actionable—key decisions were made in the first few months around resource allocation and budget that enabled future work.

The work done by City Schools and Project Evident is a good illustration of a "Next Generation" partnership. The practitioner-centric engagement was designed around the needs and timelines of City Schools. The work took an R&D approach through the iterative use of multiple data points to support decision making and by understanding that flexibility and a willingness to pivot when required are key. And the engagement intentionally involved communities throughout the process, recognizing that students, families, and teachers hold the most knowledge about what is working and what isn't in their education system.

RESOURCES AND FURTHER READING

A copy of the district's career readiness plan can be found at www.baltimorecityschools.org/sites/default/files/2021-03/CTE-StrategicPlanAppendices.pdf.

Additional resources related to this project include:

Baltimore's Promise. "Gaining Traction after High School Graduation: Understanding the Post-Secondary Pathways for Baltimore's Youth." 2018.

ESG. "Preparing All Students for Economic & Career Success: An External Assessment of Career Readiness Priorities, Practices, and Programs in Baltimore City Schools." Education Strategy Group. 2019.

Glasmeier, Amy K. "Living Wage Calculator." Massachusetts Institute of Technology. 2020. livingwage.mit.edu. 2020.

Schoenberg, Corrie, Danielle Staton, Sadie Baker, and Sydney Short. "Broken Pathways: The Cracks in Career and Technical Education in Baltimore City Public Schools." Fund for Educational Excellence. 2019.

NOTES

1. See "Career Readiness: A New Pathway Forward," Baltimore City Public Schools, March 2021, www.baltimorecityschools.org/sites/default/files/2021-03/CTE-StrategicPlanAppendices.pdf.

2. See https://go.boarddocs.com/mabe/bcpss/Board.nsf/files/BYTKY453BEC1/$file/BCPS%20CTE%20Strategic%20Plan%20with%20Appendices.pdf.

PER SCHOLAS

NAVIGATING COVID WITH PARTICIPANT FEEDBACK

PLINIO AYALA

Per Scholas is a national organization that has been advancing economic mobility for twenty-five years. Through rigorous training, professional development, and robust employer connections, we prepare individuals traditionally underrepresented in the technology workforce to enter and succeed in high-quality careers. I have been privileged to lead Per Scholas as its president and CEO since 2004, including our extensive national growth, over the past nine years.

For two decades prior to COVID-19, one hallmark of our technology career training approach was that it was rooted in immersive, classroom-based instruction. We believed we achieved our impressive outcomes—85 percent of Per Scholas learners graduate, and 80 percent of the graduates attain jobs within one year—largely because we required actual *presence* from learners.

Our classroom-based model is designed to mimic the workplace, requiring on-time attendance and professional attire, as well as facilitating group projects and encouraging communication, presentation, and collaboration skills. Learners practice on business-class hardware and software, and engage with working IT professionals who volunteer to help learners develop start-up social capital and job-search skills. Finally, in a typical Per Scholas classroom, learners form tight-knit bonds that provide added support for them to succeed. Graduates have routinely described this aspect of their training as transformative.

Twenty-eight-year-old Taiheem Wentt is just one of thousands who have benefited from this model. Taiheem overcame exceptional challenges growing up and started out on a college career. But the birth of his daughter cut short these ambitions, so he turned to Per Scholas instead. An outstanding learner, Taiheem graduated from our Network Support training and found a job paying four times as much as the security guard salary he had earned before. Today, Taiheem and his family are thriving, and Taiheem was extensively featured in 2021 on the PBS career exploration series *Roadtrip Nation*.

Moreover, from 2004 to 2007 and then again from 2011 to 2018, Per Scholas underwent long-term, random assignment evaluations, first by Public/Private Ventures (Sectoral Employment Impact Study) and then by MDRC (WorkAdvance). Both studies concluded that Per Scholas learners, all of whom attended physical classrooms, were more likely to secure jobs in tech and earned significantly more than equally qualified and motivated control group members, including those who went on to pursue other career training options.[1]

Early in 2020, though, we had just begun our first pilot of a partially remote learning model—a tech-enabled "Connected Classroom" that made it possible for an instructor teaching in New York to simultaneously teach a class of in-person learners in Dallas. We planned on testing additional hybrid and remote learning models throughout the year. Little did we know how quickly we would shift to the largest organizational experiment we had undertaken to date!

In March 2020, as the COVID-19 pandemic overtook the nation, Per Scholas reluctantly shut down all our classrooms nationally, and migrated 538 then-enrolled learners to remote instruction. We had no idea how learners might fare in a 100 percent remote framework, only that we had no other options. But we also knew we had been presented with an unusual opportunity to explore the capabilities and limits of remote learning and, perhaps, to begin to understand whether our long-standing valorization of in-person training was fact-based.

Per Scholas is fortunate to have many amazing partners. One of them, the **Bill & Melinda Gates Foundation**, helped fund our shift to remote learning, along with a **participant- and provider-centered evaluation we wanted to embed in its implementation**. Per Scholas subsequently engaged **Barrow Street Consulting (BSC)**, a Washington, DC–based independent consulting firm, to provide expertise and support for this critical evaluation.

PARTICIPANT AND PROVIDER CENTERED EVALUATION

Even prior to BSC's engagement, Per Scholas began collecting remote participant feedback on its own, from both learners and faculty members. BSC worked initially to help us organize and build on these internal feedback-gathering activities to yield better insights and to center them in a more standard evaluation design.

To these ends, BSC expanded our in-house learner satisfaction surveys, developed new instruments to assess feedback by instructors and career coaches,[2] and supplemented both survey types with focus groups. It also incorporated a learner **Net Promoter Analysis** into the overall research design. The latter is a strategy commonly deployed in the for-profit sector to formulate insights into *customer loyalty*, and has been found to correlate with outcomes such as revenue growth.

The largest challenge we confronted was the pandemic itself, since at the time we started, nearly all Per Scholas learners and faculty members were coping with COVID-19's initial economic shocks, and many also experienced health impacts—all while trying to complete an intensive, boot camp-style course. Within this context, we feared our surveys would be perceived as a nuisance, or even that we might find it difficult to recruit a truly representative group of learners for the planned focus groups.

We overcame this challenge by providing opportunities for learners and faculty members to complete the surveys during class time, and by following up persistently with non-respondents. Ultimately, 74 percent of learners (n = 259) who were enrolled when the study began completed a mid-course edition of the surveys, along with 78 percent of career coaches and 58 percent of instructional personnel (n = 46). We also successfully recruited ten diverse learners and ten instructional personnel from across Per Scholas locations to participate in the focus groups.

Another challenge was that we continued to modify many aspects of our remote program design even after the research began—often in direct response to the raw survey and focus group data as it came in. By September, for example, we had reconceived and virtualized a much larger set of in-classroom demonstrations and even some hands-on computer lab activities.

These changes meant the program model we asked respondents to rate in their end-course surveys was already becoming very different from the one midstream, an inconsistency that might be fatal to the aims of more traditional evaluation research. But, here, we precisely illustrate the distinction between traditional evaluation, which tends to measure outcomes attributable to a stable, well-defined set of activities, and a true participant- and provider-centered inquiry whose more urgent focus is to help improve program experiences and processes in real time.

RESULTS

The early participant feedback we gathered ourselves immediately helped us identify many beneficial changes to our remote learning implementation, including migration to a different video communications platform, reimagining the organization of each remote learning day, providing assistance for learners who lacked adequate technology for remote access, and supporting faculty members struggling to develop remote proficiency themselves.

BSC's analysis over the summer of 2020 helped us understand whether these earlier changes were effectively addressing learner and faculty barriers. Encouragingly, BSC's findings were quite positive. Among both learners and faculty members, our implementation of remote learning was widely perceived as a "success." Moreover, the learner Net Promoter Score for Per Scholas as a whole was strikingly high at 67 (the range is -100 to 100, but a typical score in the for-profit sector ranges between 30 and 40).

However, BSC's analysis also helped us identify several areas for improvement. For example:

- Learners and practitioners alike reported that they needed more time to deliver/complete coursework than was typical for in-person sessions. In addition, "homework" lost much of its value as a pedagogical strategy when learners already spent their entire day working at home.
- Learners and practitioners also reported that we needed to build better and more creative strategies to support hands-on skills acquisition.
- Learners felt they still lacked sufficient opportunities to develop one-to-one connections with their classmates, that aspect of Per

Scholas training so many of our in-person learners had previously told us was invaluable.
- Instructors struggled to create the same energy as they have in the classroom. One instructor noted that her remote classroom was "unnaturally quiet as everyone is on mute." This was especially difficult for career coaches.
- The initial integration of virtual IT professional volunteer engagement into remote classes had mixed success, in part because it was more difficult to manage and also because the volunteers sometimes struggled to adapt to remote interactions themselves.
- Finally, even though learner Net Promoter Scores for Per Scholas as a whole were exceptionally high, those for our remote courses were closer to the norms for this type of analysis.

We viewed these and related findings as strong confirmation that our work to develop an effective remote learning model remained unfinished. However, considered as a whole, the results persuaded us that it would be possible to provide a remote learning experience just as engaging and effective as our in-person model. **In other words, our previous bias in favor of 100 percent classroom-based training was not entirely justified.** Indeed, in one of the most revealing findings, a substantial majority of remote learners said that, in the best of worlds, they could access a *hybrid* model: one in which they attended remote sessions to learn new knowledge and physical computer labs to practice putting it to use.

IMPACT

As a result of the research findings, Per Scholas implemented improvements to our remote learning model:

- We organized a national remote training team to centralize all remote learning administration and program development across our sites.
- We provided substantially more time in each training day for learners to complete coursework and for instructors to provide individualized attention to learners.

- We developed and began distributing new "Tech Learner Kits," customized by course, so all learners can practice hands-on skills at home.
- We created virtual IT professional volunteer engagement opportunities and centralized it nationally.

The evaluation findings helped convince us to **continue offering remote learning opportunities even after we can safely return to classrooms**—and moreover, that remote learning should become one of the primary engines propelling our future national growth. This represents a momentous change vis-à-vis our pre-COVID-19 thinking, and one that can set Per Scholas on a far more ambitious growth trajectory than we had previously conceived.

REFLECTIONS

Our research methods for understanding the efficacy of remote learning were not especially innovative or unusual. But, frankly, that was its main virtue. Although we benefited very much from BSC's expertise, ours was the kind of project that nearly any practitioner might implement at a more basic level. More important was our willingness to listen, understand, and act on the information we received.

For funders and policymakers, we cannot underline enough the importance of supporting comparable efforts that may require funders reluctant to help pay for "research" or "overhead" costs to revisit their conceptions of what these terms really mean. Per Scholas was very fortunate to have funding that specifically supported its remote learning evaluation.

For researchers—especially those focused on mounting gold standard evaluations like the ones Per Scholas has hosted twice before—we would suggest that our project shows that evidence building can come in many forms. In this case, a rapidly constructed and fielded implementation analysis focused on participant and practitioner voices fostered a profound new shift in direction for Per Scholas with momentous implications for our future.

Finally, the experience I have described has reaffirmed for Per Scholas that this kind of participatory evaluation should never really end. We recently decided to extend BSC's engagement with Per Scholas so it could repeat its earlier research with a new cohort of learners to see if they view

Per Scholas and their remote training differently now that we have acted on a number of the previous findings. Evaluation is most helpful when it is coupled with improvement. Isn't that what Taiheem and so many other Per Scholas learners deserve?

NOTES

1. The MDRC researchers additionally found that the earnings difference between Per Scholas and control group participants grew larger over time, and that Per Scholas participants reported greater life satisfaction. Kelsey Schaburg and David H. Greenberg, "Long-Term Effects of a Sectoral Advancement Strategy: Costs, Benefits, and Impacts from the WorkAdvance Demonstration," March 2020, www.mdrc.org/publication/long-term-effects-sectoral-advancement-strategy.

2. The Per Scholas career training model is two-pronged. *Instructors* help learners build new technical skills. *Career coaches* teach more general job search and professional skills. They also work with learners to develop individualized career plans and serve as a primary liaison for them with job placement personnel.

SECTION 2

CONNECTING EQUITY WITH DATA AND EVIDENCE

> *The worst equity problem we're dealing with in data at the moment is that we're making prejudiced choices, but don't understand how.*
> —HEATHER KRAUSE, "ALL DATA IS BIASED"

> *Equity will continue to be elusive if we dance around the edges of racism and power dynamics and fail to address these issues in our strategies, organizations and systems.*
> —CARINA WONG, "FIVE DESIGN PRINCIPLES OF JUST PHILANTHROPY"

Decision power is ultimate power. Past norms for building evidence have imbalanced power and exacerbated inequity by creating a black box between the practitioners and communities who collect and submit data and those who evaluate it.

Authors in this section of our book point out harmful practices and propose helpful ones to bring an authentic equity lens to building evidence of social impact. Michael McAfee calls for recasting evidence as justice, and calls out current racial disparities in who is called to account to show evidence to prove their "basic humanity." ("No one ever asks a white school what evidence they have for expanding an afterschool program. . . .") Heather Krause shatters the myth that "data offers an objective, bias-free way to make decisions" and offers a roadmap for equitably using data to advance racial equity. Carina Wong, speaking to philanthropy, offers five design principles to improve a strategy's equity orientation. Chris Kingsley highlights initiatives in Los Angeles, New York City, and Cuyahoga County that take seriously the needs of agency and nonprofit practitioners and their clients. Robert Newman, Dylan Edward, Jordan Morrisey, and Kiribakka Tendo propose alternatives to the enduring tendencies in subfield evaluations in sub-Saharan Africa to extract data. Meanwhile, Amy O'Hara and Stephanie Straus describe how to de-risk civil court data through clarifying public interest and creating transparency.

Issues of inclusion link to equity, and practical cases in this section address both. Coauthors Tatewin Means, Dallas Nelson, and Dusty Lee Nelson write about Lakota data sovereignty. Carrie Cihak issues five calls to action for local governments to move toward a pro-equity approach to evidence building, born of her policy work in King County (Seattle area).

Tracy Costigan and Raymond McGhee of the Robert Wood Johnson Foundation, and Lola Adedokun, formerly at the Doris Duke Charitable Foundation, share learnings on centering evaluation norms on equity at their respective foundations. Nisha Patel flips the lens from eradicating child poverty to achieving guaranteed minimum income levels, citing cases and evidence where practitioners' cash distributions make a difference. Finally, the use case of ParentCorps describes how tapping into community voice enabled the early childhood education nonprofit to pivot to virtual programming almost overnight, amid tremendous uncertainty and fear during the pandemic.

Questions raised and addressed in this section include:

1. What does it mean to say that "all data is biased?"
2. How can evidence builders account for and diminish racial and other bias across all steps of evidence building?
3. What can practitioners, policymakers, funders, and evaluators do to support more equitable practices?

REIMAGINING EVIDENCE AS JUSTICE

MICHAEL MCAFEE

As the leader of a national organization pushing for more equitable public policies, the question I get asked more than any other is, "Where's the evidence this policy will work?"

Now, I am a big fan of evidence. Evidence, when effectively and properly marshalled, can provide vital clues toward equity. My organization, PolicyLink, relies heavily on evidence to make the case for equitable policy change in city halls, state capitals, and Congress every day. We dig through mountains of data, research, and studies to hone our policy proposals.

Evidence can, unquestionably, help improve existing programs and identify promising innovations. It can help us see where programs are falling short. It can help us challenge our own perspectives to see new ways of tackling thorny challenges.

But as important as it is to build evidence, it is just as important to ask, "Who is required to show evidence to prove their basic humanity?"

No one ever asks a white suburban town council what evidence they have for building a new park or community center. No one ever asks a white school what evidence they have for expanding an afterschool program or varsity sports. Rarely is the business association in a white town asked why they invest in improved infrastructure on Main Street.

All those investments are seen as self-evidently good for the community, so we skip right past the evidence phase to the implementation phase.

But when it comes time for truly equitable policies—policies that allow everyone to participate and prosper to their full potential—suddenly we cannot even begin to move forward without ironclad, peer-reviewed evidence. When we try to make life better for Black, brown, Indigenous, and other marginalized communities, the threshold for action is much, much higher.

This is what I call "evidence as a double standard." When policies that uplift Black and brown people are at issue, "evidence" is too often used in calculated, inhumane ways to reach some cold cost-benefit analysis completely removed from the real lives and lived experiences of the people affected by the policies. On the other hand, largely white communities too often use "evidence" as a proxy to deny poor people access to the resources and programs that wealthier people take for granted.

That is why we must face what is evident before we demand evidence. It is evident that America is built atop centuries of ingrained white supremacy and systemic racism. It is evident that the effects of our history are felt in every corner of our nation, from schools to businesses to transportation to infrastructure to banks to housing. It is evident that systemic inequities were purposefully built into the DNA of our institutions. And it is, therefore, evident that we cannot overcome these systemic obstacles without reimagining the very design of our nation.

Deployed without a deeper understanding of the roots of American injustice, evidence can become merely its own form of white supremacy by reinforcing the racist status quo. That may be a largely unintended consequence, but it is a consequence, nonetheless.

Racism has been the driving current of American public policy for 400 years. We cannot address that reality by narrowing our vision solely to "evidence" for specific policy proposals, in a vacuum of history and context. For far too long, we have treated Black people and other marginalized communities as mere test subjects in a scientific endeavor to find evidence.

When you continue to deny our nation's origin story, when our hearts are too calloused to see the humanity of others, evidence alone (as currently conceived) will never compel people to act. There are reams and reams of evidence for equitable policies that have been dutifully and painstakingly compiled by folks like the Children's Defense Fund, Urban Institute, and PolicyLink. Evidence shows we must dramatically reverse income inequality if we want to sustain our democracy. Evidence shows climate change will soon swamp states along the Gulf Coast. Yet the equitable policy solutions to those challenges remain unenacted, gathering dust on the desks of the very same elected officials who dismiss our demands for our own basic humanity with a blithe, "But where's the evidence?"

That is why it is time for us to create a new vision of evidence—evidence as justice, evidence as truth. If evidence is not leading us inexorably toward justice, we are not maximizing the use of evidence.

To create a new paradigm of "evidence as justice," we first must ask ourselves some vital questions: What does it take to reverse 400 years of systemic oppression? What does it take to undertake a truly equitable redesign of a country built upon genocide, stolen land, and slave labor? If we do not ask ourselves these questions before we set out to gather evidence, we will miss the destination. Evidence today is a microscope. We need it also to be a telescope.

When policymakers ask for evidence, they really are asking for proof that an intervention will work. They are attempting to manage risk before investing in long-underinvested communities. But equity work is risky by nature. You cannot wring the risk out of the vital work of creating a world that has never before existed.

One of the darlings of the evidence-first policy world is the Harlem Children's Zone (HCZ), a group PolicyLink has been working with for more than a decade. The basic premise of the HCZ model is that we can dramatically improve the lives of children by investing in wrap-around services—schools, mentorships, health care, parental support—that help provide children in need with the same foundational supports wealthier families take as a given.

The truth about HCZ, though, is that while the evidence is promising, it is still scarce. And it never would have made the progress it has made so far if it had not secured significant investments long before the green sprouts of "evidence" began to push through the soil.

Finding evidence for radical ideas is, by definition, extremely difficult because they are policies that would create a different world than the one in which we currently live. We have plenty of evidence for the ways white men govern. We have plenty of evidence for how patriarchies and capitalism work. We have plenty of evidence for the status quo.

But what evidence is there for how our governing institutions, public servants, laws, and regulations should act in a multiracial democracy? What evidence is there for how fiscal policy can be marshalled to lift up underserved communities? What evidence is there for how to undo the inherently oppressive structures of modern capitalism?

To be clear, many people are trying to figure out how to use evidence for justice. Participatory and community-based evaluation, in particular, provides a promising way forward for bringing the insights and voices of impacted communities into the policy process. And we can and should continue to use data when it can be marshalled in service of justice and equity.

One of the leading voices in this new movement is the Equitable Evaluation Initiative (EEI), founded by Jara Dean-Coffey to help spark "paradigm-shifting conversations" among practitioners in the philanthropic, nonprofit, and consulting communities. EEI offers five key considerations for how to reimagine evaluation and evidence:

1. Acknowledge that evaluation reflects a paradigm that cloaks privilege and racism as objectivity.
2. Explore the ways in which current practices in foundations and nonprofits and among consultants can be barriers to the adoption of equitable evaluation principles, and identify and share approaches that interrupt those habits.
3. Elevate evaluative thinking that links organizational culture, strategy, and evaluation to be a leadership competency and organizational capacity.
4. Move beyond methodological approaches and evaluator demographics to address culture and context and, in so doing, unpack our definitions of evidence, knowledge, and truth so we may create new ones grounded in this time, place, and set of intentions.
5. Continue to diversify and expand the talent pool of evaluators, and ensure that their training (both formal and informal) introduces and nurtures a myriad of new and different ways to conceptualize evidence, knowledge, and truth in service of greater validity and rigor.

Each of these steps requires us to think beyond raw numbers and spreadsheets and truly understand the historical and social context in which we are operating—and how that context requires us to think more creatively and deeply about seemingly intractable problems.

Project Evident is already moving in this direction with its goals for the Next Generation of Evidence Campaign: Practitioner Centric; Embracing an "R&D" Approach; and Elevating the Voices of Communities. The third goal, in particular, is where the researchers and policymakers can make an enormous difference almost immediately. Bringing the voices, insights, and ideas of people most affected by policies can make those policies sharper, more effective, and more equitable.

Even as we develop the next generation of evidence, though, we must constantly be asking ourselves: Evidence in service of whom? Are we requiring marginalized communities to contort themselves into the narrow boxes of the status quo, boxes they have been shut out of for hundreds of years? Or are we beginning from a place of justice and working backward to create evidence and evaluation strategies that will achieve equity—just and fair inclusion into a society in which all can participate, prosper, and reach their full potential?

As the United States becomes ever more diverse, the need for truly equitable policies becomes more urgent. But we cannot understand those policies if we continue to see them through an inequitable prism.

Evidence is not a road map; it is a flashlight. Evidence alone does not guide us to where we need to go but, rather, illuminates our path toward justice.

ALL DATA IS BIASED

HEATHER KRAUSE

In this chapter, I am going to share some stories with you that show how the worst equity problem we are dealing with in data at the moment is that we are making prejudiced choices but don't understand how. Most of us are reading this because we know that math, science, and data can improve the world. One of the reasons many people like the idea of data in the mission-driven sector is that we believe data offers an objective, bias-free way to make decisions. I have good news and bad news for you. The bad news is that this is a data myth. At every single step of a data project, we are making choices. Choices about whose lived experience to center; choices about whose worldviews get prioritized; choices about who gets reflected in the work. The good news is that, once we move past this myth, we can get to some valuable, grounded work on using data for racial equity.

Here is my favorite story about making choices in the way we use data. What is the average number of students in these classrooms? There are three students in classroom A, six students in classroom B, and nine students in classroom C.

If you said the average classroom size is 6, you are right. If you said the average classroom size is 7, you are right. These answers use the same math; they just embed a different perspective. Let's look at the math from the teacher's perspective.

FIGURE 2.2.1 **What is the average number of students across these three classrooms?**

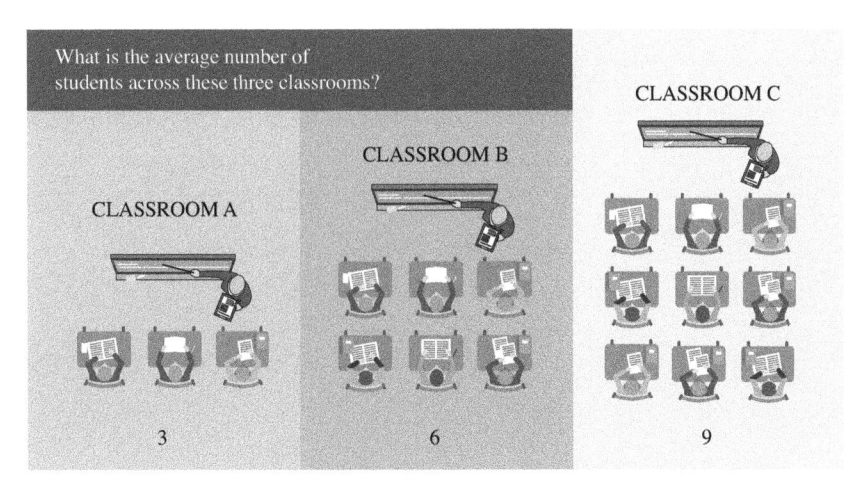

FIGURE 2.2.2 **From the teacher's perspective**

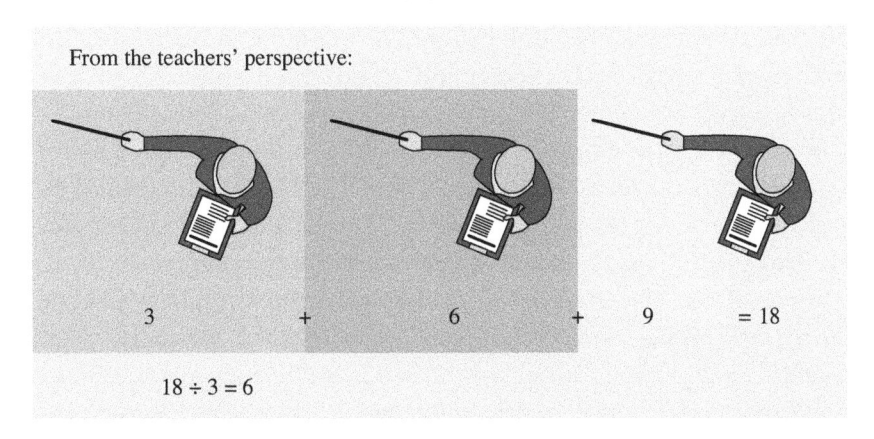

The first teacher takes a look around and sees three students, the next sees six students, and the last sees nine students. 3 + 6 + 9 is 18. Divided by the three teachers is six. The average students per classroom from the teachers' perspective is six students.

And this one is from the students' perspective.

In classroom A, each student counts three students in their classroom, including themself. In classroom B, each student sees six, including themself. And in classroom C, it is nine. Adding the total up and dividing by the

FIGURE 2.2.3 From the students' perspective

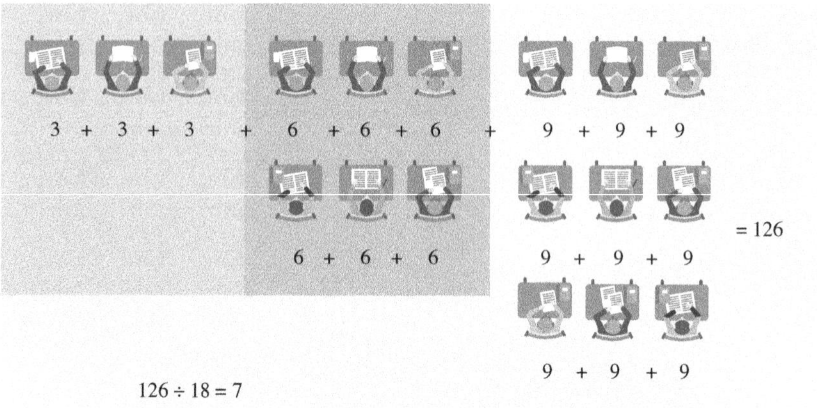

$126 \div 18 = 7$

$= 126$

number of perspectives is eighteen students. We get an average classroom size of seven.

We do the math in exactly the same way. Both processes are valid; they just center a different lived experience. The way most of us are taught to think about math can make this example, by turns, confusing, enraging, or mind-blowing. You might need to actually get out some dolls and test it.

It is important to note that we are not using a different kind of math. Both means are calculated in the same way: the sum of the units divided by the number of units. We just had to make a choice about which unit to use, where to put the locus of power, whose experience to prioritize.

Many of you (myself included for most of my life) will have felt they did not make a choice, that this is just how math works. That is the most insidious myth with all data and research. The dominant perspective is so deeply ingrained in much of data and models that it seems like the only perspective—or no perspective at all.

Let's look at another example where the math is completely correct but there is a choice to be made. This graph is looking at outcomes in an income improvement project. In this graph, we clearly see that the people in the project had a huge average increase in their income. Project success!

But the people in this project are from three different zip codes. If we are interested in the equity between these groups, we might want to measure them individually to see how they contributed to the average.

Uh oh . . . though we have increased the average income significantly, it has not been the same for everyone. We can see from these different

FIGURE 2.2.4 Measuring a project's impact on average monthly income

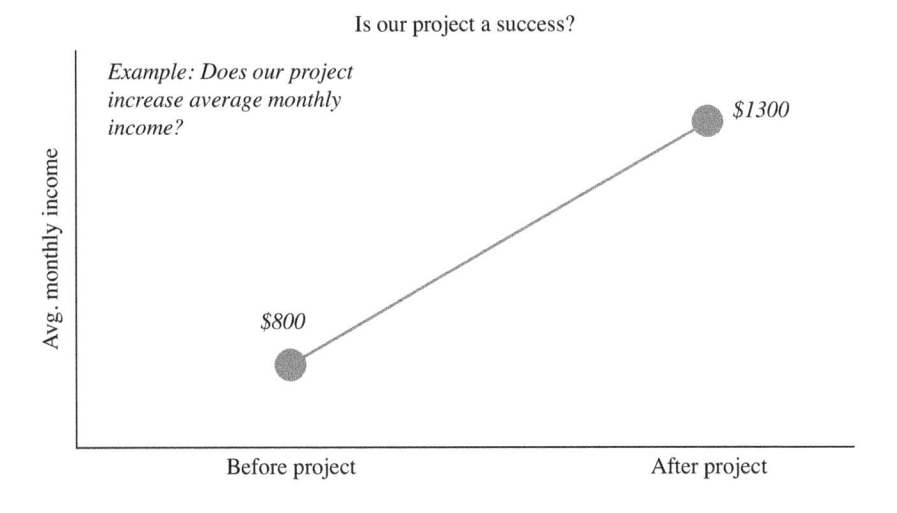

Is our project a success?

Example: Does our project increase average monthly income?

Avg. monthly income

$800 — Before project
$1300 — After project

FIGURE 2.2.5 Measuring a project's impact on average monthly income by zip code

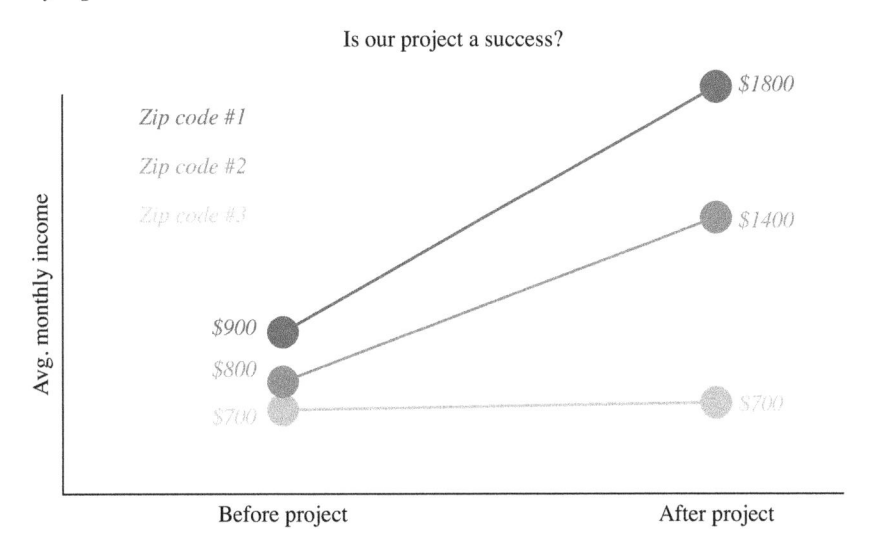

Is our project a success?

Zip code #1
Zip code #2
Zip code #3

Avg. monthly income

Before project: $900, $800, $700
After project: $1800, $1400, $700

FIGURE 2.2.6 Did we increase or decrease the income gap?

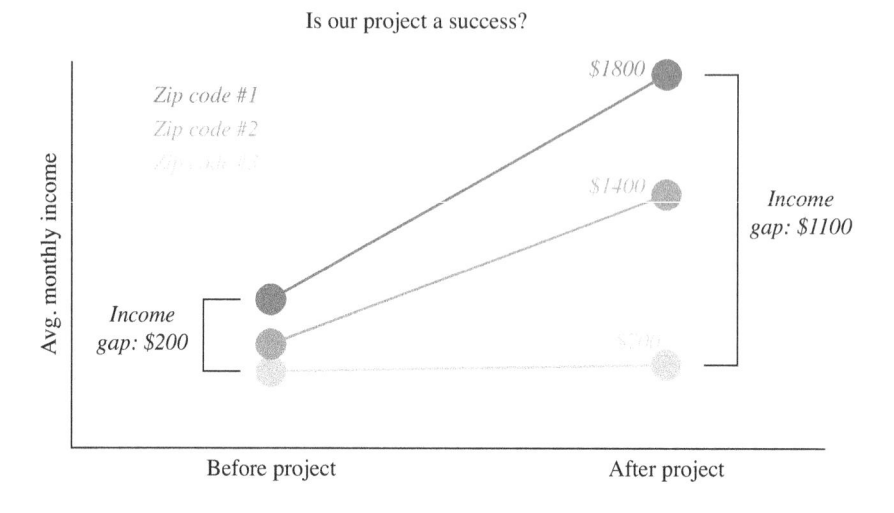

Is our project a success?

Zip code #1
Zip code #2
Zip code #3

$1800
$1400
$700

Income gap: $200
Income gap: $1100

Avg. monthly income

Before project — After project

zip code lines that the project has been a success for some people and not for others.

If we define success by "Did we increase or decrease the income gap?" then the project has failed badly. It is not that one of these is right or wrong; you may care only about the overall average, but the math is correct in both.

But it is not just a math problem. Put yourself in the shoes of someone from zip code #3. You have participated in a project, watched your neighbors' incomes rise, and heard the organization running the project using data to proclaim it a huge success. How do you feel? Probably, you are either ashamed because there must be something wrong with you or you are furious because you can see that the project did not work for you and the researchers did not include your perspective in the way they used their data.

In this case, we had a choice to make: What data should we use to measure success? All the work downstream of that choice, from the analysis to the design of the graphics, is affected by the equity implications of the initial choice.

We want to think of quantitative research as a situation in which we make one important choice, which research question to look at. And then we follow that research question through a trail of building an objective research design, collecting objective data, doing an objective analysis, and, then, hopefully, creating an objective data visualization.

Not so, though. This is a myth.

Instead, research, even, and sometimes especially, quantitative research, involves dozens and even hundreds of choices at every single step in the research process.

And there is no way to avoid making these choices. Our only option is to continue to hide behind a false narrative about "objective quantitative research" or "value-free evidence," or to figure out how to make choices in our research that better reflect the equity we want to embed.

Note that I did not say, "Now you have to learn how to make the right choice." There is no "right choice." There is no objective and bulletproof equitable data project. My clients come to me searching for that like it is the holy grail or the fountain of youth. Data projects can be intentional and transparent in their choices, but they cannot be objective or choice-free. Equity is a process, not a binary state, between equitable projects or inequitable projects.

Let's look at another time I was trying to use data for equity. I was working with a school district struggling with the way they used data about student outcomes and race. One of the issues that was particularly tense was the reporting on expulsion data. The data was being used to show that more Black and Latinx boys were being expelled from school than white boys. More often than not, this data was analyzed and displayed in a way that emphasized "the equity gap" between Black/Latinx boys and white boys.

The district wanted to improve both situations—the way they were using data about racial equity and the experiences these young men were having in school.

The district launched a project aimed at reducing the rate of expulsion of specific groups of young men. At the outset of the project, they established the research question as: "Has our initiative reduced the rate of Black and Latinx boys being expelled relative to white boys?" Unsurprisingly, this initiative, and its research, was not welcomed by the community.

When framing a research question, there are two key choices we make. The first is where we place the onus of change. The second is how we define success. In this case, the researchers had placed the onus to change on Black and Latinx boys and defined success as the rate of white boys. Neither of these choices was in alignment with the stated equity goals of the project. Essentially, this original research question can be boiled down to: "How good is our project at getting Black and Latinx boys to be like white boys."

To start making choices to align research with racial equity objectives, we needed to put the onus of change somewhere other than on the marginalized people, and define success differently. After conversation with the community and deeper reflections on the actual equity goals of the project, the research question was changed. The questions became: "Has our initiative disrupted the processes in our district that are most strongly related with us pushing out Black and Latinx boys?" and "Has our initiative improved the school characteristics that are most strongly related with creating environments that encourage Black and Latinx boys to fulfill preexisting desires to be in school?" These questions and the way they frame research were welcomed by the community. Data started to go from a weapon of disaggregation and separation to a tool that could be used to reach a common goal.

Even the smallest choices in the data process can have huge impacts. This example illustrates one of most important equity issues in research: there is a lot of power in getting to make these data choices. When we realize that data, evidence, and research are not completely objective processes, we discover they are a series of choices about whose lived experiences and worldviews we are going to center in the design, question, methodology, analysis, visualization, and more.

To equalize the power of these choices, you need to start by at least informing people in meaningful and useful ways that you are making them and explain your reasoning. This provides us all with the choice to agree or disagree, and is the gateway to getting better feedback and more nuanced perspectives.

Doing this involves vulnerability from usually privileged people and letting go of the power of the "black box" in your data process. This is a practical issue. If your data decisions are made under a veil of mathematical objectivity, "the data doesn't lie" kind of stuff, no one can even tell what your data actually means.

The truth is that even if you do not want to embed more equity in your data, it is about to be demanded of you. Research is losing its sheen of automatic objectivity. When you say this is how it is and our numbers do not lie, not everyone believes you. The kid in the crowded classroom does not agree. The project participant in the blue zip code does not think your project was a success. The Black and Latinx families do not want to participate in inequitable research. When our work does not match the lived reality of the very people the data comes from, people do not buy it, and they are right not to.

Let's talk a little bit more about feeling not seen in the data.

For example, if we are showing survey results about levels of satisfaction with our network of food banks and we have a large amount of data from white clients, a medium amount of data from Black clients, and a small amount of data from American Indian clients, we often don't even show the results from the American Indian respondents, because there are too few and, instead, we say those findings are "not statistically significant." We think we have to do this, because that is what we have been taught to do, but it is a choice. It is a choice with harmful equity consequences. It stops people from being counted, and in a data-based world, that is like saying they don't matter.

There is no math-based reason in this case that supports saying something with a small sample size is insignificant. It is a statement that is both technically and humanly incorrect. This is another data myth. It is a norm so entrenched that it feels like a rule. What we should be doing is talking about levels of uncertainty.

When we say "not statistically significant" in this case, what we mean is that we have a high level of uncertainty about this result. See how much less comfortable it is to say that? "We are uncertain" puts the responsibility where it should be, on us, the data analysts. It leads to the next natural questions of: "Why are you so uncertain about this group?" and "Could you have used a different way to be more equally certain about all the groups?"

"Not statistically significant," in this example, is a shield we hide behind instead of being transparent about our process and the meaning of our results. There are a thousand shields like it in data science. And people are figuring that out.

Data literacy and an understanding of the power structures involved in data is exploding. That is a great thing. The bar is being raised, and we need to rise up to it.

So, we have blown apart the myth that data is objective, that we can get to a "right" answer. We can get only to answers that reflect our inputs and the process we use. Our perspectives are the main shaping force behind those things. We see that our data is selective, our models are malleable, our results can be validly interpreted in more than one way, and almost all of our "data rules" are arbitrary and often unfair.

Should we abandon data and quantitative research? No. Can it still be used for good? Yes.

If we are willing to admit that we are making choices, then we can uncover them, improve them, and communicate them effectively. Then we really can use data for good. We can estimate and quantify and understand things from an equity lens.

If we value equity in our policies, practices, and systems, it is essential that the next generation of practitioners be supported with tools and training that equips them to succeed in embedding equity in their work with data.

Here are five steps to get started:

1. Include in all trainings the essential task of recognizing that we are making subjective, human choices in our data work.
2. Develop research frameworks that identify as many choice points in the research and evidence creation process as possible. Each of these choice points is the place in which a practitioner can increase the equity and center the voice they intend to represent in their work.
3. Build into emerging research best practices expectations that these choices will be made transparent—both to research subjects and evidence consumers.
4. Teach the importance of statistical methods in aligning the perspectives of the community, the learning agenda, and the world views.
5. Learn to communicate in an accessible and transparent manner about the world views, lived experiences, and quantitative choices that have been used to build the evidence.

We need to talk about what choices we are making in data and why. This is the only way.

Sometimes, it feels like people are losing trust in "science," but, actually, they are losing trust in scientists. They are losing trust in the scientists who won't even hold themselves to the standard of a high school science project: being honest about what they do know and what they don't know, and showing their work about how they are making choices.

Many researchers, practitioners, and analysts are trying their hardest to be good and just, to add valuable, truthful information to what we know about the world. But we cannot hide from criticism behind the idea that all data is objective or that numbers do not lie. Our data reflects the way we see the world, but that is a good thing.

It means that, instead of unsuccessfully trying to pretend that there is no worldview in our data projects, we can acknowledge that there are many choice points at which we embed world views and perspectives in our data projects, and then we can make these choices with purpose.

FIVE EVALUATION DESIGN PRINCIPLES OF JUST PHILANTHROPY

CARINA WONG

Building an evidence-based strategy that also centers on issues of racial equity is both art and science in philanthropy. For fourteen years, I have worked in philanthropy and tried to understand what can be done. I came to philanthropy as an educator who had worked for over a decade at the intersection of policy and practice. I also came as a student of innovation and design, with a penchant for wanting to identify an end user, gather insights, and understand motivations before jumping to solutions. I believe one must design for equity (especially in philanthropy); it is not inherent in the design process.

When we rely on quantitative evidence alone; when we ignore the experience and identified needs of those most proximate to the problem; when we prize rigor over practical application; and when we favor the machinations of philanthropy, government, and academia over what would be useful to those directly working on these problems, we are failing on equity. This is because equity requires listening to those directly affected and involved; understanding the why/how (qualitative) and not just the what (quantitative); prioritizing what is specifically helpful over what may be broadly true; and putting the needs of Black and brown people ahead of the needs of organizations and systems.

What does it mean to be successful in philanthropy (or policymaking), and whose success are we focused on? This essay wrestles with those questions and unpacks the role evidence has played in my own work, and it considers new ways of thinking about what role it might play in yours, through three projects (or acts) that I have engaged in over the last decade. Unfortunately, I won't be able to describe the richness of each project in detail, but I will illustrate how each project brought new opportunities and, ultimately, a set of design principles for me to apply in a "rinse, repeat, and relearn" way.

ACT ONE: TEACHER2TEACHER

Seven years ago, I was asked to take on a project to understand teacher narratives, networks, and needs.

Discovering the First Design Principle: Tailor Your Work for Your Partners and for Usability by Them

What was unusual about the work was the way in which we went about understanding teachers (using both a mix of qualitative and quantitative data) and what we did with that data. Ultimately, we used it to inform the development of a solution—a large online network. Teacher2Teacher, as the project still is called, was not predicated on using the network to scale a particular set of investments at the time. It truly was designed for teachers by teachers. We intentionally engaged teachers who were teaching Black and brown students and/or worked in vulnerable communities.

Uncovering the Second Design Principle: Center on Perspectives and Concerns of People Closest to the Problem

We had a key partner (teachers), but what did we know about them? We used traditional focus groups that told us teachers use social media and consume print and digital media in typical ways. We also heard them say: "Nobody knows teaching like teachers," and "We want to connect with our peers," and "We have no time to connect." We used narrative analytics, a process pioneered by Monitor 360 that combines big data and narrative analysis, to dig even deeper. From January to May 2014, we looked at over 2,400 blogs, 12,600 tweets, and 16,900 Edchats to get a sense of teachers' views of their work. This process surfaced ten key narratives—these narratives and the insights from our focus groups were then translated into a set of guiding principles for our work and continue to be a core part of how the community is still run today.

A Third Design Principle: Take a Dynamic, Interactive, and Networked Approach

Building this massive teacher network was not easy. But we exceeded our engagement goals, and the community is now a healthy and engaged network of 1.8 million educators. Teacher2Teacher has done much more than surface new ideas and disseminate best practices. It also has served as an important way for us to get continual insights in real time. When the COVID crisis broke, Teacher2Teacher was able to give us weekly insights from teachers on what they were experiencing, what would be helpful, and how they were helping each other. We see the network as about building relationships versus making transactions.

We had taken a set of clear actions: identify a partner you seek to work with who is close to the problem and seek to understand their needs; gather insights on what they care about and use those insights to inform your strategy; and build relationships and support a network that would surface what they need and let that drive how one might best support them. This seemed like a more equitable way to go about developing and surfacing solutions at the time, but could we apply those lessons to other projects? What else might we learn (or relearn) using a rinse and repeat process?

ACT TWO: ADVANCING ACTIONABLE KNOWLEDGE

About three years ago, I was asked to launch a new grantmaking portfolio focused on the use of evidence. Our initial learning questions: *If we know what works in education, why don't educators use it? How might we scale knowledge of what works beyond the places where we invest?* While these questions are frequently asked by philanthropy and policymakers, workable solutions are elusive.

Applying the Design Principles

We followed roughly the same protocols as we had in Teacher2Teacher. First, identify the partner you seek to support and get input and insights from them by developing relationships and listening to their needs. After deliberations, we chose to focus on school leaders (principals and assistant principals). We used a combination of qualitative and quantitative insights to understand and surface several insights about principal leader needs, networks, and behaviors.

One issue that emerged: improving attendance. What did the evidence base say about how to improve student attendance, and how might we share that with school leaders in ways that might optimize uptake?

We aggregated a community of over 35,000 school leaders online (known as the Principal Project) to get continued input and test our hypothesis about what would help them most. We found school leaders welcomed the connections and were eager to share their own knowledge about what works with others. When we tried to replicate this with other topics, we found that *the hard part was finding research and evidence-based practices that were actually usable or useful to their needs.*

Introducing a New Set of Problems with Evidence

There were four main reasons the evidence base was hard to find. First, the evidence base is often framed in ways that do not resonate with the problems practitioners face. For example, a principal might want to know how to develop deeper relationships in their school to reduce absenteeism, yet the evidence base is focused on dropout prevention programs. A related challenge we encountered was a mismatch between what researchers include in their published papers and what information practitioners actually want. Third, the format and distribution of the evidence base itself rarely acknowledged the busy lives of school leaders and the cadence of their day/week/year. Finally, there was a constant tension between what qualifies as evidence and how to include the modifications practitioners were making in real time to the evidence base given their local contexts.

Surfacing a Fourth Design Principle: Ask Explicitly about Equity and Make It a Condition of Success

We had identified the partner and gathered insights. We had started to build a network infrastructure to keep getting insights. But these challenges generated a new set of learning questions for us that began to reveal the importance of an equity orientation from the start. Our initial learning questions did not have an equity intention. We had framed the questions in a way that put the onus of change on the educator, and we defined success solely in terms of scale (adoption of practices).

Suddenly, we had another set of learning questions to address:

- What constitutes "evidence" and why does it seem so untimely or unhelpful to practitioners or needs expressed by families and students?
- Who is generating the evidence and how are those with lived experience influencing how the problems are framed?
- How, if at all, is the evidence base that already exists being used by those most proximate to the problem, and how is it reaching practitioners?

The Fifth Design Principle: Question Who Gets to Define Success and How It Is Measured

Heather Krause of We All Count reminds me: *If you want to have an equity orientation, you have to ask two fundamental questions: Where is the onus to change, and what is the definition of success?*

Her questions prompted a fifth design principle that had yet to be addressed. We had to reassess what success looks like. Success often is defined narrowly in terms of scale (reach or adoption of practices) versus considering other aspects, including behavior change, relationship development, power dynamics, structural change, or other leading indicators of impact. In the end, we took a field building strategy and a view of success that included distributed networks and decentralized power, as well as policy change from the top.

Finally, success depends on setting internal targets related to your evidence and equity intentions. It is one thing to start on an equity journey and another thing to actually collect data and qualitative feedback on how well you are living your values. It may take additional effort or dollars to support organizations making shifts in their orientation, and it may mean seeking out new partners and partnerships. We are trying to move beyond the usual partners and set clear targets for engaging more organizations that have high levels of equity capacity and are run by leaders of color.

ACT THREE: ASSESSMENT AND ACCOUNTABILITY

About six months into the COVID pandemic, I was asked to launch a new opportunity area. Given the pause in standardized testing in both the K12 system and the SAT/ACT, the question was raised: What opportunity might this disruption bring? I began to think about how the five design principles summarized here might apply with this very different project that was focused on policy reinvention and potential technical innovations: 1) Ask explicitly about equity and make it a condition of success; 2) Center on perspectives and concerns of people closest to the problem; 3) Tailor your work for your partners and for usability by them; 4) Take a dynamic, interactive, and networked approach, and, finally; 5) Question who gets to define success and how it is measured.

Ask Explicitly about Equity and Make It a Condition of Success

First and foremost, we started by asking an intentional question about equity. Lesson learned! We could have asked a general question as we conducted our research, such as: What was the impact of standards-based reform? Instead, we chose to ask questions in this way:

- **RQ1:** How and in what ways did standards-based assessment and accountability address structural inequities in the education system? What were the successes and challenges?

- **RQ2:** What were some of the unintended consequences (that is, negative impact) of standards-based assessment and accountability on schools and districts serving primarily Black, Latino, and students living in vulnerable communities? What pushback, if any, did standards-based assessment and accountability receive, and from whom?

- **RQ3:** Of the districts previously identified as low-performing or turnaround but are now demonstrating positive academic shifts for target students (Black, Latino, and those living in vulnerable communities), what actions were taken to address the unintended consequences of standards-based reform? Were equitable strategies and approaches used to address unintended consequences of standards-based reform? If so, what were the emerging results? What factors or conditions appear to be driving positive shifts?

We prioritized understanding the structural inequities and intentionally hired a team of diverse and equity-minded researchers to undertake the analysis.

Center on Perspectives and Concerns of People Closest to the Problem

As part of the assessment and accountability project, we conducted a landscape analysis, interviewed researchers and early architects of the standards-based reform movement, and did a lookback internally at what we had invested in and why. This is where the fact base might have ended.

But we again chose to look further by finding partners who could give us deeper insights into how those closest to the problem experience the current assessment and accountability system. We intentionally included this learning question upfront as core to our strategy: What can we learn by listening to/acknowledging the voices/views of families, educators, and students most affected by standards-based assessment and accountability since it was initiated?

Question Who Gets to Define Success and How It Is Measured

As part of our insights work, we were trying to understand how students, educators, and family members define what success looks like for their children. We still are gathering insights as this chapter goes to publication. Success from my perspective will be to surface these insights in a way that helps inform the policy conversations to come.

A Work in Progress: The Emergence of a Sixth Design Principle

One issue that remains unresolved and emerged in the Advancing Actionable Knowledge work was about who holds power and in what form. In this new project, the issue surfaced through discussions about decision making power versus those who are most impacted by the current assessment accountability system. Perhaps as this work progresses, we will have to reckon more deeply with the power dynamics between funders and other stakeholders as well. A sixth design principle? The question remains. The journey continues.

TOWARD MORE JUST PHILANTHROPY

Equity will continue to be elusive if we dance around the edges of racism and power dynamics and fail to address these issues in our strategies, organizations, and systems. While I am not an equity expert, and have a long journey ahead to be sure, I have learned that the simple act of being intentional about racial equity as a goal, and expanding our notion of what constitutes evidence is a step in the right direction. I have made the case for improving a strategy's equity orientation by starting with a set of intentional design principles. This is a starting place for a much longer and more complex journey toward using evidence in ways that lead to what I call "more just philanthropy." Just philanthropy is a mindset and a way of approaching strategy development that involves engaging stakeholders in new ways and acknowledging that the solutions to the most wicked problems lie in the hearts and minds of those most proximate to the problem. It is a discovery process, not a solution, and I am still learning.

CAN THE NEW DATA ECONOMY GIVE BACK TO COMMUNITIES?

CHRIS KINGSLEY

Data has earned a bad reputation within the social sector, with the most acute complaints coming from the front-line workers and communities it is meant to benefit. Teachers disparage "flawed, unfair and incomprehensible" new uses of statistics to measure their performance.[1] Social workers are "drowning" in data they are required to collect but lack the training to use.[2] And some communities, citing long histories of having been over-researched but underserved, are organizing to reassert rights over how their data are used and to insist on broader definitions of what *kinds* of data and evidence matter.[3] While it has been promoted as a tool to help organizations continually prove and improve the value of their work, data is more often associated with production management philosophies, narrow registries of evidence-based programs that come with mandatory certification from remote experts, and complex matrices of indicators imposed by different funders with competing theories of change.

It is not as though proponents of data and evidence have been running down a blind alley here. These activities have value, and a focus on measurement and outcomes imposed from the top down is *one* part of the answer to what we might call the Practitioner's Prayer ("God grant me the courage to fix programs and policies that don't work, the resources to expand those that do, and the data to know the difference"). The education and social sectors direct more than $1 trillion each year, and it is eminently reasonable for taxpayers, practitioners, students, and other participants to know what their contributions are achieving. Moreover, there is a strong moral case for collecting the data necessary to understand and begin to correct the effects of decades of discriminatory practices and policies on the well-being of communities of color.

But there are consequences to teachers, social workers, and nonprofits relating to the use of data almost exclusively as a tool to define, limit, and control their programs and organizations rather than to interrogate, explore, and strengthen their work. A top-down approach to measurement causes expensively developed, quickly abandoned systems to proliferate in the back offices of agencies and nonprofits. It breeds cynicism among frontline staff about new data collection activities that detract from doing their jobs without returning anything of obvious practical value. And, over decades, it has eroded relationships with students, clients, and communities who are too frequently required to sign over access to so much personal data without being invited into conversations about how it's being used. (In the words of Chicago Beyond: "Why am I always being researched?"[4])

These are the complaints of results-oriented people, many of whom would agree with Mark Friedman's rallying cry that "trying hard isn't good enough" but are keenly aware that the data-driven regime that has been built around them is serving somebody else's purpose.[5]

A NEW DATA ECONOMY

The social sector's leaders have a tremendous opportunity to overhaul this broken information economy and, in so doing, put data in the service of innovation, systems reform, and rebuilding cooperation between agencies, nonprofits, and their communities. And the starting point for that transformation is to restructure the market for data and evidence.

Why do so few tools exist for families to manage their own social services profiles or to compare the efficacy of different providers? How can it be that the high-stakes testing systems built to evaluate schools and teachers do not return timely, useful management information to principals and superintendents? Much of the data infrastructure supporting services to children and families was built to the specifications of public and private funders to facilitate payment, auditing, and outcomes reporting. The occasions when these investments in better data also result in local innovation and improvement are more the result of happy accidents than design; funding incentives are rarely aligned to sustain and scale them.

So, what does the alternative look like?

We can look first to organizations that are serious about designing tools from the perspective of clients and practitioners, incorporating—or at least emulating—the missing "market demand" of communities and nonprofits. For example, when Code for America launched its Integrated Benefits Initiative,[6] staff began as clients would, by applying to different public benefits and documenting the impediments they would have to correct to create more friendly, uniform services for families.

More than capital "R" research and evaluation, public leaders need data partners that can contribute to rapid-cycle analysis and problem solving. This is the kind of service the University of North Carolina's Charlotte Regional Data Trust provided Charlotte-Mecklenburg Public Schools when they identified hundreds of students receiving housing and homelessness services,[7] unknown to the district, and qualified them through the McKinney-Vento Homeless Assistance Act for additional funding available as well as resources like transportation services and expedited enrollment.

The need for this kind of responsive analysis is particularly acute during moments of crisis, as Colorado discovered in the early days of the COVID-19 pandemic when it turned to the state's Evaluation and Action Lab at the University of Denver to quickly connect licensed childcare workers to centers that urgently needed them[8] to serve the children of essential workers. This kind of disaster response illustrates a more broadly generalizable lesson that funders should take to heart: By the time this kind of data infrastructure becomes mission critical, it is too late to build it from scratch. It is *exploratory* analysis within the social sector, not summative reports on the result of a program, that can create the space for new thinking and different responses. A data economy that supports this kind of exploratory analysis is one that can help build, test, and scale innovative solutions. Some examples:

- During the pandemic, Los Angeles County, California, leveraged its long established integrated data infrastructure[9] to support people who were experiencing homelessness and at a greater risk of contracting COVID-19. By linking information from healthcare and homeless management information systems (HMIS) datasets, county researchers working with the University of

Pennsylvania and UCLA were able to assess discrete levels of vulnerability among the aging homeless population and propose housing and service models that matched their level of risk. They also were able to estimate potential cost offsets to Medicaid and the county that would help recapture funds needed to help stabilize people in housing.

- New York City's experience during Hurricane Sandy prepared them to be a reliable, community-engaged partner when NYC was the epicenter of the COVID-19 outbreak in 2020. The city's Center for Innovation through Data Intelligence's (CIDI) used its existing cross-agency workgroup to quickly map vulnerable populations,[10] drawing on integrated client data from NYC Health and Human Services and overlaying information on public housing, retirement communities, and shelter sites. NYC's immediate aid targeted these most at-risk populations and contributed to a more equitable response and recovery to the pandemic.

- Cuyahoga was the first county in the United States[11] to receive social impact financing after building their project on cross-agency data analysis that suggested there could be tremendous benefits—both to families and to the county budget—to providing coordinated housing and social supports to mothers with children in the foster care system, to more rapidly stabilize and reunite them.

These are examples of work that take seriously the needs of agency and nonprofit practitioners and their clients, and that use data to interrogate problems, explore new solutions, and put authority in the hands of decision makers who are closer to the point of service. They are initiatives that use data as a flashlight and not as a hammer. Much more of this is possible.

WHAT IT WILL TAKE

This chapter has been critical of a data economy that revolves around the planning decisions of large government systems and private funders rather than one that reacts to market forces reflecting the needs of nonprofit practitioners and communities. That result was not inevitable, however, and creating a different economy for data and evidence will require infrastructure and new capacity within communities. All of us have a role to play in laying that foundation—funders included, and especially.

Infrastructure

The kinds of data projects that can strengthen the decision making of communities and practitioners share common elements. They are developed by data intermediary organizations that center on practitioner needs and have built trusting relationships with their agency and nonprofit partners, often with formal governance arrangements that include business and legal agreements. When these projects use external technical expertise, that expertise is martialed through organizations like Code for America and the U.S. Digital Service with deep knowledge of the pain points, incentives, and limitations of their public and nonprofit partners. These organizations are vehicles of a more responsive and innovative economy for data products and tools. The networks that connect them are the roads by which new tools, policy analyses, and initiatives propagate.

These intermediary organizations and networks are chronically undersupported parts of the sector. Several of the projects described here were created by social entrepreneurs operating outside the bounds of their professional responsibility, sometimes against the incentives of their funding. To create and sustain this kind of adaptive data capacity requires more than project-oriented grants and capital dollars for modernizing technology. It takes patient support for the crucial "soft" work necessary to understand the priorities of agency and community leaders, negotiate terms of access to their information, and prove that this kind of data infrastructure can solve real problems. Once established, these data intermediaries—whether university-based policy labs, state offices like Kentucky Stats, or local nonprofits like the New Orleans Data Center—tend to persist and expand into new domains where they can rapidly and cost-effectively build projects. It is in the enlightened self-interest of government and philanthropic funders to help develop these practitioner- and community-oriented organizations within the counties and states where we work, and to start *before* there is an urgent need for evidence.

Sharing and Building Power

Fans of data and evidence should recognize that our goals of using data for good depend on earning social license, described in Amy O'Hara's chapter as something that exists when the public trusts that data will be used responsibly and for societal benefit. A caution flag has been flying for several years that parents and communities—particularly communities of color—are dissatisfied with their place in the development of this data economy. For parents and educators, it was the 2013 public launch of a centralized data sharing platform, inBloom, that catalyzed three years of protest and hundreds of pieces of privacy legislation[12] aimed at curbing the collection and use of information on students. Recent and more pointed arguments from civil rights organizations like Data for Black Lives and the Leadership Conference for Civil and Human Rights have focused on the need to renegotiate limits on the use of data and technology tools,[13] and to foreground issues of race and racism. The data of communities of color is often collected and used, but communities are rarely included in framing the field's research priorities.

Organizations like Actionable Intelligence for Social Policy have responded by collaborating with some of these critics (colloquially, "frenemies" of data) to develop a roadmap for centering race in data use, integration, and governance.[14] And a few local and state data intermediaries are making earnest attempts to give parents, nonprofits, and communities of color real power at those governance tables where decisions are made about what kind of evidence is important to build. But these efforts are nascent and more difficult to manage the further organizations get from neighborhoods and schools. The National Secure Data Service envisioned by the 2018 Evidence Act, for example, has tremendous potential to contribute to racial equity analyses of U.S. programs and policies at all levels, but questions remain whether or not Americans will tolerate the federal government directing such a powerful tool.

The Annie E. Casey Foundation argues that the social sector should lean into this debate[15] about data infrastructure and data innovation, and that, rather than trying to "abolish big data,"[16] funders and civil rights advocates alike should create and enforce standards for the fair and good use of data. The good uses are potentially vast: from AI tools that already triage crisis calls to the Trevor Project[17] and target lead remediation efforts[18] to new talent screening models[19] that promote more diverse technology workforces by rewarding aptitude rather than educational pedigree.

These "fair and good" uses are not inevitable, however. Civil rights critics of these technologies—and of data's use in the social sector—are right to point out that, in some ways, the more *likely* outcome is the opposite, that data science will be deployed in ways that systematically disadvantage poor and minority communities through greater surveillance and actuarial discrimination. This, too, is already happening, as anyone who reads Upturn's weekly newsletter[20] knows.

The difference between these two competing visions parallels the problem this chapter began with, and its fault line is the willingness of our data economy's most powerful actors to cede some control over whose questions take priority and whose decisions new data tools interrogate. By giving nonprofits and affected communities a greater stake in the creation and ownership of this kind of evidence, we enable a much more dynamic and fair market for new ideas and solutions. This is the right moment for the social sector to recommit itself to uses of data that are not only useful to practitioners but also *empowering* to the communities that the Great Society, which inaugurated so much of this kind of policy analysis, was created to help.

NOTES

1. American Federation of Teachers press release, "Federal Suit Settlement: End of Value-Added Measures for Teacher Termination in Houston," October 10, 2017, https://www.aft.org/press-release/federal-suit-settlement-end-value-added-measures-teacher-termination-houston.

2. Putting Data to Work to Improve Child Well-Being: A Post-Convening Report. Casey Family Programs and the National Governors Association for Best Practices: Washington, DC, 2006.

3. See "Why Am I Always Being Researched," at the Chicago Beyond website, https://chicagobeyond.org/researchequity/, as well as the work of groups like Our Data Bodies, https://www.odbproject.org/.

4. Ibid.

5. Mark Friedman's seminal book *Trying Hard Is Not Good Enough: How to Produce Measurable Improvements for Customers and Communities* (Santa Fe, NM: Parse Publishing, 2019) helped popularize results-based accountability, .

6. Code for America, "Reshaping the Safety Net: The Integrated Benefits Initiative State Cohort," Medium, August 28, 2018, https://medium.com/code-for-america/reshaping-the-safety-net-the-integrated-benefits-initiative-state-cohort-32ea621e40d5.

7. See Child and Youth Integrated Homelessness Data Report: Part 2 page at Charlotte Urban Institute website, https://ui.charlotte.edu/story/child-and-youth-integrated-homelessness-data-report-part-2.

8. See "Critical Data Ensures Child Care for Essential Workers," Colorado Evaluation and Action Lab blog, June 24, 2020, https://coloradolab.org/blog-2020-06-24-child-care-for-essential-workers/.

9. See AISP Network: Los Angeles County Enterprise Linkages Project 2.0 page at AISP website, www.aisp.upenn.edu/network-site/la-county-2/.

10. See "Identification and Mapping System for Vulnerable Populations," at NYC Center for Innovation through Data Intelligence website, https://mcusercontent.com/07732221ebbe01a8b8186e8ef/files/b1d68383-43bc-49b8-a7f8-e4b8d5724f08/Vulnerable_Populations_Description_and_Process.pdf.

11. See Integrated Data Is Key to "Pay for Success" page at the Annie E. Casey Foundation website, www.aecf.org/resources/integrated-data-is-key-to-pay-for-success/.

12. "The Legacy of InBloom, Working Paper, Data Society, February 2, 2017, https://datasociety.net/pubs/ecl/InBloom_feb_2017.pdf.

13. See Principles page on Civil Rights Privacy and Technology Table website, www.civilrightstable.org/principles/.

14. "A Toolkit for Centering Racial Equity Throughout Data Integration," Actionable Intelligence for Social Policy, https://aisp.upenn.edu/centering-equity/.

15. The Annie E. Casey Foundation, "Four Principles to Make Advanced Data Analytics Work for Children and Families," The Annie E. Casey Foundation, October 27, 2020, www.aecf.org/resources/four-principles-to-make-advanced-data-analytics-work-for-children-and-famil/.

16. Yeshi, "Abolish Big Data," Medium, July 8, 2019, https://medium.com/@YESHICAN/abolish-big-data-ad0871579a41.

17. Josh Weaver, "The Trevor Project Launches New AI Tool to Support Crisis Counselor Training," The Trevor Project, March 24, 2021, www.thetrevorproject.org/blog/the-trevor-project-launches-new-ai-tool-to-support-crisis-counselor-training/.

18. "Machine Learning Models Identify Kids at Risk of Lead Poisoning," Science News, September 21, 2020, www.sciencedaily.com/releases/2020/09/200916113523.htm.

19. Michele Cunningham, "Time to Use Data for Social Good," Catalyte, www.catalyte.io/data-for-social-good/.

20. See About page on the Upturn website, www.upturn.org/about/.

STOP EXTRACTING

OUR DATA, OUR EVIDENCE, OUR DECISIONS

ROBERT NEWMAN WITH DYLAN EDWARDS, JORDAN MORRISEY, AND KIRIBAKKA TENDO

In the 1990s, I was working in central Mozambique as the country coordinator of an international NGO. My team worked closely with district health management teams, the branch of local government responsible for health service delivery in the communities where we operated. We found that these teams spent significant amounts of time collecting data and submitting it to their bosses at the provincial or national level of the health system. The whole system was very opaque for the district teams. As far as they were concerned, they collected data and waited for decisions to be made elsewhere. The teams themselves did not have an appreciation for the potential power of these data to catalyze immediate action and drive local public health improvement.

The teams we were working with spent a lot of time collecting data on immunization rates, for example. This information was painstakingly captured at health facilities, which, in post-war Mozambique, lacked even the most basic infrastructure, like electricity and running water. These paper records were sent to provincial health offices, but it was not clear to the teams at the health facilities what happened next, or what happened to the paper forms they submitted to their bosses. It was important to them that they collected the data because they knew they would be in trouble with their bosses when they did not submit their forms on time. But beyond that, what happened to all that information was a bit of a mystery.

In response, we designed and implemented a program to support these teams to work with the data they were collecting before passing it up the chain to provincial level. We taught the teams how to look at the data, how to perform relatively simple analyses, and how to identify potential local actions *they* could take in response to those data without waiting for feedback from provincial or national health officials, which could take over a year or might never happen at all. If the data showed immunization rates were falling at one specific health facility, we taught them to ask the kinds of questions that might allow them to solve problems themselves. Were there sufficient supplies of vaccines and syringes? Were there enough appropriately trained staff on duty? Could it be a transport issue? Or, perhaps mothers are preoccupied with harvest time. Are there any social or cultural reasons people might not trust vaccines?

At the time, this program, which we called using data for decision making, seemed at once a simple and radical concept. Now, more than twenty years later, there is an enormous focus on "big data" in global health. Unfortunately, much of this discourse has played out in the conference rooms of wealthy countries, far from the halls of the ministries of health that are, ultimately, responsible for the analysis and use of public health data, and even farther from the front-line district health teams collecting those data. In fact, "data" has become big business and, in many ways, has come to resemble an extractive industry. Large and powerful organizations fund and push for the collection (or extraction) of data from lower-resource settings, which are then collated, analyzed, and published, often in prestigious international journals and with much fanfare and celebration about the power of big data to drive evidence-based programming.

The disconnect we saw in Mozambique all those years ago is too often still at play: data collection is something local healthcare workers do. Using data to make decisions is something that happens in a boardroom somewhere else. *We collect. We submit. They analyze. They decide.* By treating the generation of evidence as an extractive industry, we risk entrenching patterns of exploitation that have been in place since colonial times. As long as we continue to do so, we will reinforce the divide between health care workers who collect data and the academics, funders, governments, and companies that use those data to make decisions.

How, then, can we make the process of collecting data and building evidence better serve the needs of the people affected by the decisions these data are used to inform? In the remainder of this chapter, we focus on two key areas in need of reform: *building grassroots capacity* to more effectively analyze data and build evidence, and *increasing transparency* of the process of transforming data into evidence.

BUILDING GRASSROOTS CAPACITY

The front-line staff collecting data at peripheral levels of health systems continue to lack opportunities to learn and develop data analysis skills, and, therefore, see their role as one of submitting the data to the next level of the system before getting scolded for failing to do so. They do not see it as within their remit to use those data to inform *their* decision making in the service of setting and advancing *their* objectives. By failing to foster and enable the transformation of data into evidence for informed decision making at the front lines of health systems, the development community perpetuates this unfortunate cycle. The sort of work we were doing in Mozambique remains largely unfunded and unfinished.

While data and digital technologies for health have strong potential to catalyze improvement in health systems and health outcomes, the people tasked with managing these technologies often do not fully understand the potential of these systems to inform *their* decision making. Too often, well-meaning providers of technical assistance have started with a new tool, or have shown up with evidence for a particular intervention, and expected the receiving team to respond promptly and positively to either adopt the tool or create new policies and programs. They come with a solution to a problem they believe they understand rather than coming to the table seeking first to understand the specific contexts, challenges, and opportunities present in that country, region, district, or community. Those local contexts, challenges, and opportunities are precisely what Ministry of Health professionals working at various levels in the system are best poised to provide expertise in.

We believe there is a missed opportunity to bolster a fundamental understanding of the intrinsic value and potential of data, evidence, and digital tools. Specifically, we think it is critical that Ministry of Health officials and cadres *first* appreciate the importance of timely and accurate data for decision making in managing their work, *then* the possibilities presented by digital systems and tools to use that data and evidence to drive better decision making and more effective management.

While we should not be Luddites, we should be skeptical of individuals and groups selling tools and technologies that promise to provide near-magical solutions to problems. Instead, we believe that if Ministry of Health teams at all levels of the system develop an appreciation and understanding of the power of data and evidence, the skills to analyze and transform data into evidence, and how to use that evidence to drive programmatic improvement, then they will be able to broker the sort of partnerships and request the types of tools needed to support *their* efforts *in their* contexts.

When front-line staff have a greater appreciation of the fundamental value of data to support their work, they will have a greater stake in ensuring that accurate, timely information is captured. This will create a virtuous circle, where data leads to better decisions, which leads to better outcomes, which increases the demand for good quality data.

One example of how this can work has emerged from Ethiopia's Community-Based Data for Decision Making (CBDDM) strategy. Under this program, community health workers collect data to create maps of the households they support. These maps also display information relating to each household's reproductive, maternal, newborn, and child health needs. Meetings are then held at community health facilities to review the data to identify barriers to access to services and implement solutions. This allows health workers to set targets, plan and prioritize more effectively, and monitor progress. An evaluation of the program found that the intervention led to significant improvements in the uptake of maternal and child healthcare services.[1]

INCREASING TRANSPARENCY

There also needs to be greater transparency concerning the transformation of data into evidence. While we often refer to data and evidence in the same breath, there is not much transparent discourse on the process by which, for example, epidemiological data are turned into evidence. And, generally, that transformation is taking place far from where the data are collected. In some cases, the statistical methodologies and modelling being used are so complex that even specialists are not capable of understanding the process and, therefore, cannot question it or the underlying assumptions used.

This evidence is then used to set targets that countries are expected to meet, further disempowering public health leaders at national and peripheral levels.

In addition, not all sources of data are given equal weight. Randomized controlled trials have emerged as the "gold standard" of evidence for health and many other domains. And while well-designed RCTs can, indeed, be powerful sources of evidence, overly focusing on them risks ignoring other highly relevant and more easily (and inexpensively) collected data that could allow for more local hypothesis generation and testing to drive programmatic decision making.

This work of increasing transparency in evidence building can be time consuming and resource intensive, and does not produce the sort of quick wins generally attractive to large development funders. Therefore, despite all the talk about the importance of data in recent years, not much has changed. Only in the last few years have we seen a surge in discussions about the power of data (especially big data) and its translation into evidence to drive public health programming and accelerate achievement of ambitious global health goals.

Meanwhile, back in the countries and communities from which the data were collected, little action is likely to have been taken in response to the data. As I learned in Mozambique, feedback takes such a long time to reach the initial source (if it ever does get there) that these data-turned-evidence may seem irrelevant.

There are, however, some notable examples of approaches that use inexpensive, community-based approaches to data collection and use these data to hold service providers accountable. In South Africa, for example, a coalition of civil society organizations analyzed the local government's budget for sanitation in a major city's informal settlements and compared it to actual services received on the ground by carrying out a "social audit." The civil society organizations mobilized residents of a poor neighborhood to take stock of public sanitation infrastructure projects in their community and compare their findings to the official figures provided by the city government.

The social audit concluded the city was failing to monitor contractors, leading to wasteful expenditures and human rights violations.[2] After a presentation of the coalition's findings, the local municipality agreed to repair and better maintain sanitation facilities for 5,000 informal settlement residents, including the installation of new doors, taps, and drains.

Partnerships like this demonstrate what is possible when citizens are given access to information and provided with the skills to interrogate it. It also shows how government transparency might lead to improved levels of trust in government.

DATA FOR DEMOCRACY

The examples of health workers in rural Mozambique or a civil society organization in South Africa may seem removed from the daily reality of many, particularly in wealthier countries. But the broader point is relevant whether you live in Chicago or Chimoio: being able to engage with data has the potential to enable practitioners to make better decisions. More than that, it enables practitioners to translate data into meaningful information and evidence.

Access to information is critical for a functioning democracy. It allows citizens to participate in decisions that shape their lives, to influence the way those in power make decisions, and to hold them accountable for those decisions. Quality information is an indispensable tool in advocating for equal treatment and enabling people to fully participate in civic life. It allows health workers to make timely, informed decisions on where to focus their resources, and enables civil society organizations to hold governments to account for their spending.

Democracy is about more than holding regular elections. At its core, it is about giving people greater control over the decisions that affect their lives. This chapter has argued that greater transparency combined with concerted effort to build data literacy skills of local practitioners will allow people to take greater ownership of their evidence and make more informed decisions. It also gives people the skills they need to hold governments accountable. Ultimately, this will contribute to a deepening of democracy.

Achieving this, however, will require both significant investment and a shift in mindset. Funders looking to support development projects that focus on the use of data are often interested in high-tech tools and innovative technologies with the potential to disrupt old ways of doing things. However, if the appropriate data skills are not in place, these projects are unlikely to build traction over the long term. This implies a need for a longer-term and more practitioner-centric approach, recognizing that building data skills locally is essential to ensuring the sustainability of any investments. Practitioners, for their part, must recognize their own role: *Our data. Our evidence. Our decisions.*

NOTES

1. A. M. Karim, N. Fesseha Zemichael, T. Shigute, and others, "Effects of a Community-Based Data for Decision-Making Intervention on Maternal and Newborn Health Care Practices in Ethiopia: A Dose-Response Study," *BMC Pregnancy Childbirth* 18, no. 359 (2018), https://doi.org/10.1186/s12884-018-1976-x.

2. Social Justice Coalition, "Report of the Khayelitsha 'Mshengu' Toilet Social Audit," International Budget, 2013, www.internationalbudget.org/wp-content/uploads/Social-Justice-Coalition-Report-of-the-Khayelitsha-Mshengu-Toilet-Social-Audit.pdf.

DE-RISKING DATA

EQUITABLE PRACTICES IN DATA ETHICS AND ACCESS

AMY O'HARA AND STEPHANIE STRAUS

INTRODUCTION

Data on individuals are collected on almost every facet of our lives: our location, well-being, purchases, and interests. Despite this fact, many of us do not understand the ways in which our data are being used (see art installation in Designboom 2018[1]). Whether working for pay, using social media, or scrolling our phones, we often mindlessly agree to terms of service and data use, not rifling through the reams of legalese or considering the benefits of the data use at hand. Similarly, data are highly valued for many secondary uses, including research and evaluation in the government and nonprofit space. Yet these governments, nonprofits, and philanthropy organizations that enable this secondary data analysis do not always communicate with their data subjects and the greater public *why* they are using individuals' data. Individuals may not have been given an informed choice about their data being used or considered how their data will be repurposed for program evaluation, trend analysis, strategic planning, or predictive modeling. This is an issue because this lack of transparency undermines the public's trust that data will be used for greater good, which hampers future data efforts and precludes proper community engagement.

In addition to the lack of public transparency, there are no adequate, widely accepted guardrails for responsible data use in a big data world focused on evidence building. Ethical review from the biomedical world may be a poor fit for assessing responsible data use by government agencies, requiring principles more pertinent for program participants and communities represented in surveys. For example, the Menlo Report (2012) affirmed that the Belmont principles of beneficence, justice, and respect for persons, from the medical ethics world, were a sound fit for information and communication technology research, and added a fourth principle (respect for law and public interest), and encouraged development and implementation of ethical impact assessments. From these principles sprung the Institutional Review Boards (IRB) that govern all ethical data use on human subjects. However, for secondary data uses, IRBs do not always apply, and data users are left to their own devices to ensure their data subjects are being properly protected and their communities properly informed.

There are many points of discussion relating to private sector, administrative, and research uses of data. This document focuses on administrative and research uses of data not motivated by concerns about monetization. *This chapter summarizes the landscape of ethical, trusted data use as it currently exists in the research and evaluation ecosystem in the United States to discuss the current blind spots and what they mean for equitable data practices. We suggest that fair and equitable practices around data ethics and access are essential to the sustainability of administrative research uses of governmental, private and public data—and the risk of not using data for these purposes far outweighs the risk of using them. We recommend ways to improve the usage, access to, and provisioning of these datasets, highlighting real-world examples that, although promising, represent isolated instances and so must be properly scaled to produce true high-level impact.*

ROOM TO IMPROVE ON PUBLIC INTEREST, TRUST, AND TRANSPARENCY

The Administrative Data Research Network in the United Kingdom (ADRN-UK), the primary government data intermediary for their Office of National Statistics, found that the public is broadly supportive of their data being used as long as: 1) the work is in the public interest; 2) data privacy and security needs are being met; and 3) there is trust and transparency (Waind 2020). The United States often is quite strong in data privacy and security, but has lagged in establishing what the public interest is, and is equally weak in creating trust and transparency for its data subjects and stakeholders. The National Institute of Standards and Technology (NIST) routinely establishes the baseline that federal information technology systems must meet to prevent unauthorized access. NIST continually monitors the needs of the government, updating its standards (NIST 2020a), and develops new frameworks, such as the Research Data Framework (NIST 2020b). The National Institutes of Health launched All of Us[2] in 2018, aiming to build a massive database containing the electronic health records, biomarkers, and survey responses from 1 million participants to improve precision medicine. All of Us has privacy and trust principles, as well as data security policy principles. These principles, along with their data security framework and certificates of confidentiality, aim to protect privacy for the people in this longitudinal study.

Across these initiatives, however, there has been less a focus on *why* this data must be used. This gap in explaining why using administrative data is in the public interest, and a parallel lack of transparency about current and planned uses, is significant. These initiatives are needed to create an environment of trust between data owners and data subjects, which feeds directly into the concept of social license. *Social license* exists when the public trusts that data will be used responsibly and for societal benefit. Social license pertains to the reuse of government records, as well as data held by other organizations, such as healthcare systems, post-secondary institutions, and private sector companies. It requires an understanding of what safe use would be, belief that data security terms and conditions will be met, and trust that enough value will be created through data uses (Data Futures Partnership 2017). Data users must earn *and* maintain trust. This requires continuous communication and engagement to align user intentions and data subject preferences (O'Hara 2019). However, there often are power imbalances between data subjects, controllers, and users. Additionally, there are not always opportunities for direct communication with or consent from research subjects about secondary data uses. Some key questions in building social license include:

Public Interest	Trust	Transparency
• How do researchers and evaluators make clear what they are going to *do* with the data?	• How do we shift from disempowered users (who feel their data are out of their control) to empowered data users?	• How do we increase knowledge about data use across a wide range of subjects of various ages and cultures, and who communicate in a multitude of languages?
• Can they explain their findings, why they matter, and what they plan to do next?	• How can data use be seen as contributing to bettering health, communities, and society?	
• Does the audience understand how their data were used? How are the learnings applied to real-life issues?	• How do we make sure data collection and retention are not exploitative?	

INVESTING IN SOCIAL LICENSE

To address public interest, trust, and transparency, we need a balance of norms that apply to all evidence-building data uses, with sensitivities specific to each type of data and how it is used. This could involve a combination of government regulations and standards, as well as norms about data use and public involvement. Like the American Humane organization, with its "No Animals Were Harmed"® certification in film productions that meet a rigorous standard of care for animal actors, we need standard-setting followed by compliance monitoring. The film industry knows that allegations of noncompliance will be investigated and that productions failing to meet standards will be sanctioned. Similarly, in government and philanthropy, oversight bodies can ensure regulations and standards are met, and gather input on evolving concerns of the communities contributing data and affected by the data uses. Supporting groups include the federal government, state and local governments, and philanthropy.

Federal Government

Government can aid in the development and introduction of standards and policies that boost transparency, and can help define the public interest through deeper use of data for evidence building. Under federal regulations, data users already must abide by standards from the National Institutes of Health and inform Institutional Review Boards (IRB), which often act as the only ethical checkpoint before big data analyses. IRBs are helpful for research on human subjects in outlining proper informed consent to reflect respect for persons and in applying the ethical principles of justice and beneficence.

However, many uses of secondary data are exempt from IRB review, and there is no federal standard for assessing data ethics.[3] We can do better. Looking to our international peers, New Zealand has a digital government strategy that produces standards and guidance for online engagement, and the Information Commissioner's Office (ICO) in the UK has a data sharing code of practice that informs researchers on what they need to tell data subjects (New Zealand government 2020; Information Commissioner's Office 2019).

State and Local Governments

State and local governments can pass laws and incorporate transparency and trust building into policies. For example, in 2008, Seattle, Washington, passed an executive order on inclusive outreach and public engagement (Nickels 2008). An outreach and public engagement liaison from each city department now helps community members with the translation and interpretation of policies using data and with understanding study specifics and broader public health issues. They have tools to support public engagement, including an evaluation template to gauge the effectiveness of their engagement efforts. The city of Fort Saskatchewan developed a public engagement framework centered on the representation of diverse voices and encouraging dialogue with its citizens to develop solutions for issues affecting their lives (The Praxis Group 2012). The Actionable Intelligence for Social Policy (AISP) at the University of Pennsylvania serves networks of state and local governments using data to improve service delivery. Their learning cohorts benefit from their Toolkit for Centering Racial Equity Throughout Data Integration,[4] which highlights best practices in advancing racial equity through data sharing and integration. With activity templates, it also guides users in identifying which stakeholders they should engage from within their community.

Efforts to work across governments, with academic support, are growing, as well. For example, the Societal Experts Action Network,[5] a collaboration between the National Science Foundation and the National Academies of Science, Engineering, and Medicine, connects expert researchers with mayors and city officials to develop evidence-based recommendations to support local, state, and national responses, particularly in the wake of the COVID-19 pandemic. However, there are over 19,000 cities in the United States, and there are 1.5 million nonprofits. What networks must form to scale such activities? A challenge is that community engagement, sharing and opening data, exploring consent issues, and—most importantly—clearly communicating how and why data are used are activities spread across teams, seldom falling to one role to do this important trust-building work. Additionally, many of these efforts are jurisdiction-specific. While these state and local data ethics/public engagement efforts are excellent examples, they represent isolated instances across the U.S. data use landscape and, coupled with the lack of standards and incentives to build trust and transparency, amount to limited progress overall.

Philanthropy

What can philanthropists do to encourage better practices? They can require attention to trust-building and transparency, just as they require accountability for expenditures, evaluation after convenings, and data archiving. Philanthropists also can encourage norms and systems that hold researchers accountable for appropriate data use and clear communications with stakeholders, learning from less successful endeavors (Carter and others 2015; Dahl and Saetnan 2009). Philanthropic organizations also can influence the public perception of data use through targeted messaging, such as the Data Saves Lives[6] campaign led by the European Patients' Forum (EPF) and the European Institute for Innovation through Health Data (i~HD). Data Saves Lives is an initiative that shares relevant information and best practices on the use of health data to help both health patients and the general public understand the importance of health data use and what safeguards the health community has in place. In addition to targeted messaging, nonprofit organizations can work in collaboration with government organizations to create more application-based recommendations to facilitate data users in changing their practices. The UK Anonymisation Network (UKAN), a nonprofit organization that works with the UK's ICO, designed an operational method of planning called the Anonymisation Decision-Making Framework, which data users can reference to anonymize data and remain compliant to UK legislation. In the United States, the nonprofit organization Thrive! uses data to facilitate equity audits and help local governments identify programs to invest in that will reduce disparities, thus disrupting the generational cycle of poverty. Thrive! currently is launching local government pilot programs in Massachusetts, Vermont, and New York (Gardizy 2021).

Philanthropy can insist on translational work, communicating what was learned and its relevance. Grantees can be required to communicate their methods to safeguard data during and after use, and stress the benefits gained relative to the minimal risks of using the data. The community of funders, data controllers, and data users must articulate how data use is less risky than non-use; that is, that the risk of not knowing whether a treatment is effective outweighs the managed risks of using data. That message must be heard by policymakers, regulators, stakeholders, the media, and the general public. Beyond grant reports and scholarly outputs, this message can be delivered through Hill briefings, development of draft legislation, and op-eds.

EQUITABLE DATA PRACTICES

Marginalized and vulnerable communities face data equity challenges. Organizations are acknowledging and addressing current inequities, including institutional, financial, and technical barriers that prevent these communities from accessing data or conducting analyses of interest, as well as the actual and potential harms that stem from misuse of data, even in efforts for evidence-based policymaking.

Through the Urban Institute's Elevate Data for Equity project,[7] briefs and resources are available that encourage researchers and communities to manage data through its life cycle. These briefs and reports contain actionable items for researchers to incorporate, such as seeking communities' interests in research design elements, accounting for the potential social risk of research publications in reinforcing inequities, and returning research results to community members in open-access journals. As described above, more tools are available from AISP's Toolkit for Centering Racial Equity throughout Data Integration[8] projects. Pew Charitable Trusts is engaged in a Civil Legal System Modernization project,[9] focusing on open, efficient, and equitable courts—with the individuals involved in the court system at the center. Equitable courts encourage transparency and access to justice, regardless of representation status, race, ethnicity, economic status, disability, and language spoken.

The Civil Justice Data Commons,[10] part of Pew's Civil Legal System Modernization project (see text box below), has shaped its product to protect the marginalized individuals present in civil court data. By collaborating with nonprofit and community advocacy organizations, social service providers, and courts, who lend voices to those involved in the civil legal system, we have built a technical infrastructure and data governance model with an equity lens, with elements such as a systematic research proposal approval process for desired users of the Commons, and thorough de-identification and disclosure avoidance protocols to guard against re-identification of data subjects.

Philanthropic organizations also are pursuing equitable data practices. The Robert Wood Johnson Foundation (RWJF) established the National Commission to Transform Public Health Data Systems,[11] which soon will publish recommendations to improve health equity. The commission used a framework of truth, racial healing, and transformation, analyzing how current and historic institutional racism and discrimination (for example, against people of color, of those with different abilities, or based on sexual orientation and gender identity) impact laws and policies. Another way in which philanthropy can help lead equitable data practices is through community-based participatory research, which treats community members as research partners, not just data subjects, involving them in the entire process from research question development to data analysis (Lief 2020). American Indian tribes have successfully used this participatory research, partnering with local universities and research institutions in Texas and South Dakota, for example, to take ownership of the economic development data collected on their communities, correct inaccuracies in existing federal government data, and produce actionable solutions tailored to their on-the-ground needs.

CIVIL JUSTICE DATA COMMONS

As part of the Pew Legal System Modernization initiative, we have founded a Civil Justice Data Commons[12] that applies the best practices of data governance to civil court data. We aim to create a

(continued)

> secure, robust repository for civil legal data, gathered from courts, legal service providers, and other civil law institutions, that will enable stakeholders, researchers, and the public to better understand the civil legal system in the United States. By working with stakeholders in legal aid, social services, and advocacy organizations, we are building fair and equitable access to court data. We also are working with the courts to address their knowledge gaps, particularly surrounding fairness, equity, and access to justice. Our project relies on philanthropic support from the Alfred P. Sloan Foundation and Pew Charitable Trusts, as well as the National Science Foundation, to develop capacity to build the evidence—in a system where resources are lacking in individual courts or within state court systems. This project will have implications beyond courts alone, as researchers, nonprofits, and government organizations alike can apply for access to the CJDC to examine the connections of civil court involvement to economic, labor, health, and other social outcomes.

RETURNING TO "NO SUBJECTS WERE HARMED"

Clear, ubiquitous messaging is needed to explain that we can build evidence without harming people, creating social license. Evidence is a public good, and building it comes with broad societal benefit. Individuals, groups, and communities should discuss the harms, actual and perceived, that could come from data use. Discussing these harms should be a dialogue, not a conversation ender. Could a No Harm certification work with personal data? Only if the public recognizes and believes in it. We must work together to incentivize data controllers and users to adopt practices the public can recognize. We must strive for social license, producing evaluations showing that data use can be additive to our knowledge, not just extractive from the data subjects and their communities.

NOTES

1. See "Artist Visualized the Lengthy Terms of Services of Large Corporations like Facebook and Instagram," Designboom, May 7, 2018, www

.designboom.com/readers/dima-yarovinsky-visualizes-facebook-instagram-snapchat-terms-of-service-05-07-2018/.

2. See NIH, https://allofus.nih.gov/.

3. GSA released a framework to support federal leaders and data users in 2020. See https://resources.data.gov/assets/documents/fds-data-ethics-framework.pdf.

4. See https://aisp.upenn.edu/wp-content/uploads/2022/07/AISP-Toolkit_5.27.20.pdf.

5. See Societal Experts Action Network website, www.nationalacademies.org/our-work/societal-experts-action-network.

6. See Data Saves Lives website, https://datasaveslives.eu/.

7. See Marcus Gaddy and Kassie Scott, "Principles for Advancing Equitable Data Practice," Urban Institute, June 2020, www.urban.org/sites/default/files/publication/102346/principles-for-advancing-equitable-data-practice_0.pdf.

8. See https://aisp.upenn.edu/wp-content/uploads/2022/07/AISP-Toolkit_5.27.20.pdf.

9. See Project Civil Legal System Modernization, Pew Trusts, www.pewtrusts.org/en/projects/civil-legal-system-modernization.

10. See Georgetown Law website, www.law.georgetown.edu/tech-institute/.

11. Robert Wood Johnson Foundation, Better Data for Better Health, www.rwjf.org/en/library/collections/better-data-for-better-health.html.

12. See Georgetown Law website, www.law.georgetown.edu/tech-institute/.

REFERENCES

AISP. A Toolkit for Centering Racial Equity Throughout Data Integration, https://aisp.upenn.edu/wp-content/uploads/2022/07/AISP-Toolkit_5.27.20.pdf.

Bailey, M., Dittrich, D., Kenneally, E., and Maughan, D. "The Menlo Report." *IEEE Security & Privacy Magazine* 10, no. 2 (2012): 71–75. https://ieeexplore.ieee.org/document/6173001.

Carter, P., Laurie, G. T., and Dixon-Woods, M. "The Social License for Research: Why Care Data Ran into Trouble." *Journal of Medical Ethics* 41 (2015): 404–09. https://jme.bmj.com/content/41/5/404.citation-tools.

Dahl, J. Y., and Sætnan, A. R. "'It All Happened So Slowly': On Controlling Function Creep in Forensic DNA Databases." *International Journal of Law, Crime and Justice* 37, no. 3 (2009): 83–103.

Data Futures Partnership. A Path to Social License: Guidelines for Trusted Data Use. 2017. www.aisp.upenn.edu/wp-content/uploads/2019/08/Trusted-Data-Use_2017.pdf.

Designboom. "Artist Visualizes the Lengthy Terms of Services of Large Corporations like Facebook and Instagram." May 7, 2018. www.designboom.com/readers/dima-yarovinsky-visualizes-facebook-instagram-snapchat-terms-of-service-05-07-2018/

Gardizy, A. "Building an Escape Route from Poverty." *Boston Globe*. January 17, 2021. www.bostonglobe.com/2021/01/17/business/building-an-escape-route-poverty/.

Information Commissioner's Office. Data Sharing Code of Practice: Draft Code for Consultation. 2019. https://ico.org.uk/media/2615361/data-sharing-code-for-public-consultation.pdf.

Lief, L. How Philanthropy Can Help Lead on Data Justice. *Stanford Social Innovation Review*. February 6, 2020. https://ssir.org/articles/entry/how_philanthropy_can_help_lead_on_data_justice#.

Mackey, E., Elliot, M., and O'Hara, K. The Anonymisation Decision-Making Framework. UKAN Publications. 2016. https://fpf.org/wp-content/uploads/2016/11/Mackey-Elliot-and-OHara-Anonymisation-Decision-making-Framework-v1-Oct-2016.pdf.

National Institute of Standards and Technology. SP 800-53 Rev. 5: Security and Privacy Controls for Information Systems and Organizations. 2020a. https://nvlpubs.nist.gov/nistpubs/SpecialPublications/NIST.SP.800-53r5.pdf.

National Institute of Standards and Technology. Research Data Framework (RDaF): Motivation, Development, and a Preliminary Framework Core. 2020b. January 1, 2020. www.nist.gov/system/files/documents/2020/12/01/Preliminary%20RaF%20final_12-01-2020.pdf.

New Zealand Government. Standards & Guidance: Engagement. October 20, 2020. www.digital.govt.nz/standards-and-guidance/engagement/.

Nickels, G. J. *Executive Order 05–08: Inclusive Outreach and Public Engagement*. Office of the Mayor, Seattle. April 4, 2008. http://clerk.seattle.gov/~CFS/CF_309282.pdf.

O'Hara, K. *Data Trusts: Ethics, Architecture and Governance for Trustworthy Data Stewardship*. WSI White Paper #1. 2019. https://eprints.soton.ac.uk/428276/1/WSI_White_Paper_1.pdf.

The Praxis Group. Collaborative Governance Initiative: City of Fort Saskatchewan Public Engagement Framework. 2012. www.fortsask.ca/en/your-city-hall/resources/Documents/Public-Engagement-Framework.pdf.

Waind, E. Trust, Security and Public Interest: Striking the Balance. ADR-UK Strategic Hub & Economic & Social Research Council. May 2020. www.adruk.org/fileadmin/uploads/adruk/Trust_Security_and_Public_Interest-_Striking_the_Balance-_ADR_UK_2020.pdf.

LAKOTA PERSPECTIVE ON INDIGENOUS DATA SOVEREIGNTY

DALLAS M. NELSON, DUSTY LEE NELSON, AND TATEWIN MEANS

The Thunder Valley Community Development Corporation envisions a liberated Lakota nation through our language, lifeways, and spirituality. From our homelands on the Pine Ridge Reservation in South Dakota, we have been working persistently to create opportunities for our youth, dismantle systemic oppressive systems, foster an authentic Lakota regenerative community, and, most importantly, carry on our language and lifeways for future generations.

Our origin story began with our relatives challenging and empowering us with the questions: "How long are you going to sit back and let others decide the future of our children? Are you not warriors?" From inception, prayer has been the guiding force that has allowed us to grow and refine our effort of liberation. From inception, we have taken that challenge of not sitting back and letting others tell us what is best for us and our children.

Throughout our history of living on the Pine Ridge Reservation—also known as Prisoner of War Camp #334—our families, communities, and ancestors endured and continue to endure an all-out genocidal attack by the federal government and churches to remove our languages, land, history, and way of living. Yet we still are actively living through our language and lifeways in our homelands because our ancestors and families never swayed from them. As we continue in our journey toward liberation, we have to challenge Western settler colonialism and work to control the narratives that surround our work—including issues, ideas, and practices around data sovereignty and data governance.

Why is it important we change the narrative of data collection, data acquisition, data storage, and data access? As this chapter is being written, our Indigenous nations across the world are at the brink of losing their languages and lifeways. Indigenous Data Sovereignty is directly tied to the Indigenous language and lifeway reclamation and revitalization efforts being carried forth by the very people it has been extracted from. It is important to understand, that as Indigenous nations revitalize and reclaim our languages and lifeways, we have to work in concert with reclaiming our data to ensure our language movements can sustain themselves far into the future. Access is an ongoing barrier for our children, families, and communities. Access to language, whether in person, via the internet, a book, or a recording, is limited. To sustain the efforts to create a movement where our language is normalized, we have to create sustainable efforts around protecting and safeguarding our data.

WHAT IS INDIGENOUS DATA AND DATA SOVEREIGNTY?

According to the University of Arizona Native Nations Institute, Indigenous sovereignty is the right of Native nations to govern themselves (Rainie and others 2017). The Te Mana Raraunga–Māori Data Sovereignty Network defines [Indigenous] data as "the digital or digitazable information or knowledge that is about or from [Indigenous] people, our language, lifeways, resources or environments"; [Indigenous] data sovereignty as referring to "the inherent rights and interests that [Indigenous people] have in relation to the collection, ownership, and application of [Indigenous] Data"; and [Indigenous] data sovereignty as referring to "the principles, structures, accountability mechanisms, legal instruments, and policies through which [Indigenous peoples] exercise control over [Indigenous] Data" (Te Mana Raraunga 2018). And as Stanford University professor Matthew Snipp put it, "Quite simply, data sovereignty means managing information in a way that is consistent with the laws, practices, and customs of the nation-state in which it is located" (Snipp 2016).

Furthermore, Indigenous data contain knowledge about our environments, cultures, and community members at both an individual and collective level. The concrete boundaries between data, information, and knowledge are more fluid in an Indigenous context than in a traditional Western context, which also has implications for the governance of Indigenous data (Carroll and others 2019).

The United Nations in September 2007 developed the Declaration on the Rights of Indigenous Peoples (UNDRIP). Within this international document, under Article 18, it reads: "Indigenous peoples have the right to participate in decision-making in matters which would affect their rights, through representatives chosen by themselves in accordance with their own procedures, as well as to maintain and develop their own indigenous decision-making institutions." While the case law is limited with respect to Indigenous data sovereignty issues and rights, the Indigenous data sovereignty movement grew in 2015 at an international convening in Australia to determine Indigenous rights under the International Declaration on the Rights of Indigenous Peoples. It was determined at that time that Indigenous nations owned the rights to their citizens' data and also had the ability to determine how that data would be used.

Here in South Dakota, specifically in Pine Ridge, Standing Rock, and other Indigenous reservations, we are facing a moment in history where we have to reclaim, revitalize, then sustain our language movements. Questions that must be answered or used as a guide toward creating sustainable Indigenous data sovereignty practices and systems include: Who owns the data? Whose data is it? Who controls it? Who benefits from it? Who benefits from it financially?

INDIGENOUS EDUCATION

In our Lakota Language & Lifeways Initiative at Thunder Valley Community Development Corporation, we see language as our education. Everything radiates from our language; it contains our connection to the land, blueprints on how to live, and thousands of years of knowledge and teachings—but, more importantly, our language is our liberation. Anton Treuer writes in the *The Language Warrior's Manifesto*, "Language revitalization is nothing short of a pathway to liberation. When we shake off the yoke of colonization, we no longer have to be defined by that history. We do not become decolonized. We become liberated—unconquered. That should be our goal for every one of our children and all the children yet to be born over the next seven generations" (Treuer 2020, 168). Therefore, language and education are at the forefront of Indigenous data reclamation and data sovereignty.

Our Lakota Immersion Montessori (preschool to elementary), adult education programming, and elder philosophy and language preservation programming are front-line efforts creating safe environments free of oppression and grounded in our belief system. By centering our languages, our effort is radically shifting the narrative of what education for Indigenous peoples should be. Through our Indigenous education efforts, we are enacting data sovereignty at its purest form, because when we create anything from our language, we are not only reviving our language but positioning ourselves as stewards of that information moving forward.

Like with our language, our data is not owned by a specific Lakota person or persons but is guarded and protected by all Lakota people. As Te Reo Irirangi o Te Hiku o Te Ika (Te Hiku Media), an organization dedicated to language preservation and learning, explains via their website: "Indigenous people do not have a concept of private ownership of land and resources, that's a Western construct by which many of us are required to abide by. We see ourselves as the caretakers of our environment and society. Likewise, when we gather data to improve our services, we're taking care of the data given to us, and we follow tikanga (cultural protocols) when we need to make decisions around using data or providing access to data" (Te Reo Irirangi o Te Hiku o Te Ika 2017).

THE NEAR FUTURE

Liz La quen náay Kat Saas Medicine Crow writes: "Information, data, and research about our peoples—collected about us, with us, or by us—belong to us and must be cared for by us" (United States Indigenous Data Sovereignty Network). This is still a new and emerging idea to all the Indigenous communities around the world and specifically in the United States. Indigenous nations are still grappling with the effects and aftermath of the federal governments' and church systems' effort to take their languages and lifeways.

Indigenous peoples across the world have been collecting, analyzing, and aggregating data for thousands of years. The National Congress of American Indians (2018) writes that we, Indigenous peoples, "have always been data creators, users, and stewards." Pre-reservation days, the wild wild west was considered untamed and open for all in terms of acquiring land and conquering peoples. Today, the wild wild west is still present—but the free-for-all is taking place with our language, our education, and our data.

"As the Indian Wars concluded and American Indians were relocated to reservations, much of the data gathering on which they depended for generations was also forcibly seized," UCLA professor Desi Rodriguez-Lonebear writes. "Removal from their ancestral homelands, coupled with the decimation of wild game, population decline, and the boarding school system, stripped Indians of their traditional sources of knowledge and survival" (Rodriguez-Lonebear 2016, 258).

Indigenous people have not been in a position to be able to control the data and information that has been collected from them since European contact. From the moment the camera was introduced in the mid-1800s to today's advanced technology, we as Indigenous nations have been studied, recorded, photographed, sterilized, measured, and displayed by the colonizer and the non-Indigenous people who are infatuated with a romanticized misrepresentation of our living cultures. We have continuously been portrayed as savages, and the plains Titunwan people have served as an image for pan-Indianism in mainstream media. Movies, documentaries, studies, books, and dictionaries have been made about us, for us, and in the name of allyship. The narrative shift begins with actively working toward understanding that our data is just that: ours.

Understandably, our grandparents and great-grandparents were well aware of the significance of losing our language and ways, so they took to recording (audio/video) to ensure future generations would have access. With good intentions, our grandparents openly gave of their knowledge, history, and language to anthropologists, linguists, scientists, authors, etc. This created a large database of information that spans universities, colleges, private libraries, nonprofits, and other digital and hard storage platforms. We now are facing the issue of access, control, and guardianship. Non-Indian white institutions are actively working against this effort of language and data reclamation and data sovereignty (Niyake Yuza 2021).

The very reason this is an issue is because of the genocide, the taking, the termination, the relocation, the boarding schools, the mining, and the broken treaties. This is violence. Make no mistake, however it is said—whether it be through eloquent think pieces, intellectual terminology, or adding a linguistic spin to it—it is violence. The calls to toxic positivity, the calls to spirituality and being a good relative, whatever presents itself as a guard to the ongoing excavation of our elders' knowledge, life experiences, and the paradigm of being Lakota is still, at the end of the day, violence. We will not hold hands and sing in brotherly harmony the songs you took from us and recorded with your foreign voices. Taking our data—our language, our sacred songs, stories, words—whitewashing it and then selling it back to the very people to whom it belongs is a violent act, especially to our people who have experienced multitudes of loss and genocide throughout recent history.

The answer to colonization is not better colonization or a diet version of colonization. Indigenous data sovereignty is the final frontier in which we find ourselves in a vulnerable position once again, defending our natural resources from exploitation. Data collection is an unchecked process in which linguists and researchers freely take, analyze, and form solutions that fit their narratives. "In the indigenous world, data has a contentious history tied to the survival of native peoples on one hand, and to the instruments of the colonizer on the other," Rodriguez-Lonebear writes. "Indigenous data engagement in the United States is inextricably tied to the subjugation of American Indians and federal policies of Indian extermination and assimilation" (Rodriquez-Lonebear 2016, 257). We must take a stand and construct safeguards as Indigenous language and education activists and spiritual beings who are on a mission to ensure our traditional lifeways and connection to our identity remains authentic for generations to come.

It is our hope that, within our homelands, we can continue to advocate for our inherent sovereign right to protect and honor our data, create systems founded in the philosophy of guardianship, and, ultimately, reclaim our grandmothers' words and history so our children in the future have access to sustain the movement.

SOURCES

Carroll, S. R., Rodriguez-Lonebear, D., and Martinez, A. "Indigenous Data Governance: Strategies from United States Native Nations." *Data Science Journal*, July 8, 2019. https://datascience.codata.org/articles/10.5334/dsj-2019-031/.

Kipp, D. R. "Encouragement, Guidance, Insights, and Lessons Learned for Native Language Activists Developing Their Own Tribal Language Programs." Co-Founder of the Piegan Institute of the Blackfeet Nation Sat with Twelve Native American Language Activists for Most of the Day on March 11, 2000, at the Piegan Institute's Cut-Bank Language Immersion School, Browning, Montana. Piegan Institute's Cut-Bank Language Immersion Schools. 2000.

National Congress of American Indians. Resolution KAN-18-011: Support of US Indigenous Data Sovereignty and Inclusion of Tribes in the Development of Tribal Data Governance Principles. June 4, 2018. www.ncai.org/attachments/Resolution_gbuJbEHWpkOgcwCICRtgMJHMsUNofq YvuMSnzLFzOdxBlMlRjij_KAN-18-011%20Final.pdf.

Nickerson, M. "First Nation's Data Governance: Measuring the Nation-to-Nation Relationship" Discussion Paper. May 2017. https://static1.squarespace.com/static/558c624de4b0574c94d62a61/t/5ade9674575d1fb25a1c873b/1524536949054/NATION-TO-NATION_FN_DATA_GOVERNANCE_-_FINAL_-_EN.DOCX.

Niyake Yuza, C. "Ochethi Shakowin Data Sovereignty." Lakota Language Reclamation Project, January 22, 2021. https://lakotalanguagereclamationproject.com/blog/2021/1/22/ochethi-shakowin-data-sovereignty.

Rainie, S. C., Rodriguez-LoneBear, D., and Martinez, A. Policy Brief—Indigenous Data Sovereignty in the United States. 2017. https://static1.squarespace.com/static/5d3799de845604000199cd24/t/5d6f93c9c5442b00013e4b69/1567593418017/Policy%2Bbrief%2Bindigenous%2Bdata%2Bsovereignty%2Bin%2Bthe%2Bunited%2Bstates%2BV0.3%2Bcopy.pdf.

Raraunga, T. "Principles of Māori Data Sovereignty." October, 2018. www.temanararaunga.maori.nz/nga-rauemi.

Rodriguez-Lonebear, D. "Building A Data Revolution In Indian Country" In *Indigenous Data Sovereignty toward an Agenda*, pp. 253–274. Australian National University Press. 2016.

Te Reo Irirangi o Te Hiku o Te Ika. Kōreromāori.io. 2017. https://korero maori.io/.

Treuer, A. *The Language Warrior's Manifesto: How to Keep Our Languages Alive No Matter the Odds*. Minnesota Historical Society Press. 2020.

United States Indigenous Data Sovereignty Network. n.d. Retrieved May 11, 2021, https://usindigenousdata.org/.

BUILDING EVIDENCE AND ADVANCING EQUITY

A CALL TO ACTION FOR LOCAL GOVERNMENT

CARRIE S. CIHAK

For the past twenty years, I have been immersed in local government decision making in King County (Seattle, Washington) government. My training as an economist has naturally led to an interest in applying evidence-based practice to my work in the public sector. But my most important education and work has come through engagement with the diversity of communities in King County on advancing racial equity.

King County government's intentional focus on equity and social justice has been grounded in data and evidence from its inception about a dozen years ago.[1] At first glance, King County's metrics depict a flourishing region—one that has weathered even the effects of COVID-19 better than many other places. Yet, our communities attest and a deeper look at the data show that our region suffers from large disparities by race and place. Despite some important gains, many disparities have persisted and even worsened.

All local governments must recognize that we have contributed to racialized disparities and have a responsibility to eliminate them. Building an anti-racist pro-equity future requires local governments to work with community to deeply challenge the status quo, innovate, be willing to fail, and try again. Often, local governments assume that community-based and evidence-based practices don't mix. But my experience is that communities are eager to build, interpret, and use data and evidence. It is not that local governments need to set aside data and evidence to work with community; it is that we need to do the hard work of challenging our data and evidence practices to be more driven by, inclusive of, and responsive to communities.

Fundamentally, King County and other local governments cannot become anti-racist organizations that contribute to building a pro-equity future without co-creating and innovating with community, and that includes how we use data and evidence. We need to work with community to create the conditions under which the next generation of evidence flourishes in our organizations, and apply a constant vigilance so data and evidence are used in service of equity.

Here are *five calls to action for local governments* to support evidence building and use for a pro-equity future.

1. LEVERAGE LOCAL GOVERNMENT'S MULTIPLICITY OF ROLES

Local governments are directly accountable to the community for outcomes and the use of data and evidence to increase and demonstrate impact. We are at once policymakers, funders, practitioners, and implementers of evidence-based practice, contributing to both the supply of and demand for evidence building and use. Local governments, therefore, have an opportunity and a responsibility to model the way. It is time to get our act together. Here are a few ways to leverage the roles we play.

As Funder—Set-Asides for Data and Evidence: Data and evidence need to be recognized as foundational practices that contribute to impact, not as "overhead" that easily can be cut. Local policymakers can signal their commitment by setting aside a percentage of their budgets specifically for data and evidence building.

In 2015 and in 2021, King County voters approved an annual property tax levy (estimated at $132 million in 2022) called Best Starts for Kids (BSK) to help ensure every child here grows up happy, healthy, safe, and thriving.[2] BSK includes a 5 percent set-aside dedicated to data and evaluation. Those funds have allowed King County to develop a Child Health Survey, giving us data on the health and well-being of our youngest residents and their families for the first time.[3] The funds also help build capacity in hundreds of community organizations to use data and evidence to contribute to BSK results.

As Policymaker—Learning Agendas: Local governments cannot complain that evidence generated by researchers doesn't meet our needs if we are not clear about what questions are our highest priorities. We can do that through development of learning agendas,[4] now required of federal executive agencies by the Foundations for Evidence-Based Policymaking Act.[5]

At King County Metro Transit, we see the opportunity to move beyond learning agendas, which we have used for some programs, to develop an agency-wide strategic evidence plan.[6] Working with community, a strategic evidence plan will help set our learning priorities and build capacity and skills for continuous evidence generation and use over the long term.

As Implementer—Act on the Results: Local governments need to be clear and work with community partners on what we will do with results from evidence we build together. Too often, we leave programs showing positive results in the "pilot" phase because we have not considered how to scale them up. And, often, ineffective programs limp along for too long. Acting on results more quickly allows us to invest funds where they have the biggest impact.

That does not mean every promising finding results in big new investments, nor does it mean every null finding results in overturning a policy or eliminating a program. Evidence building takes time, and we should strive for evidence-informed decisions that balance many other considerations local governments face. For example, through a randomized control trial study, the Lab @ DC[7] found that body-worn cameras[8] had no statistically significant impact on police use of force and other outcomes measured in the study. The Lab @ DC provided a thoughtful analysis of several possible reasons for this result.[9] The city continues to use body-worn cameras for their important transparency and evidentiary value while focusing on rigorously evaluating other innovative efforts to improve police-community interactions.[10]

As Convener—Partner across the Regional Evidence Ecosystem: Local governments can convene other organizations, like universities, philanthropies, nonprofits, and the private sector, to partner on evidence building and use. For example, King County was instrumental in bringing partners together to form HealthierHere, a regional nonprofit driving and testing innovations to advance equity and improve health and wellness as part of Washington State's Medicaid Transformation.[11] HealthierHere's collective action model allows us to better link and interpret data, much of which is maintained by King County, to catalyze and test innovations across the healthcare system.

2. CENTER EQUITY AND INVOLVE COMMUNITY FROM THE BEGINNING

To build a pro-equity future, local governments need to center equity and involve community in all of our processes, including evidence building. People closest to the issues also are closest to the solutions, and we need to be continuously engaged with community so they drive priorities and innovations.

An exciting example of this is the co-creation of a new Mobility Framework by King County Metro Transit and the community-based Mobility Equity Cabinet.[12] The framework, adopted in early 2020, provides overarching policy guidance for how Metro Transit can advance mobility, particularly for communities "where the needs are greatest." Working closely with communities at the earliest stages of policy development produced a much stronger and innovative policy response. A commitment to continuously engage with communities as we implement and build evidence will also produce better, more durable solutions.

3. BUILD EVIDENCE FROM THE BEGINNING

Like equity considerations, local governments often treat evidence building as an afterthought. Decision makers often fail to ask about the learning objectives or establish success criteria until well into a project or until something isn't working well. When we do not consider equity from the beginning, rather than advancing equity, we often end up having to mitigate negative impacts. Likewise, when we do not consider evidence building from the beginning, it may be impossible to build strong evidence. Equity and evidence work together from the outset to support stronger pro-equity impact and outcomes.

In 2015, King County was one of the first jurisdictions in the country to implement a discounted transit fare for people with low incomes.[13] While we built an equity and performance measurement focus into implementation, we failed to consider how we would build causal evidence about the impact of fare discounts on mobility and quality of life outcomes among different communities. The program enrolled 60,000 people at its peak and while we have measures on how much people use the benefit, retrospectively we do not have the ability to rigorously demonstrate how the lower fare created changes in mobility and other life outcomes for people in the region.

Learning from this, when, in 2020, King County implemented a fully subsidized annual transit pass,[14] available at no cost to our residents with the lowest incomes, we considered evidence building from the beginning. This has led us to stronger partnerships, data infrastructure, and other mechanisms for program improvement, and will allow us to demonstrate the impact of the pass on quality-of-life outcomes for the diversity of communities in King County.

4. INVEST IN INTERNAL AND EXTERNAL CAPACITY AND RELATIONSHIPS

Local governments often approach the generation of evidence as something they contract for through external researchers. Often, there is little interaction between researchers and government staff, with results of the research being delivered in a report several months later, which then sits on a (proverbial) shelf.

Local governments need to be more engaged in the production of evidence if that evidence is to be put to use. Even where government brings staff with evaluation expertise in house, teams like the Lab @ DC demonstrate that the most useful evidence building occurs when those researchers are continuously engaged with program staff. This ensures that evidence is highly tuned to program needs, that continuous learning and improvement occurs, and that program staff build knowledge and skills that help interpret evidence.

In King County, we have benefited greatly from "matching services," such as through the State & Local Government Innovation Initiative at J-PAL,[15] to pair us with researchers on specific evidence-building projects. From there, we have invested in building those into long-term partnerships, such as with the Wilson Sheehan Lab for Economic Opportunities at Notre Dame (LEO)[16] and the Regulation, Evaluation, and Governance Lab at Stanford Law School (RegLab),[17] where we now are involved in several evidence-building projects together. We are able to generate useful research much more quickly with each project as these researchers build their expertise and relationships in King County, and we integrate data across more projects.

Local governments also need to recognize the value of the knowledge, expertise, and time that communities bring. We need to pay community members and community-based organizations for this expertise, just as we pay for the expertise of consultants with whom we regularly contract. As well, when we require community-based organizations to participate in data or evidence-building activities, we need to fund and support their capacity to do so.[18]

These relationships among people dedicated to the same goals while holding different perspectives and roles benefit us in countless ways and are much more nimble, durable, rewarding, and—frankly—fun than the transactional interactions that come with a contract, which tend to vanish when the work is complete.

5. SHARE RESULTS AND LEARNINGS TRANSPARENTLY AND BROADLY

Too often, the results of a research project do not make it beyond the program being examined. We need better mechanisms in local government to share evidence across programs and agencies, with community and the public, and with other local governments. The value of sharing evidence is not just about the results but also about the thought process that went into establishing our hypotheses, what failures of implementation we recovered from, and how we were able to build process equity.

Evidence building in King County is strengthening through sharing across departments, with local communities, and with other governments. Evidence on the impact of case management in a homelessness prevention project has directly influenced an initiative to reduce barriers to transportation through community navigators. In Best Starts for Kids, we have established a regular practice of data deep dives to enlist communities' expertise in the interpretation of data and results.[19] Outreach to other jurisdictions started a few years ago regarding evaluation of income-based transit fares has led to the establishment of an Interjurisdiction Transit Equity Research Collaborative, a monthly convening of over twenty major transit agencies to share learnings, challenges, and research.

If local governments can make progress on the five areas above, we will be well poised to make the following *three requests of the research and evidence-building community.* At the Causal Inference for Social Impact Lab at the Center for Advanced Study in the Behavioral Sciences at Stanford University,[20] which I co-direct, we are taking up these considerations:

1. Treat Government and Community as Equal Partners: The most productive evidence-building processes are where government staff, community members, and researchers work together as equal partners. We ask research partners to value the expertise of our staff and community partners and respect the priorities of community and the multiple constraints and pressures of the environment in which local government operates. These constraints don't always make for the perfect research project, but they often are more likely to reflect the actual conditions in which policy and program innovations are implemented.

2. Innovations in Causal Inference: We also need innovation in the methodologies and practices used to build rigorous evidence. While randomized controlled trial experiments are one important tool, we should prioritize evidence building for our most important questions and promising interventions that advance equity, regardless of whether we can randomize. We also need research practice innovations that center equity while challenging the definitions of core elements of our evaluative practice (validity, rigor, and objectivity) so that inquiry better reflects the multiplicity of experiences within multiple cultural contexts, as envisioned by the Equitable Evaluation Initiative[21] with their Equitable Evaluation Framework™.[22]

3. Collaboration across Researchers: The decisions individual researchers make can have enormous effects on the results generated and, subsequently, on the policy responses that impact residents' lives. Local governments need researchers who are willing to consult, collaborate, and act as "critical friends" with one another and with us and our community partners. We need researchers who know that any one study does not provide definitive answers, understand that the best studies provide some answers and more questions, and are willing to work across disciplines so local governments and communities can make evidence-informed decisions based on the best imperfect information.

Just imagine the progress we could make to advance racial equity if local governments, community, and our research partners were co-conductors of this evidence train!

NOTES

1. See King County Equity and Social Justice website, https://kingcounty.gov/elected/executive/equity-social-justice.aspx.
2. See Best Starts for Kids page on the King County Department of Community and Human Services website, https://kingcounty.gov/depts/community-human-services/initiatives/best-starts-for-kids.aspx.
3. See Best Starts for Kids Health Survey page on the King County Department of Community and Human Services website, https://kingcounty.gov/depts/community-human-services/initiatives/best-starts-for-kids/survey.aspx#:~:text=The%20Best%20Starts%20for%20Kids%20Health%20Survey%20is%20a%20survey,Washington%20to%20collect%20this%20information.
4. See the Learning Agendas page on the Evaluation.gov website, https://www.evaluation.gov/evidence-plans/learning-agenda/.
5. See "Achieving the Promise of the Evidence Act," Results for America, https://results4america.org/evidence-act-resources/.
6. See "Supporting Effective Policymaking through the Development of Strategic Evidence Plans," Project Evident, https://www.projectevident.org/updates/2020/9/2/supporting-effective-policymaking-through-the-development-of-strategic-evidence-plans.
7. See The LAB @ DC website, https://thelab.dc.gov/.
8. The LAB @ DC, "Do Body-Worn Cameras Influence Police-Community Interactions?," https://thelabprojects.dc.gov/body-worn-cameras.
9. See The LAB @ DC, "Do Body-Worn Cameras Influence Police-Community Interactions?" conclusions, https://bwc.thelab.dc.gov/conclusions.html.
10. See The LAB @ DC, "Can Knowledge of Historical and Cultural Context Have an Impact on Policing," https://thelabprojects.dc.gov/historic-cultural-training.
11. See "An Overview and Highlights from Our Current Work," Healthier Here website, www.healthierhere.org/our-work/.
12. See the Mobility Framework page on King County Metro's website, https://kingcounty.gov/depts/transportation/metro/about/policies/mobility-framework.aspx.
13. See the ORCA LIFT page on the King County Metro website, https://kingcounty.gov/depts/transportation/metro/fares-orca/orca-cards/lift.aspx.

14. See the Subsidized Annual Pass page on the King County Metro website, https://kingcounty.gov/depts/transportation/metro/fares-orca/subsidized-annual-pass.aspx.

15. See the State and Local Innovation Initiative page on the J-PAL website, www.povertyactionlab.org/initiative/state-and-local-innovation-initiative.

16. See the Wilson Sheehan Lab for Economic Opportunities website, https://leo.nd.edu/.

17. See Stanford University's Regulation, Evaluation, and Governance Lab website, https://reglab.stanford.edu/.

18. "New in 2020, More Data and Evaluation Support," Best Starts for Kids blog, January 7, 2020, https://beststartsblog.com/2020/01/07/new-in-2020-more-data-and-evaluation-support/.

19. "What's a Data Dive?" Best Starts for Kids blog, June 12, 2018, https://beststartsblog.com/2018/06/12/whats-a-data-dive/.

20. See the Causal Inference for Social Impact Lab page at the Stanford University website, https://casbs.stanford.edu/programs/causal-inference-social-impact-lab#:~:text=In%20Spring%202021%2C%20CASBS%20will,Social%20Impact%20Lab's%20Data%20Challenge.&text=Unlike%20most%20data%20challenges%2C%20the,the%20questions%20posed%20to%20them.

21. See the Reimagining the Purpose and Practice of Evaluation page of the Equitable Evaluation Initiative website, https://www.equitableeval.org/.

22. See the Equitable Evaluation Framework page of the Equitable Evaluation Initiative website, www.equitableeval.org/framework.

HOW FUNDERS CAN CENTER EVALUATION NORMS ON EQUITY

TRACY E. COSTIGAN AND RAYMOND MCGHEE JR.

The purpose of this chapter is to provide a field-level look at how philanthropy can support more equitable evaluation practices to produce evidence that is relevant to community and practitioner interests, as well as funder goals. We start with a description of practices at the Robert Wood Johnson Foundation (RWJF), a field leader in philanthropic evaluation, describing how the foundation has shifted its approaches to center equity in evaluation processes and outcomes. We then turn to the history of evaluation in philanthropy more broadly and describe what it will take for the field to move toward equity-centered approaches in evidence generation, offering examples of steps RWJF has taken toward this goal.

EVALUATION AT RWJF: AN EVOLUTION TOWARD EQUITY

During its nearly fifty years in operation, RWJF has experienced an evolution in its vision and strategies, originating with a focus on improving health and health care, progressing to addressing the social determinants of health, and then to further the achievement of health equity in the context of building a Culture of Health. In 2020, RWJF sharpened its strategies, emphasizing the role of structural racism as a barrier to health equity, magnified in the contexts of the COVID-19 pandemic and anti-Black violence (RWJF 2020).

Over this period, the foundation developed its definition of health equity. As an outcome, equity is defined as *everyone having a fair and just opportunity to live a healthier life*. This requires removing obstacles to health such as poverty and discrimination, and their consequences, including powerlessness and lack of access to good jobs with fair pay, quality education and housing, safe environments, and health care. For research and evaluation, health equity is measured as *reducing and ultimately eliminating disparities in health and its determinants that adversely affect excluded or marginalized groups* (Braveman, Arkin, Orleans, Proctor, and Plough 2017). Like other philanthropies, it took several iterations to get to these definitions because of the challenge of clearly articulating a measurable outcome that can be sensed at a visceral level and yet is filled with nuance, multiplicity, and complexity. While many in philanthropy have articulated verbal and written affirmations of equity, the greater challenge has been implementing real change to embed equity into strategy and organizational values that lead to actions consistent with these declarations. Through this evolution, RWJF's commitment to building evidence has held steady, articulated in its first guiding principle: *We seek bold and lasting change rooted in the best available evidence, analysis, and science, openly debated.*

RWJF is considered a pioneer in philanthropic evaluation, and is known for using evaluation to build evidence about program impacts: to support program improvement, scale, and spread, and to guide decision making. The foundation engages with evaluators to design fit-for-purpose evaluations to inform its own work as well as that of others. RWJF does not subscribe to one type of methodology; rather, it supports evidence generation across a continuum of methods that respond to the unique research and evaluation questions of each body of work. Although the commitment to evidence remains unchanged, specific approaches to evaluation have progressed to keep pace with RWJF's increasingly focused commitment to equity. This parallels the evolution of evaluation across philanthropy as the sector struggles with challenging questions around the roles of validity, rigor, and relevance.

THE ROLE OF EVALUATION IN PHILANTHROPY

In philanthropy, rigorous evaluation of social programs has been central to evidence generation, beginning in the 1970s, as a way to measure program impact, usually at the individual grant level. In the decades that followed, evaluation shifted to measuring broader outcomes across clusters of grants and programs. More recently, as the sector has shifted its focus to solving more complex systems-level problems, including advancing equity, evaluation has, again, shifted its focus toward informing strategic progress (Coffman and Beer 2016; Coffman 2016). Over time, philanthropic evaluation has turned more inward to examining foundations' own progress, losing sight of the communities they aim to serve. This calls to question the role of evaluation in evidence generation for practitioners and communities.

In the last few years, philanthropy is again evolving evaluation practices, in response to internal and external influences. Internally, foundations are examining operations and approaches with respect to centering equity. Concurrently, they are rethinking strategies to tackle the complexity of the systems preserving inequities. Furthermore, nonprofit sector leaders are pushing philanthropic institutions to examine their roles in perpetuating white-dominant narratives and culture despite trying to advance equity.

Externally, social justice movements challenging the structures of inequity in society have accelerated, particularly in light of the events of 2020, including the COVID-19 pandemic, which illuminated health disparities, and the groundswell of protests against long-standing racial inequity and police brutality. This has brought into focus the role of institutions and systems in preventing equitable outcomes. Communities and practitioners are asking questions about evaluation: What is its value relative to its historical origins? Who is it meant to serve? What is its relevance to advancing equity? As a result, philanthropy has been challenged to support communities in new ways, including addressing how evaluation supports evidence generation.

Finally, concepts like Critical Race Theory (CRT)[1] are appearing in the sector. CRT has had a significant influence in challenging philanthropy and those they fund to reimagine forms of evidence. Philanthropies have adopted new value and mission statements, with intentions to implement new practices that make equity a reality in strategies and practices. There is an urgent need for the next wave of philanthropic evaluation to center equity in design and measurement. Evaluation as a form of evidence must be in service to the communities and people most affected by the systems philanthropy is seeking to change. This shift also requires philanthropy to consider historical context, root causes, and status quo of the systems that drive inequities.

FIGURE 2.9.1 Equitable Evaluation Framework Principles

1	2	3
Evaluation and evaluative work should be in service of equity:	Evaluative work should be designed and implemented commensurate with the values underlying equity work:	Evaluative work can and should answer critical questions about the:
• Production, consumption and management of evaluation and evaluative work should hold at its core a responsibility to advance progress toward equity.	• Multi-culturally valid, and • Oriented toward participant ownership.	• Ways in which historical and structural decisions have contributed to the condition to be addressed, • Effect of a strategy on different populations, on the underlying systemic drivers of inequity, and • Ways in which cultural context is tangled up in both the structural conditions and the change initiative itself.

Source: Dean-Coffey, J. (2017). Equitable Evaluation Framework™. Retrieved from Equitable Evaluation Initiative, https://www.equitableeval.org/framework

WHAT DOES IT MEAN TO CENTER EQUITY?

The Equitable Evaluation Initiative (EEI 2017) has called into question philanthropic approaches to evaluation, encouraging the sector to transform evaluation to better fit with these newfound equity commitments. The equitable evaluation framework (EEF) offers three principles that have the potential to produce rigorous and relevant evidence that takes into account historical, structural, systemic, and cultural drivers related to decisions and outcomes (see figure 6.4-1).

Centering equity does not mean abandoning rigor. Rather, strong evaluation design driven by EEF principles achieves both. Analysis of equitable evaluation approaches that pits rigor versus equity is simply wrong ("A 'Mischaracterization' of the Movement Toward More Equitable Evaluation" [Letter to the Editor] 2020). Rather, the aim is to disrupt historic philanthropic orthodoxies around evaluation and replace them with a framing that is in service to equity. These old orthodoxies included centering on the foundation, who defined success and was the primary user of evaluation results. They also centered on the evaluators, based on traditional academic credentials, as objective experts who have the final say about meaning and impact. And they emphasized quantitative and experimental methods, which were usually the only approaches deemed sufficiently rigorous.

There is an opportunity now to shift the field away from these old orthodoxies to a new set of guiding principles that center equity in the work, while maintaining the standards of evidence generation. This includes first recognizing expertise in equal measure across the ecosystem, particularly privileging community and practitioner voices, designing and embracing continuous evidence building driven by their evidence agendas. It also means expanding the sector's thinking about rigor, encouraging fit-for-purpose mixed methods designs in the work.

Centering equity also requires evaluative work to reconsider validity, identifying the multiplicity and complexity of truth and moving away from the white-dominant culture frame that prioritizes funders' questions. The work needs to move toward expanding perspective to consider questions and test assumptions from all parts of the ecosystem. Evaluation needs to lift up the voices and perspectives of community members, organizations, local leaders, practitioners, and decision makers, and account for the context, culture, and power structures in the system. Producing valid evidence often requires expanding the scope of design and analysis.

What does this look like in practice? The EEF is not a tool, method, or rubric. Rather, it is a set of principles for reflection and learning about how evaluation practices can create the conditions to deeply examine and understand the work. It emphasizes the need to continually check beliefs, assumptions, and approaches and to continually recalibrate approaches throughout the process. It is possible to shift foundation norms and expectations around evaluation to support equity in process and outcomes while maintaining the rigor of high-quality methods and producing insights valuable to the interests of various stakeholders (EEI and GEO 2021). Moreover, given philanthropy's interests in advancing equity, not shifting in these ways creates a false sense of comfort in the evidence and will do harm to those most affected by structural inequities.

In recent years, a number of resources have been published that describe ways in which the sector is progressing to incorporate EEF principles and center equity in the work. Various examples describe concrete ways in which groups have transformed evaluation practices (for example, Annie E. Casey Foundation 2020; WestEd 2019; Forum for Youth Investment 2020, Public Policy Associates 2020; TCC Group 2021; Community Science 2021). An example of a community-generated framework designed to build meaningful evidence is the Chicago Beyond Initiative (2018), which articulates seven barriers to equity and impact perpetuated by the long-standing power and control of funders. These include the lack of: *access* to wisdom that is missing because communities are not at the table; *information* about and *accountability* to the communities who are the subject of the research; *ownership* by and *value* to the community because funders and evaluators are centered in the work; and *authorship* credit to the community. Evaluation design must address these barriers to address equity. These and other resources offer ideas for the sector to translate these principles into practice (EEI and GEO 2021) to produce evaluative evidence that informs decisions (Lynn 2021).

RWJF's progress toward centering equity in evaluation, learning, and evidence generation has resulted in reexamining often long-standing approaches once considered best practices. This has included designing learning and evaluation plans that advance both community-practitioner interests and funder goals. It has included consideration of how opportunities are shaped, including scope and selection criteria, along with how these are shared, reviewed, and awarded. New approaches also include setting budgets for evaluations that support the effort necessary to center equity. It also means working with evaluators who are shifting their approaches: constantly checking biases and assumptions; using more mixed-methods approaches with iteration; repeatedly bringing grantees into design, implementation, analysis, and communication. And, it is imperative to clarify what equity means for each effort, both in terms of design processes and in measurement. Finally, as RWJF moves to develop evaluation around more complex strategies focused on systems change, we are being more deliberate in how we center the voices of those most affected by inequity, by giving community members the opportunity to help select the evaluators working in their communities, as well as co-design activities. In doing so, communities and practitioners are developing important lines of inquiry and measures in the evaluations. Throughout the work, communities, funders, and evaluators must feel empowered to hold each other accountable to these evaluation efforts, coming to agreements about how to raise questions at times when equity seems to be losing its place at the center of the work.

The demands of equity require philanthropy to be responsive in a variety of ways. Evolving evaluation practice to center equity is in the collective best interest, especially for the communities that have been most harmed by extractive practices of researchers and evaluators. It is an opportunity for philanthropy, through its grantmaking and field building, to expand our vision to embrace the next generation of evidence. Integrating rigorous methods with a comprehensive design process that includes and amplifies the perspectives of those most affected by the systems under study will produce more rapid program improvements, further insight into what is necessary to produce systems change, and, ultimately, more robust and meaningful study of impact at all levels. Philanthropies individually can do this work; they can, as well, build partnerships to create more coherent funding packages and processes that support this next generation of evidence. Taken together, these actions can help philanthropy embrace more equitable evidence practices going forward.

NOTE

1. CRT views racism as a pervasive and systemic phenomenon that functions on many levels, necessitating the centering of the voices of people of color and seeking to highlight their lived experiences. K. Bridges, *Critical Race Theory: A Primer.* Concepts and Insights Series (Washington, DC, Foundation Press: 2019); and D. Stovall, "A Challenge to Traditional Theory: CRT, African-American Community Organizers, and Education," *Discourse: Studies in the Cultural Politics of Education* 26, no. 1 (2005), pp. 95–108.

REFERENCES

Annie E. Casey Foundation. Step-by-Step Guide on Using Equity Principles in Social Science Research. 2020. www.aecf.org/blog/step-by-step-guide-on-using-equity-principles-in-social-science-research/.

Braveman, P., Arkin, E., Orleans, T., Proctor, D., and Plough, A. *What Is Health Equity? And What Difference Does a Definition Make?* Princeton, NJ: Robert Wood Johnson Foundation. 2017.

Bridges, K. *Critical Race Theory: A Primer.* Concepts and Insights Series. Foundation Press: Washington, DC. 2019.

Chicago Beyond. *Why Am I Always being Researched?* Chicago Beyond Equity Series, Vol. 1. 2018. https://chicagobeyond.org/researchequity/.

Coffman, J. "Oh for the Love of Sticky Notes! The Changing Role of Evaluators who Work with Foundations." April 28, 2016. https://medium.com/@jcoffman/oh-for-the-love-of-sticky-notes-the-changing-role-of-evaluators-who-work-with-foundations-66ec1ffed2e4#.ifphr6o4a.

Coffman, J., and Beer, T. "How Do You Measure Up? Finding Fit between Foundations and Their Evaluation Functions." *The Foundation Review* 8, no. 4 (2016): 27–43.

Community Science. *What Have We Learned about Evaluating Equity-Promoting Efforts?* 2021. www.communityscience.com/news-detail.php?news=201.

Forum for Youth Investment. *What's the Role of Equity in Evaluation Policy?* Washington, DC: 2020. https://forumfyi.org/wp-content/uploads/2020/02/Equity-in-Evaluation-Policy-Winter-2020.pdf.

EEI. Center for Evaluation Innovation, Institute for Foundation and Donor Learning, Dorothy A Johnson Center for Philanthropy, and Luminare Group. "Equitable Evaluation Framing Paper." Equitable Evaluation Initiative. July 2017. www.equitableeval.org.

Equitable Evaluation Initiative & Grantmakers for Effective Organizations. *Shifting the Evaluation Paradigm: Equitable Evaluation Framework^{TM}*. 2021. www.geofunders.org/resources/shifting-the-evaluation-paradigm-the-equitable-evaluation-framework-1332.

Lynn, J. *Strengthening the Philanthropic Evaluation Field: The Walton Family Foundation's Initial Exploration.* 2021. Walton Family Foundation and PolicySolve.

"A 'Mischaracterization' of the Movement toward More Equitable Evaluation, Letter to the editor." *The Chronicle of Philanthropy.* February 20, 2020. www.philanthropy.com/article/A-Mischaracterization-of/248085.

Public Policy Associates. *Using an Equity Lens in Evaluation. 2020.* https://publicpolicy.com/using-an-equity-lens-in-evaluation/.

Robert Wood Johnson Foundation. *Racism and Health: Inequities across Our Nation Have Their Roots in Discrimination.* Princeton, NJ. 2020. www.rwjf.org/en/library/collect ions/racism-and-health.html.

Stovall, D. "A Challenge to Traditional Theory: CRT, African-American Community Organizers, and Education." *Discourse: Studies in the Cultural Politics of Education* 26, no. 1 (2005): 95–108.

TCC Group. *Equity and Evaluation: Models of How Equity Can and Does Impact Evaluation.* 2021. www.tccgrp.com/insights-resources/insights-perspectives/equitable-evaluation-part-1-equity-as-a-leading-principle/.

WestEd Justice & Prevention Research Center. "Reflections on Applying Principles of Equitable Evaluation." 2019. www.wested.org/wp-content/uploads/2019/07/resource-reflections-on-applying-principles-of-equitable-evaluation.pdf.

LEADERSHIP IS CAUSE; EVERYTHING ELSE IS EFFECT

LOLA ADEDOKUN

As a first-generation millennial Black woman, I have been heartened by the rare but important professional spaces where leaders of color are equally valued for their work skills and life experience. But generally, I have found senior leadership in philanthropy and the social sector to be overwhelmingly white-led and Eurocentric in its values, priorities, and vision. This trend, in turn, influences who receives funding and who does not. To buck this trend is not easy, but during my eight-year tenure as Director of the Child Well-being Program at the Doris Duke Charitable Foundation (DDCF), it was essential to our mission of creating a world where all children and families have the opportunity to thrive.

To do so, I worked with colleagues, peers, and grantees explicitly to draw attention to the need for investments in the leadership and professional development of social sector leaders of color who bring both personal understanding and natural affinity for the needs of residents, and will put their needs at the center of decision making related to policies, practices, and programs. Often, such leaders of color will shoulder the responsibility of the expectations of their job while also enlightening their white counterparts as to where systemically racist practices exist and how they can be disrupted. In this way, hiring leaders of color creates positive ripple effects across the entire organization, including the way it goes about gathering evidence of social impact.

These ripple effects remain in short supply at a time when they are most needed. The COVID-19 pandemic and the interrelated financial and social justice crises have further reinforced the need to develop a cadre of entrepreneurial-minded social service leaders of color who can realize their visions to transform the social service sector in ways that can truly better the lives of the individuals, children, and families they serve.

IMPLICATIONS FOR THE NEXT GENERATION OF SOCIAL SECTOR LEADERS

We need to get better at addressing root causes of social ills that reinforce multi-generational poor health and well-being outcomes. The COVID-19 pandemic and related crises have exposed, yet again, deep fissures in our social fabric. Social and health crises like homelessness, opioid misuse, gun violence, obesity, and so many others continue to disproportionately affect children and families of color or those living in low-income environments. Sadly, our social and human service systems are not designed to address the root causes of these ills nor the complex contextual factors that continue to trap communities and families in unjust and unhealthy circumstances for generations.

We need to think globally and learn from other countries. The global nature of the pandemic also has reminded us that no country's GDP can inoculate its government and citizenry from the need to invest in responsible, equitable, and empathetic leadership. We need such visionary and collaborative leadership to build a strong social fabric, resilient to adverse conditions and actors. And we need to look beyond our own borders to find and make connections with exceptional social sector and community leaders around the world tackling similar issues.

For example, through the DDCF *African Health Initiative*, we have observed our colleagues in Africa commit to and exercise evidence-driven decision making and leadership, while also investing in the next generation of leaders who are better prepared to transform social service cultures and norms. They continue to serve as a refreshing resource on how to truly support leaders who envision changing a system. The *Child Well-Being Program* has applied important leadership lessons from our work in Africa to our U.S.-focused work in the social and human services sector to support root-cause solutions that will speed child and family well-being.

We need to serve needs as we also work to solve problems. This takeaway is particularly important as leaders now have two jobs—to respond to increasing needs and demand for services while also responding to calls to dismantle and/or transform systems that perpetuate racist and discriminatory practices. Few leadership and professional development programs provide adequate support to prepare leaders to meet these challenges.

For example, in 2018, our Child Well-Being Program invested in a portfolio of national leadership and professional development programs to expand opportunities and positively reinforce networks of racially and ethnically diverse mid- and senior-level social service leaders. These programs aim to increase the visibility of leaders committed to transforming policies and programs, building and strengthening community and agency partnerships and making sustained improvements in the well-being of children and families in the United States.

Quickly and radically improving well-being for children and families in the United States requires transforming systems and strengthening local, state, and federal supportive safety nets that better and more equitably serve them. The complexity of this undertaking requires visionary, diverse, collaborative, adaptive, and entrepreneurial leaders and teams to implement and sustain new ways of thinking and working. Unfortunately, there is a lack of coordination and opportunities for skills building for leaders in the social and human services sectors, primarily because there are limited resources and incentives in place to support them.

We need to invest in the leadership, professional development, and networks for leaders of color. Making explicit investments in the leaders of color is key. Though they have suffered and continue to carry the brunt and the burden of bringing attention to their work and their communities, they persist. They, in fact, exemplify the type of resilience essential to navigate the complex social challenges our country faces today. Furthermore, recognizing the emotional and social toll of working to serve children and families in need, direct investments need to be made in supporting networks for leaders of color.

We need evidence-building methods that center a gender and racial lens. Poor health and well-being outcomes disproportionately disadvantage children and women of color. These outcomes reflect both the symptoms and results of historical and ongoing systemic racism. In recent decades, policymakers and practitioners have increasingly embraced and demanded evidence-based solutions that center equity in improving outcomes for children and families, yet the research methodologies and practices that generate the evidence do not adequately account for race and its role in driving these outcomes.

While researchers now routinely collect data that disaggregates by race and ethnicity, a growing body of researchers recognize the critical need to lead with a racial and gender equity lens in other dimensions of the research process. This will enable them to strengthen evidence-based interventions and provide much needed data for policymakers to reduce racial disparities and advance equity in child and family outcomes.

HOW A NEXTGEN LEADER BUILDS NEXTGEN EVIDENCE

To quote Angela Jackson, "Leaders who arise from the communities and issues they serve have the experience, relationships, data, and knowledge that are essential for developing solutions with measurable and sustainable impact."[1]

To build next generation evidence, the next generation of leaders need intellectual curiosity and a clear *vision*; they are *creators and dreamers* on the constant quest to achieve the ideal world—where every human is cherished, beloved, and enabled to lead self-determined lives.

They are *resilient* and impervious to naysayers. They have the data and evidence on their side, so they must be able to persist. And they have a willingness to mentor and develop the next generations.

They must be *storytellers*—not simply good communicators—who can weave a story to compel others. This is a rare skill, but one of the most important. Relatedly, they must be credible messengers—often those with lived experience and/or those who directly reflect the racial and ethnic diversity of the people they are serving.

They must be *collaborative*—willing to engage others in their thinking—particularly the suspects not in their regular circles. Collaboration is an underrated and essential skill. It takes a sense of *confidence and humility*; it requires patience and thoughtfulness. When collaborating well, there is also a need to support conversations around success and failure.

They must be *courageous* and willing to disrupt narratives, leaning into data and evidence while also making that evidence accessible.

Where to find these leaders? Community organizers tend to have these skills. We need to invest in them explicitly, along with social entrepreneurs who know how to move from vision to building on and adapting that vision as needed.

These leaders also must excel at systems thinking. There is often an assumption that bureaucrats or those working within systems automatically have the skills, bandwidth, or interest to tackle complex systems challenges. Just as we recognize physician scientists as practitioner scholars—with academies in place to recognize and preserve their leadership (for example, National Academies of Sciences, Engineering, and Medicine)—the same standards and expectations should be set for practitioners leading the way in building evidence in the social service sector.

There is an essential need to support the next generation of practitioner-scholars in the human and social service sectors. They will be the ones best positioned to inform conversations and apply evidence to advance equitable transformation and strengthen the systems meant to serve people. They bring a human-centered and scholarly lens that is often overlooked and, consequently, underapplied.

EXEMPLARS OF NEXT GENERATION LEADERSHIP

I can name a plethora of evidence-driven leaders whose impact and visibility has grown due to our commitment to their development and investment in all parts of their leadership:

Dr. Clinton Boyd[2] refers to himself as an activist researcher. In addition to a number of other awards, he earned a competitive DDCF-funded fellowship through the Doris Duke Fellowships for the Promotion of Child Well-Being. He continues to advance in his career and is gaining much-deserved recognition as a next generation leader promoting supportive services for Black fathers—an often ignored and excluded population. He is bringing his expertise to the University of Chicago-Chapin Hall as a faculty member while also taking on a community leadership role at Fathers, Families, and Healthy Communities.

Aisha Nyandoro,[3] CEO of Springboard To Opportunities, which supports residents of affordable housing, and innovator of the organization's Magnolia Mother's Trust. Not only does Aisha lead the first ever direct cash assistance program that directly applies a gender focus, but also influences policies at the state and national level to ensure that the experiences and voices of Black Mothers are authentically valued and incorporated in the design of policies and programs.

Dr. Koku Awoonor Williams[4] began his career working at the district level in Ghana and then serving as a regional health director. Because of his leadership, vision, and positive impact on the lives of millions of Ghanaians, he was placed to serve as director of policy, planning, monitoring, and evaluation at the Ghana Health Service to strengthen the community-based system of care. Throughout his leadership trajectory, he has mentored and supported several generations of burgeoning Ghanaian leaders at all levels of the health system.

With his vision and DDCF support, Dr. Manzi Anatole[5] was able to pursue his graduate studies at the University of Rwanda and at Harvard Medical School but continues to focus his work and vision in the African context, and in Rwanda specifically. He recently was awarded the prestigious Aspen Institute New Voices Fellowship to enable him to deliver on his passion for mentorship of the next generation of leaders.

In these examples, I think it important to note that we not only invested in their training and professional development, but also championed and gave greater visibility to their work. This is critically important, as the bulk of formal investments in leadership focus solely on the individual, but we believe it is equally important to fund and facilitate their work and the communities they are focusing on as well. Further, we found that leaders of color in this space often feel lonely or isolated in their work. We intentionally work to respond to their requests to connect them to other powerful networks of leaders, where they are not simply observers but are leaders in the conversation.

These leaders are just a few examples of the next generation of leaders whose influence and impact will have ripple effects in communities around the world. Despite their brilliance as individuals, they still struggle to gain the deserved visibility, funding, and access to global platforms that their white colleagues receive. It is our collective responsibility as leaders to trust them, invest in them, champion them, and work in solidarity with them.

WHAT SHOULD/COULD FUNDERS DO TO SUPPORT THE NEXT GENERATION OF EVIDENCE-DRIVEN LEADERS OF COLOR?

There is an urgent need to support the growing cadre of accomplished scholars of color who work to expand the perspectives reflected in research and to design more equitable and racially and ethnically representative policies and practices based on program evidence. For example, the National Center for Education Statistics reported that in fall 2018, just 4 percent of full-time professors at degree-granting postsecondary institutions were Black and 3 percent were Latinx. Among assistant professors, 8 percent were Black and 6 percent were Latinx. These data indicate that 50 percent of Black and Latinx scholars fall out of academia before achieving the highest rank. Additionally, Native American faculty represented less than 1 percent of all faculty from degree-granting institutions.

Researchers of color are more likely to account for racial disparities in their research design and analysis. Researchers from racially and ethnically diverse backgrounds and lived experiences are uniquely equipped to partner with under-researched and underserved communities to develop and implement research approaches that reflect and elevate the backgrounds, needs, and cultural and linguistic practices of Black, Indigenous, and other diverse populations. Researchers of color are more likely to drive research and implement research methodologies that take into account racial disparities and cultural context. Furthermore, they are more likely to identify, value, and understand the protective value of culture and community and seek authentic partnership with communities to inform their work. Authentically engaging community stakeholders—including parents, young people, and community residents who are experiencing the challenges the researchers are interested in understanding—is critical to defining locally relevant research questions, designing inclusive data collection and analysis tools, and interpreting and equitably disseminating results.

We need to build the capacity of research institutions and/or networks to increase funding and direct support to researchers of color and the teams that they lead and nurture. Researchers of color are more likely to face bias and discrimination and less likely to receive adequate recognition and support from their academic institutions and funders, making it difficult for them to enter and remain in academic settings as well as to promote and increase the use of their important work. As a component of advancing their research agendas, researchers of color also need clear financial and nonfinancial incentives to ensure that they are supported in their academic journeys. For DDCF, this has meant funding the individual researcher, their research, and the development of networks of other researchers of color. We also incent, with matching funds, the development of research centers of excellence or institutes within their university or organization that emphasize a focus on the lived experience of people of color as well as offer support for executive vouching and mentoring to strengthen their leadership and advance their professional development

For example, in recent years, the DDCF *Child Well-Being Program* has contributed to efforts to support researchers of color through direct investments in their research; promoting training, mentorship, and pipeline building; and facilitating a space and platform for funders to discuss and collaborate on increased funding opportunities. Grants have supported the National Indian Child Welfare Association to evaluate a parenting program developed by and for Native communities, and the Scholar Development Program through the Society for Research in Child Development to help researchers of color successfully navigate the complex process of obtaining NIH awards. In partnership with the William T. Grant Foundation, the Child Well-Being Program also co-founded a funders learning community of more than fifteen public and private funders committed to reducing racial gaps in research funding and improving career advancement for scholars of color.

CONCLUSION

In the not-so-distant past, it has felt unsafe to speak truth to power, to name the things that are wrong and the roles race and racism have played in social ills. Today, I am proud to acknowledge that I feel more welcomed and more confident to be vocal on these matters. Because of the next gen leaders of color around me, I am refueled, reignited, and energized, even as I share in feelings of deep exhaustion from responding to calls for input and wisdom at a time when the country and the world still feels it is rallying against people of color in so many intentional and entrenched ways. Still, I am privileged to be part of a loud and proud army of leaders and champions who are demanding systems transformation that authentically values the expertise and experience of leaders of color in the social service and research sectors.

NOTES

1. Angela Jackson, John Kania, and Tulaine Montgomery, "Effective Change Requires Proximate Leaders," *Stanford Social Innovation Review*, October 2, 2020, https://ssir.org/articles/entry/effective_change_requires_proximate_leaders.

2. See the Clinton Boyd Jr. page at the Zero to Three website, www.zerotothree.org/our-team/clinton-boyd-jr.

3. See the Aisha Nyandoro page on the Springboard To Opportunities website, https://springboardto.org/about/leadership/.

4. See the e-Health Africa Conference page at the Anadach Consulting Group website, www.anadach.com/blank-t7407.

5. See the Anatole Manzi page at the University of Global Health Equity website, https://ughe.org/meet-the-team/anatole-manzi.

THE CASE FOR CASH

NISHA G. PATEL

INTRODUCTION

A limitation when developing solutions to social problems is the tendency to start with existing government programs and work incrementally to improve them rather than anchoring on bold goals and working backward to design solutions. Using the latter process encourages more expansive thinking to tap into the full body of existing and emerging evidence. When it comes to addressing child poverty—or poverty more broadly—*what if we started with a bold goal like a minimum level of income for all?*

CONTEXT

Whether children have the chance to thrive is linked to their families' opportunities to access adequate income, and that typically has led policymakers to focus on jobs, education, and training for parents as the solution to child poverty. However, evidence shows that parents with low incomes often face considerable obstacles to getting and keeping jobs that pay enough. These challenges include: limited formal education and work histories; caring for young children; lack of stable, affordable, high-quality child care; children with special needs; domestic violence; physical and mental health issues; trauma and toxic stress; and lack of stable housing. The more of these challenges parents face, the less likely they are to be employed.[1]

The primary federal policy response intended to support parents with access to jobs over the past twenty-five years has been the Temporary Assistance for Needy Families (TANF) block grant. The evidence from programs that informed the development of TANF policy in the mid-1990s found that they increased *labor force attachment* but failed to increase *income* for families.[2] Today, only a small share of families with children experiencing poverty receive access to income support[3] through TANF, and states spent only about 10 percent of TANF funding on work, education, and training in FY 2019. So, it is unsurprising that TANF has done little to sustainability reduce child poverty or increase social and economic mobility.

A CHANGING ECONOMY AND CHILD POVERTY

Much has changed since TANF was designed in the twentieth century, before the internet was widely used and before Google, the iPhone, LinkedIn, Uber, or Amazon existed. Availability of jobs with good wages, benefits, and advancement opportunities for people with less formal education has declined.[4] Contract, temporary, on-call, and gig economy jobs have surged,[5] and the workforce has fissured across employers by wages and education.[6] The way people get jobs has shifted rapidly to online job search and recruitment tools, and the potential for displacement of jobs by automation and artificial intelligence has increased.[7]

In the midst of these changes, in 2019, 17 percent of all children were living in poverty. Despite declines since 2010, Black (30 percent) and Native American (30 percent) children were still about three times as likely as Asian and Pacific Islander[8] (10 percent) and white (10 percent) children to be living in poverty. Hispanic and Latino (23 percent) children were more than twice as likely than Asian and Pacific Islander or white children to be living in poverty.[9]

ENTRENCHED STRUCTURAL RACISM AND GENDER INEQUITY

Prior to COVID-19, TANF and other safety net programs failed to address the structural issues that keep many families of color trapped in poverty. As just one example, for the past five decades, the Black unemployment rate almost always has been double the white unemployment rate,[10] even in tight labor markets. Racial discrimination by employers[11] and occupational segregation by race and gender are factors.[12] Prior to the pandemic, only

20 percent of white men were working in low-wage jobs versus almost 40 percent of Black women and 46 percent of Hispanic women.[13]

The bottom line is that even in a "good" economy, lots of people were locked out of opportunity—and we always have needed ways to supplement or replace income from employment.

A GLOBAL PANDEMIC AND THE POLICY RESPONSE

Then came the pandemic, an economic crisis, and the federal policy response. The initial policy response in 2020 was not universal, but was designed to provide direct cash to the majority of people in the United States through two primary mechanisms: 1) Economic Impact Payments of up to $1,200 per individual and an additional $500 per child without regard to parental employment status; and 2) Pandemic Unemployment Assistance, which added $600 per week in federal benefits to state unemployment benefits. These payments dramatically reduced poverty for as many as 13 million people in the early months of the pandemic.[14] This evidence reinforced the importance and efficiency of direct cash in helping families both survive a crisis and thrive over the long term when they have the ability to save. Nearly 8 million people slipped into poverty when the cash assistance ended.[15]

TWENTY-FIRST-CENTURY EVIDENCE

The policy responses to the pandemic built on the evidence from guaranteed income programs and strengthened the case for providing direct cash to children and their families who would otherwise struggle to make ends meet. Guaranteed income exists at the state level in the form of the Alaska Permanent Fund, which provides a direct cash payment of $1,000 to 2,000 to every person in the state annually. The evidence shows that it does not discourage full-time employment—and, in fact, increases part-time employment.[16] Guaranteed income also has been tested in tribal communities. For example, the Eastern Band of Cherokee "casino dividend" provides eligible people approximately $4,000 in unconditional cash payments. It does not reduce labor force participation and improves educational outcomes for children, with better attendance and more years of education completed.[17]

Guaranteed income has recently been tested via a randomized control trial at the municipal level through the Stockton Economic Empowerment Demonstration (SEED), which provided direct cash payments of $500 per month over two years to residents living in low-income neighborhoods of Stockton, California. SEED's first-year findings revealed that people who received cash payments went from part-time to full-time employment at more than twice the rate of the control group, saw unemployment drop, and were better able to handle unexpected expenses and make payments on their debt. And, people who received cash payments were healthier, showing less depression and anxiety and enhanced well-being.[18] The Stockton pilot has spurred and influenced many other local-level pilots around the country, several of which are being designed with principles that align with the Next Generation of Evidence Campaign.

THE NEXT GENERATION OF EVIDENCE

In stark contrast to the now dated evidence that informed the development of TANF and its complex rules and work participation rates, numerous efforts that embody Project Evident's principles of being practitioner-centric, embracing an R&D approach, and elevating voices of communities are building evidence across the country. Examples of efforts that align with each principle are outlined below.

Being Practitioner-Centric: THRIVE East of the River

THRIVE East of the River (THRIVE) provided emergency cash to nearly 600 households in Ward 8 of the District of Columbia. The collaboration among four community-based practitioner organizations (Bread for the City, Far Southeast Family Strengthening Collaborative, Martha's Table, and Building Bridges Across the River) was designed to address the disproportionate economic impact of the COVID-19 pandemic on the individuals and families they serve. The practitioners designed THRIVE and its evaluation in close partnership with the Urban Institute.

THRIVE sought to: 1) alleviate crisis by providing families with immediate access to cash, healthy food, and dry goods; 2) stabilize families by connecting them to the full range of government resources for which they were eligible, and 3) foster mobility by assisting families to secure a more resilient future. The program provided $5,500 in direct cash to participants, with the option of either lump sum or monthly payments of $1,100 for five months. The first payments began in July 2020, and the program continued to recruit participants and provide payments through January 2022.[19]

In addition to working closely with the practitioner organizations to provide continuous data and reporting for program management, the Urban Institute engaged residents of Ward 8 as community-based researchers. A summary report on THRIVE outcomes and implementation found that participants most commonly used the cash payments for housing and food costs. Additional uses included transportation, debt reduction, and professional goals, such as investments in small businesses. After receiving payments, participants reported better mental health and lower rates of food insecurity compared to other people with low incomes.[20]

Embracing an R&D Approach: UpTogether

An R&D approach involves a disciplined process for learning, testing, and improving to enable timely and relevant continuous evidence building. UpTogether, which serves families across the country, embodies such an approach. For twenty years, UpTogether has continuously collected outcome data demonstrating that families can increase economic and social mobility when self-determination and mutual support are fostered. Their strength-based approach includes *capital*, in the form of direct payments to families; and *choice*, in that families have agency to use the money as they see fit. Their UpTogether Community®, an online platform, delivers cash to families via direct deposit or prepaid card and has features families can use to build and strengthen their social networks. UpTogether listens to families, learns from the actions families take to improve their lives, and uses that data to influence the ways philanthropy and government invest in communities.[21]

Building on their R&D approach, UpTogether is partnering with the Massachusetts Department of Transitional Assistance (DTA, the state's TANF agency) on an evaluation on the effects of social and financial capital on economic mobility and well-being. The two-year RCT includes families with income below 200 percent of poverty and/or who receive economic assistance through DTA. The study is tracking the impact of direct cash combined with social capital building.[22]

Elevating the Voices of Communities: Magnolia Mother's Trust

Communities must have the power to shape and participate in the evidence-building process of practitioners and the field. Springboard to Opportunities, which serves Black mothers living in subsidized housing communities in Jackson, Mississippi, embodies this philosophy, which it describes as "radically resident driven." As part of co-designing a program with mothers striving for social and economic mobility, the staff conducts regular focus groups with residents. A key insight from one of these focus groups was that families had very little, if any, access to discretionary cash, which led to both economic and emotional stress. The voices of the mothers made clear that, as they tried to create a better life for their children (for example, returning to school for more education and training), income volatility interfered with their goals.[23]

Based on the women's insights, in the fall of 2018, Springboard to Opportunities launched the Magnolia Mother's Trust, with a pilot cohort of twenty Black mothers who received $1,000 unconditional direct cash payments for twelve months. A larger study of a second cohort of 110 women began in March 2020, and a third cohort launched in 2021. Results from the second cohort, which started as the nation began to shut down due to the COVID-19 pandemic, found that, in contrast to the comparison group, recipients were less likely to report debt from emergency financing, more likely to have children performing at or above grade level, more likely to seek professional help for chronic illness and sickness, and able to budget more for food and household costs, resulting in lowered food insecurity and better access to basic needs.[24] The evaluation findings from the third cohort of 95 mothers were released in August 2022 and revealed that among participants, 98 percent felt somewhat or extremely supported to meet their family's needs, 79 percent felt more hopeful about their future, 82 percent felt more hopeful about their children's futures, and 70 percent felt capable of caring for their own emotional, physical, and mental health needs.[25]

In addition to the quantitative data, narrative change that elevates community voices is a key aspect of the program. Springboard to Opportunities has created a Storytelling Lab to support participants to share their stories with a wider audience, including oral stories geared for podcasts, storytelling events, town halls, and policy conferences; as well as written stories for publication as Op-Eds.

WHERE DO WE GO FROM HERE?

The outcomes of the 2020 federal direct cash measures in response to the pandemic and the evidence of the effectiveness of direct cash payments from state, tribal, and municipal programs over the past several years, along with other bodies of evidence about the effectiveness of tax credits[26] and child allowances,[27] helped lay the groundwork for the March 2021 American Rescue Plan. The American Rescue Plan contained several direct cash components, including Recovery Rebate payments of up to $1,400 and expanded, advance refundable child tax credit (CTC) payments of up to $3,600 for children under six and $3,000 for children six to seventeen. Columbia University found that the monthly CTC payments ($250 to $300 per month per child) kept 3.7 million children out of poverty in December 2021 and reduced monthly child poverty by nearly 30 percent.[28] Evidence of the impact of these direct cash payments on poverty over the past year should inform longer-term policy development, such as the possibility of creating a permanent child allowance, as well as other forms of recurring direct cash payments to populations facing financial hardship.

NOTES

1. Sheila Zedlewski, "Welfare Reform: What Have We Learned in Fifteen Years?," Urban Institute, 2012.

2. For example, evidence from the NEWWS study often is cited as supporting "labor force attachment," but the models that were evaluated failed to increase income for families. National Evaluation of Welfare-to-Work Strategies, U.S. Department of Health and Human Services, December 2001, www.mdrc.org/sites/default/files/full_391.pdf.

3. In 2019, for every 100 families in poverty, only twenty-three received cash assistance from TANF—down from sixty-eight families in 1996. Laura Moyer and Ife Floyd, "Cash Assistance Should Reach Millions More Families to Lessen Hardship," Washington, DC: Center of Budget and Policy Priorities, November 30, 2020, www.cbpp.org/research/family-income-support/cash-assistance-should-reach-millions-more-families-to-lessen.

4. David H. Autor, "Skills, Education, and the Rise of Earnings Inequality among the 'Other 99 Percent,'" *Science* 344, no. 6186 (2014), pp. 843–51; Jae Song, David J. Price, Fatih Guvenen, Nicholas Bloom, and Till von Wachter, "Firming Up Inequality," draft, Wordpress, October 22, 2016, https://fguvenendotcom.files.wordpress.com/2014/04/fui_22oct2016_final_qje_submit.pdf.

5. Lawrence F. Katz and Alan B. Krueger, "The Rise and Nature of Alternative Work Arrangements in the United States, 1995–2015," Working Paper 22667, 2016, Cambridge, MA: National Bureau of Economic Research.

6. David Weil, *The Fissured Workplace* (Cambridge, MA: Harvard University Press, 2014).

7. David H. Autor, "Why Are There Still So Many Jobs? The History and Future of Workplace Automation," *Journal of Economic Perspectives* 29 (2015), pp. 3–30.

8. Poverty among Asian and Pacific Islander children varies widely among racial and ethnic groups. For example, in 2016, among Asian subgroups, the percentage of children living in poverty ranged from 6 to 37 percent. The percentages of children living in poverty were higher in some subgroups, ranging from 15 percent for Vietnamese children to 37 percent for Bangladeshi children. The percentages of Cambodian, Chinese, Korean, Laotian, Nepalese, and Thai children living in poverty were not measurably different from the overall Asian percentage. The percentages of Asian Indian, Filipino, and Japanese children living in poverty (6 percent each) were lower than the overall Asian percentage. National Center for Education Statistics, "Status and Trends in the Education of Racial and Ethnic Groups," February 2019, https://nces.ed.gov/programs/raceindicators/indicator_rads.asp.

9. Kids Count Data Center, "Children in Poverty (100 Percent Poverty)," Annie E. Casey Foundation National KIDS COUNT: Population Reference Bureau, analysis of data from the U.S. Census Bureau, 2005, 2008, 2010, and 2013 through 2019 American Community Survey, https://datacenter.kidscount.org/data/tables/8447-children-in-poverty-100-by-age-group-and-race-and-ethnicity?loc=1&loct=1#detailed/1/any/false/1729,37,871,870,573,869,36,133,35,16/2757,4087,3654,3301,2322,3307,2664|17,18,140/17079,17080.

10. Olugbenga Ajilore, "On the Persistence of the Black-White Unemployment Gap," Center for American Progress, February 24, 2020, https://www.americanprogress.org/issues/economy/reports/2020/02/24/480743/persistence-black-white-unemployment-gap/.

11. Marianne Bertrand and Sendhil Mullainathan, "Are Emily and Greg More Employable than Lakisha and Jamal? A Field Experiment on Labor Market Discrimination," Working Paper 9873, July 2003, NBER, https://www.nber.org/papers/w9873.

12. Kate Bahn and Carmen Sanchez Cumming, "Factsheet: U.S. Occupational Segregation by Race, Ethnicity, and Gender," Washington Center for Equitable Growth, July 1, 2020, https://equitablegrowth.org/factsheet-u-s-occupational-segregation-by-race-ethnicity-and-gender/.

13. David T. Ellwood and Nisha G. Patel, "Restoring the American Dream: What Would It Take to Dramatically Increase Mobility from Poverty?" US Partnership on Mobility from Poverty, January 2018, www.mobilitypartnership.org/restoring-american-dream.

14. Ben Zipperer, "Over 13 Million More People Would Be in Poverty without Unemployment Insurance and Stimulus Payments," Economic Policy Institute, September 17, 2020, www.epi.org/blog/over-13-million-more-people-would-be-in-poverty-without-unemployment-insurance-and-stimulus

-payments-senate-republicans-are-blocking-legislation-proven-to-reduce-poverty/.

15. Zachary Parolin, Megan Curran, Jordan Matsudaira, Jane Waldfogel, and Christopher Wimer, "Monthly Poverty Rates in the United States during the COVID-19 Pandemic," Center on Poverty and Social Policy at Columbia University, October 15, 2020, www.povertycenter.columbia.edu/news-internal/2020/covid-projecting-monthly-poverty.

16. Damon Jones and Ioana Elena Marinescu, "The Labor Market Impacts of Universal and Permanent Cash Transfers: Evidence from the Alaska Permanent Fund," February 5, 2018, https://ssrn.com/abstract=3118343 or http://dx.doi.org/10.2139/ssrn.3118343.

17. Ioana Elena Marinescu, "No Strings Attached: The Behavioral Effects of U.S. Unconditional Cash Transfer Programs," Working Paper, NBER, February 2018, www.nber.org/papers/w24337.

18. Stacia West, Amy Castro Baker, Sukhi Samra, and Erin Coltrera, "Preliminary Analysis: SEED's First Year," Stockton Economic Empowerment Demonstration, March 2021, https://static1.squarespace.com/static/6039d612b17d055cac14070f/t/6050294a1212aa40fdaf773a/1615866187890/SEED_Preliminary+Analysis-SEEDs+First+Year_Final+Report_Individual+Pages+.pdf.

19. Program description provided by THRIVE East of the River.

20. Mary Bogle et al., "An Evaluation of THRIVE East of the River: Findings from a Guaranteed Income Pilot during the COVID-19 Pandemic" Urban Institute, February 2022, https://greaterdc.urban.org/sites/default/files/publication/105445/an-evaluation-of-thrive-east-of-the-river_1.pdf.

21. See A Community-Centered Approach to Socioeconomic Mobility page at the UpTogether website, https://www.uptogether.org/approach/.

22. See UpTogether, "Trust and Invest Collaborative," https://info.uptogether.org/tic.

23. Rachel Black, "Centering the Margins: A Framework and Practices for Person-Centered Financial Security Policy," Aspen Institute Financial Security Program, December 15, 2020, www.aspeninstitute.org/publications/centering-the-margins-a-framework-and-practices-for-person-centered-financial-security-policy/.

24. "The Magnolia Mother's Trust: The Invaluable Benefits of Investing in Black Mothers," Springboard to Opportunities, March 2021.

25. Asia Moore, Christyl Wilson Ebba, Nidal Karim, and Sashana Rowe-Harriott, "Magnolia Mother's Trust 2021–2022 Evaluation Report," Social Insights, August 2022, https://springboardto.org/wp-content/uploads/2022/08/MMT-Evaluation-Full-Report-2021-22-website.pdf.

26. See, for example, *Tax Policy Center Briefing Book*, Tax Policy Center, 2021, www.taxpolicycenter.org/briefing-book/how-does-earned-income-tax-credit-affect-poor-families.

27. National Academy of Sciences, Engineering, and Math, "A Roadmap to Reducing Child Poverty," 2019, www.nap.edu/catalog/25246/a-roadmap-to-reducing-child-poverty.

28. Zach Parolin, Sophie Collyer, and Megan A. Curran, "Sixth Child Tax Credit Payment Kept 3.7 Million Children Out of Poverty in December," Columbia University Center on Poverty and Social Policy, January 18, 2022, https://www.povertycenter.columbia.edu/publication/montly-poverty-december-2021.

PARENTCORPS

A COLLABORATIVE EVIDENCE-BUILDING STRATEGY GUIDED BY FAMILY VOICE

LAURIE MILLER BROTMAN, SHANIKA GUNARATNA, ERIN LASHUA-SHRIFTMAN, AND SPRING DAWSON-MCCLURE

Housed at NYU Grossman School of Medicine's Center for Early Childhood Health & Development, ParentCorps is a family-centered early childhood intervention designed to enhance the pre-K experience in historically disinvested neighborhoods. Its mission: to help schools partner with families to build a future where all children thrive. ParentCorps combines professional development for educators (to support school staff to form strong, culturally responsive relationships with families and promote children's social-emotional development), a group-based family program (to support families to promote children's healthy development), and a classroom-based social-emotional learning program (to help children learn to identify and communicate their feelings, develop a positive sense of self, build healthy relationships, and more).

From starting with one pilot in 2000 to significant expansion as part of New York City's Pre-K for All, ParentCorps has invested in a range of evidence-building strategies, including randomized controlled trials, mixed methods studies, and data feedback loops to evaluate impact, inform continuous improvement, and ensure alignment with practitioners' and families' lived experience.

Today, ParentCorps—fueled by a multidisciplinary team including researchers, social workers, and educators—serves more than 3,000 families annually, primarily families of color, in New York City, Detroit, Michigan, and Corpus Christi, Texas. In this case study, we focus on our evidence-building experiences in partnership with the NYC Department of Education's Division of Early Childhood Education, considering the needs and perspectives of policymakers who invest in and support programs at scale, and practitioners, including school leaders, social workers, pre-K teachers, and family support staff.

PARENTCORPS' APPROACH TO EVIDENCE BUILDING

The value ParentCorps places on data-informed decision making is evidenced by both our history of evidence building through RCTs and our responsive programmatic data collection. Led by a dedicated team, we have continuously strengthened our capacity to employ a broad range of methodologies to address critical questions and to fit the phase of inquiry.

2000–2002

An initial pilot in partnership with the Harlem Children's Zone illuminated the promise of a unique approach to bringing together families of pre-K children, honoring culture and affirming parents' autonomy to choose for themselves which evidence-based parenting strategies fit with their values, beliefs, and goals for their children.

2003–2010

We carried out two RCTs to evaluate impact in eighteen schools in high-poverty Brooklyn neighborhoods with more than 1,200 families. The first demonstrated short-term impact on children's social-emotional learning, evidence-based parenting practices, and greater parental involvement in children's learning. The second replicated these findings and demonstrated long-term impact on children's mental health (e.g., emotional and behavioral problems), academic achievement (e.g., reading at grade level), and physical health (e.g., obesity).

2010–2015

Using a range of evidence-building strategies, we expanded our understanding of ParentCorps' potential, including:

- Feasibility across settings (i.e., pre-K programs housed in elementary schools, in community-based organizations, and in Head Starts)
- Resonance and impact across diverse populations (i.e., immigrant families, Black, Latinx, and Asian American families)
- Theory of action
- Return on investment across the lifespan

It is important to note that, until this point, our team of researchers and practitioners largely drove our learning agenda, securing funds to answer questions *we* deemed critical.

2015–Present

The launch of our partnership with the NYC Department of Education in the context of the city's ambitious new universal pre-K initiative represented an inflection point. Though ParentCorps' initial RCTs yielded rich learnings, they taught us little about how to achieve scaled impact; how to work in partnership with a school district and pre-K programs to develop and operationalize evidence building; and how to prioritize the district's learning questions in addition to our own.

This phase of our work—to scale ParentCorps in more that fifty pre-K programs prioritized by concentrated poverty—generated key learnings:

- ***Shared understanding of evidence is critical.*** For instance, in early discussions with the NYC Department of Education, we invested deeply in policymakers' understanding of ParentCorps' theory of action, illustrating the relationship between our programs (the "what"), aspects of facilitation (the "how"), and proximal and long-term outcomes (the "why"). These discussions laid the foundation for ongoing dialogue about ParentCorps at scale, informing decisions on monitoring progress, evaluation, and evidence-based investments for children and families.
- ***Multiple strategies are needed to understand what lands for practitioners.*** In working with educators, ParentCorps aims to

build authentic relationships and create learning environments where shifts in beliefs are possible. For instance, in professional development, we invite educators to share candidly when they disagree with us or a colleague ("Tell us when you have a different perspective.") or have negative or unspoken emotions ("permission to feel") so that facilitators can tailor the space for these reflections. In data collection, we aim to create ample room for educators' experiences and ideas for improvement. At multiple timepoints in programs, we seek ratings and open-ended feedback via surveys and coaching conversations. Each data collection strategy brings benefits (e.g., anonymity of online surveys) and limitations (e.g., discomfort with technology), and so we challenge ourselves to avoid a one-size-fits-all approach and consider whose perspectives may be excluded by any one method. Insights into practitioners' experiences are critical to inform facilitation and content improvements.

- *Scale and evidence building must evolve in tandem.* Early on, policymakers expressed a priority to unlock ParentCorps' model to serve as many children as possible across the city's pre-K population. Thus, we co-created a strategy to "un-bundle" professional development to reach hundreds of pre-K programs (and thousands of pre-K teachers and leaders) prioritized by concentrated poverty and developed a new social-emotional learning tool to distribute universally (to 70,000 pre-K children and their families annually), helping create a shared language for children's feelings at home and school. Importantly, we then developed new evidence-building plans in partnership with the NYC Department of Education to assess these innovations implemented at scale.

We worked to bridge the worlds of evidence and practice internally. With consultation from Project Evident (2018–2019), we bolstered our team's capacity to articulate and apply iterative feedback loops and co-develop measures based on our coaches' conceptualization of the essential elements through which ParentCorps promotes adult behavior change. By increasing our own internal curiosity, harnessing constructive frustration with status quo data processes, and strengthening our coaches' comfort with and value for data, we made progress in building a learning culture that facilitates agile evidence building and program adaptation.

CHALLENGES AND RESPONSES

Over more than two decades, much has changed in both our program and our evidence-building approach. We have evolved from a strong focus on sharing the science of early childhood development with educators and families to adopting a broader approach (that also includes building authentic relationships, honoring culture, understanding race and racism, and practicing self-reflection) to transform the pre-K experience; from locating the problem in poverty alone to a much more expansive, evolving view of structural racism; and from leading a learning agenda ourselves to co-creating with the nation's largest school district.

Perhaps the biggest challenge for ParentCorps was in March 2020, when COVID-19 engulfed NYC as its early epicenter and forced abrupt school closures. Seemingly overnight, amid tremendous uncertainty and fear, we worked to take ParentCorps programs and evidence-building virtual.

In the earliest months of the pandemic, we considered every point of contact with families as an opportunity to assess need and inform rapid adaptation, both immediately and in what was sure to be an unpredictable school year to come. Touchpoints included trauma-informed phone surveys with parents participating in two ongoing RCTs and conversations with families attending group-based family programs (that had quickly adapted to virtual to preserve community). In parallel, we listened to teachers in the context of post-school closure coaching conversations and virtual facilitated professional development. These touchpoints helped us take the pulse of families and teachers at a hectic moment and, serendipitously, also convey that information to the school district, which urgently sought better understanding of family and teacher needs to inform resource allocation and other decisions.

The needs became clear, including support with grief and loss; finding predictability during uncertain circumstances; managing anxiety in self and others; connecting through empathy; and taking care of one's own mental health. These needs informed the adaptation of ParentCorps content and the development of new programs, including a four-session virtual program for parents (Parenting through the Pandemic), training of NYC's early childhood social worker workforce to facilitate this new program with families in historically disinvested neighborhoods most impacted by the pandemic, and both self-guided and virtual facilitated professional development for teachers and leaders. Alongside this rapid program adaptation and development came nimble data collection to establish meaningful feedback loops around all program elements, prioritizing non-burdensome strategies (e.g., reducing survey length and frequency; increasing brief interim reporting to content creation teams).

In particular, supporting more than one hundred early childhood social workers to deliver Parenting through the Pandemic across NYC highlights the practitioner-led feedback loops we strive for. In a plan co-developed with the NYC Department of Education, we trained and coached the social workers to facilitate Parenting through the Pandemic and—after programs launched—utilized surveys to capture their facilitation experiences and their evolving understanding of families' needs. These data then informed each subsequent coaching session's design and content (e.g., including large and small group discussion; topic-driven coaching; opportunities for self-reflection; and modeling of facilitation). By leveraging the skills and infrastructure we had built for ongoing learning, we were able to support large-scale program delivery through crisis, rooted in responsive coaching that met a critical workforce's needs.

REFLECTIONS AND CONCLUSIONS

ParentCorps' sustained commitment to evidence building of over twenty years is rooted in our core value for learning. By embedding a dedicated data team in our structure, leaning into our home institution's mission of discovery and available resources (e.g., expertise in population health and health equity; qualitative and quantitative methods; community based participatory research; data capture and visualization tools) and seeking funding for evidence building from philanthropic partners who share this value, we have become resilient to contextual challenges that might otherwise lead to the deprioritization of evidence building.

Especially through the ongoing crisis of COVID-19, a next-generation evidence approach has proven critical, enabling us to prioritize the needs of educators, children, and families over strict adherence to fidelity protocols, flexibly adapt programming, and avoid shutting down programs at a time when the home-school connection was vital. Years of building muscle around collaborative evidence building also meant that partners trusted us to make these adaptations, knowing we would accompany these changes with nimble data collection and collaborative, critical analysis of implementation and impact. Moving forward, this collaborative evidence-building approach is central to our strategy to scale Parent-Corps nationally.

NOTES

L. M. Brotman, E. Calzada, K. Y. Huang, S. Kingston, S. Dawson-McClure, D. Kamboukos, A. Rosenfelt, A. Schwab, and E. Petkova, "Promoting Effective Parenting Practices and Preventing Child Behavior Problems in School among Ethnically Diverse Families from Underserved, Urban Communities," *Child Development* 82, no. 1 (2011): 258–276, https://doi.org/10.1111/j.1467-8624.2010.01554.x.

L. M. Brotman, S. Dawson-McClure, E. J. Calzada, K. Y. Huang, D. Kamboukos, J. Palamar, and E. Petkova, "Cluster (School) RCT of ParentCorps: Impact on Kindergarten Academic Achievement," *Pediatrics* 131, no. 5 (2013): e1521–e1529, https://doi.org/10.1542/peds.2012-2632; S. Dawson-McClure, E. Calzada, K. Y. Huang, D. Kamboukos, D. Rhule, B. Kolawole, E. Petkova, and L. M. Brotman, "A Population-Level Approach to Promoting Healthy Child Development and School Success in Low-Income, Urban Neighborhoods: Impact on Parenting and Child Conduct Problems," *Prevention Science: The Official Journal of the Society for Prevention Research* 16, no. 2 (2015): 279–290, https://doi.org/10.1007/s11121-014-0473-3; L. M. Brotman, S. Dawson-McClure, D. Kamboukos, K. Y. Huang, E. J. Calzada, K. Goldfeld, and E. Petkova, "Effects of ParentCorps in Prekindergarten on Child Mental Health and Academic Performance: Follow-Up of a Randomized Clinical Trial through 8 Years of Age," *JAMA Pediatrics* 170, no. 12 (2016): 1149–1155, https://doi.org/10.1001/jamapediatrics.2016.1891; L. M. Brotman, S. Dawson-McClure, K. Y. Huang, R. Theise, D. Kamboukos, J. Wang, E. Petkova, and G. Ogedegbe, "Early Childhood Family Intervention and Long-Term Obesity Prevention among High-Risk Minority Youth," *Pediatrics* 129, no. 3 (2012): e621–e628, https://doi.org/10.1542/peds.2011-1568.

N. Hajizadeh, E. R. Stevens, M. Applegate, K. Y. Huang, D. Kamboukos, R. S. Braithwaite, and L. M. Brotman, "Potential Return on Investment of a Family-Centered Early Childhood Intervention: A Cost-Effectiveness Analysis," *BMC Public Health* 17, no. 1 (2017): 796, https://doi.org/10.1186/s12889-017-4805-7.

L. M. Brotman et al., "Scaling Early Childhood Evidence-Based Interventions through RPPs," *The Future of Children* 31, no. 1 (2021): 57–74, https://futureofchildren.princeton.edu/sites/futureofchildren/files/foc_combined_5.3.21.pdf.

SECTION 3

ELEVATING COMMUNITY VOICE

> *Many . . . have begun to raise the alarm that big data . . . will compound and exacerbate racial inequality . . . accompanied by calls for community involvement. But it is crucial that community engagement is not an add-on or window dressing for programs long in the making.*
>
> —MARIKA PFEFFERKORN, "DATA JUSTICE AND RISKS OF DATA SHARING."

Shaping questions, collecting and synthesizing information, and sharing results are critical parts of an equitable evidence-building journey. Each of these steps is made more relevant by gathering input and feedback from communities served, both practitioners and their participants, and elevating what they have to say, their "community voice." Moreover, inviting participation and sharing results with the people one seeks to serve also may infuse the process of building evidence with respect and dignity, critical to growing community trust.

The essays in this section feature Dan Cardinali, formerly with Independent Sector, who writes about building trust through listening to communities, and building evidence in service of what communities express as their own goals. Marika Pfefferkorn's subsequent chapter tells a story of data justice after the St. Paul, Minnesota, school and police districts broke community trust through an ineffective process for getting sign-off for a data sharing agreement that could lead to racial profiling. In the face of public outrage, the districts pivoted to community consultation and, spurred by community leaders, found a better way. John Brothers of T. Rowe Price Foundation calls on philanthropy to "find evaluative approaches that help communities use their own data for their own self-determination," and to invest in the measurement systems of community-based organizations. And Rhett Mabry of The Duke Endowment writes about unintended consequences when philanthropy fails to listen to community.

Use cases that follow include The Duke Endowment's Summer Literacy Initiative, New York University's Criminal Justice Lab, and girls empowerment nonprofit Pace Center for Girls. All show how gathering input and feedback from program participants has advanced social missions and improved participants' experiences and results.

Questions tackled in this section include:

1. Where in the cycle of evidence building is it best to draw in community input and loop back to communities with findings?
2. What tools and processes are useful for doing so?
3. How do these practices strengthen decision making and implementation?

THE POWER OF COMMUNITY VOICE

DANIEL J. CARDINALI

We are in the midst of a national reckoning of confronting the racial inequities that permeate every aspect of our nation. Given our nation's history of using evidence as a powerful tool for driving change, we have a serious opportunity and challenge before us: attending to how evidence is used as we confront and understand our nation's history of racial inequity. Also, how evidence is used as we work toward a more racially equitable and sustainable nation becomes essential.

For evidence to be truly useful in taking up this work, it needs the appropriate context, relational framing, and community grounding. At Independent Sector, we hold true that individual and collective flourishing are inextricably linked. Because of this, evidence—and the production and analysis of it—is beneficial to all when viewed in a framework of progress; not just on an individual basis but also a collective one.

Evidence and the building of it needs to include people and communities who are not flourishing by developed methodologies, as well as partners and institutions that can provide support, reflection, and insight for positive systemic change. For stakeholders—including foundations, government, practitioners, and researchers—this is an important, and at times a complicated, shift toward making democratic spaces to incorporate community voices in their institutional missions, grantmaking, and evidence-building work. They are doing this with full knowledge of the power, advantages, and insight that philanthropy and their institutions bring to the nation. By building the appropriate structures to unearth more relevant evidence and bringing this resolute awareness to our work, we know data can inform policies and solutions that lead to progress and, ultimately, help people thrive.

Two examples from my career support this idea. For seventeen years, I worked at Communities in Schools, a national nonprofit that supports students in overcoming barriers and staying in the education system for long-term success. I watched and learned of Black and brown children pipelined into the juvenile justice system because school officials used data—grades, absentee rates, and the number of behavioral issues—punitively. The corresponding response to this evidence from school officials was institutional and systemic punishment, which harmed a student's long-term future.

We, in the Communities in Schools movement, worked with students, families, practitioners, school personnel, and board members and realized these data were showing us what many authorities were missing—that a young person was experiencing deep distress. To avoid sending a young person into the juvenile justice system, we partnered with specialists and community members and knew the education system needed to build a constructive environment for student resilience and long-term flourishing. We used the same evidence as school officials and the juvenile justice system, but we realized this: It is *how* people in and with power in these systems use it that matters. We also took this important step: We centered the lives and future of Black and brown students and their families in our work. We continuously asked what it would take for them to flourish, and then used data and evidence to create the conditions for individual and collective flourishing.

Our movement, too, was aided by disaggregated data by race, class, and geography, as well as the explosion of the integrated student support field and the national awakening to racial inequities. In both cases, officials used evidence to release large amounts of public money. In the punitive context, dollars went to the juvenile justice system, school-based security officers, and metal detectors. When we centered our efforts on resilience and thriving, money helped fund better social service support, holistic family supports, and shifting a school environment into a place in which all students could thrive.

The other example involves our Upswell Summit, which Independent Sector powers each year to bring together changemakers for sense-making discussions about communities, collaboration, the social sector, and our nation. To help us gauge feedback, we use a variety of evaluation methods, including the Net Promoter Score (NPS) system, which is a number rating for each overall summit. We have one required question of people who attend each summit: "Would you recommend Upswell to a friend or colleague?"

From 2016 to 2018, our NPS rating hovered in relatively the same range, with an increase or decrease that was not extraordinary. Still, based on feedback and continuously testing hypotheses, we made changes to the next summit. But from 2018 to 2019, our NPS rating dropped dramatically. We took a deeper look at our evidence and audience feedback, and from a design approach, we asked: Given our data, how can we turn the curve of user experience, given our mission and what people say they care about? For 2020, we settled on two key Upswell Summit themes instead of several: racial justice and all aspects of health. These two themes still allowed us breadth and depth in our sense-making, social sector work. From 2019 to 2020, our NPS rating increased by more than twenty times—an indication that we built, collected, and used evidence to be more focused, adaptive, relevant, and forward looking. Our Upswell Summit audience told us that our anti-racism discussions gave them tools to use in their social sector work and everyday life—giving us, in 2020, our highest "strongly agree" ranking, of 92 percent. We are applying all our lessons to our everyday work because we want to do better.

For the social sector, recognizing the influence and role of evidence in our missions is pivotal. As an engine of renewal, the social sector plays a unique and powerful role in America, especially on our much-coveted path to progress and inclusion. The social sector listens and works with people in communities. We build trust. We help implement public policies. We are partners in crafting public policy solutions, as well.

What is key is that we agree on and accept ways in which people in communities, especially those that are structurally marginalized, define what individual and collective human and environmental flourishing looks like for themselves, their loved ones, and their neighborhoods. Then, we build evidence in service of progress toward that human, environmental, and community flourishing. This framework will lead to deeper answers and policy and community solutions because people will have a vested interest in all of it. It will accurately reflect their lives and communities. **We have come to believe that building evidence like this is a deeply**

authentic way to help people steward the environments where they live. It also allows for adaptation that accounts for the rich diversity across communities nationwide.

In no way am I advocating that methodologies and tools that hard and social sciences have to offer be left out of this evidence-building process. Rather, these methodologies and tools should be put in service to communities engaging in development of their own pathways forward to authentic human flourishing. Critical to the application of these methodologies and tools will be the trust of people in communities and the relationships to institutions designated to serve them. Trust is one of the most pressing adaptive challenges of our day.

Given how COVID-19 swept across the United States and world in 2020, we face an opportunity to rebuild a stronger social sector and healthier communities. So, establishing what trust looks like in communities is of paramount importance. We need to take care and be thoughtful in how we frame our steps and goals, and how we collect and build evidence.

In our journey, this means listening well, interrogating our own steps, and devoting enough time to get beyond a transactional relationship with communities and their members. It means being a partner because trust, by definition, is relational and characterized by vulnerability. It extends beyond individual agency by vesting or sharing your own ability to flourish with other people. It spotlights critical questions that Independent Sector and many in the social sector are asking to ensure we center equity on people at the margins of society.

The United States has so much important work before us. We have so much to gain to support our array of rich and vibrant communities and collective potential. We cannot ignore this question about evidence and how we build, disseminate, codify, apply, and accept it. We use evidence to disperse power and money. It can liberate, illuminate, and inspire. It also can stifle. Broadening our ideas around evidence can lead to making more of us in America whole people and restore some semblance of balance to the intersecting systems in which we live.

One unanswered part of this important calculus of equity and healthier communities for all remains—particularly for all types of leaders: Are you and your organization willing to build new evidence, analyze existing data in the context of a racial equity analysis and resilience, or consider the changing and relational nature of both, so everyone in the United States can thrive? Are you willing to ask new questions and accept new or overlooked answers as evidence to make sense of it all? If so, how can we collaborate to support the common good for people in our communities? What new lessons are you learning? What are you seeing? If not, are you interested in a conversation about the role of evidence in life, the social sector, and making greater progress for everyone?

DATA JUSTICE AND THE RISKS OF DATA SHARING

MARIKA PFEFFERKORN

Many data scientists, activists, and others have begun to raise the alarm that big data, algorithms, and a lack of transparency in AI development will compound and exacerbate racial inequality. Often these alarms are accompanied by calls for community involvement. But it is crucial that community engagement is not just an add-on or window dressing for programs long in the making. That is exactly what happened in Minnesota's Saint Paul Public Schools. An unexpected twist in what was supposed to be a community-engaged process to improve services for "at risk" students resulted in a proposal for data sharing with the potential for discrimination and worse. Twin Cities Innovation Alliance (TCIA) brings together people, institutions, organizations, and communities to generate an educate-engage-equip model to activate community members more broadly on big data, predictive analytics and algorithms, and the engineered complexity of data-centric technology. Two of our biggest priorities are building partnerships to promote data literacy and agency and facilitating community action against harmful practices like data entrapment. Here is how it worked in Saint Paul, and what happens next.

A DATA SHARING STORY

At first, it seemed like a great idea. In the winter of February 2015, the Ramsey County prosecutor's office pulled together Saint Paul Schools and the city of Saint Paul to discuss how agencies could better coordinate resources. They then contracted with community hosts to hold engagement sessions on how to more efficiently and effectively deliver services. The response from Ramsey County residents participating in the year-long community engagement sessions was clear; they wanted to implement restorative practices rather than punitive models; they wanted reduced dependence on school resource officers and police intervention, and they wanted systems to proactively engage in solutions with the community. But when the official community engagement process report was released, many who participated were shocked. The report focused on using technology and data sharing between schools, counties, and municipal entities. Many community members had never heard a word about data sharing or other technological interventions raised in the community engagement sessions. Instead of getting a report that summarized community ideas and contributions to re-thinking school discipline, they got a plan for data sharing that seemed completely disconnected from the topics they all had discussed during the engagement sessions.

Soon after, Ramsey County, the city of Saint Paul, Saint Paul Police, and Saint Paul Public Schools announced their plans for a joint powers agreement (JPA) to begin a data sharing process. Their stated goal was to improve communication between schools, juvenile justice, prosecutors, public health, and child protection agencies through data sharing. The agreement included use of artificial intelligence and predictive analytics to identify students "at risk."

Community members were skeptical that the JPA would have benevolent impacts. While they acknowledged the need for better coordination between social welfare agencies and schools, having predictive analytics and law enforcement agencies in the mix was disturbing given the long history of racism in the criminal justice system. Our concerns increased when we got our hands on a copy of the JPA. It was full of technical jargon, lacked clarity about who was responsible for student data, and left many questions unanswered about how data would be used ethically to drive predictions about "risk." We also were alarmed at how quickly the county was moving toward this technical fix without consulting the community about the proposed data sharing practices. It felt like a bait and switch.

A COMMUNITY SUMMIT

TCIA gathered with other concerned community partners to organize. In contrast to the city and county's opaque decision making and lack of community inclusion around data practices, we drew on a host of authentic relationships and community-centered practices to engage folks in a deep dive into what the JPA was and how it could impact students and their families. The full story is outlined in the report, *Defeating the JPA: A Story of Community Empowerment through Education & Coalition Building*,[1] but here I want to take a moment to describe the culminating event of our process, the Cradle to Prison Algorithm Community Summit, which took place on November 10, 2018.

This summit met community members where they were in their understanding of the JPA. It featured experiential learning and included interactive, fun workshops like algorithmic improv and making a human algorithm poem. We hosted a tech talk featuring Yeshimabeit Milner, co-founder of Data for Black Lives; a "Dare to Data Clinic," and an activity called "One Mic" where parents could share their learning in bite-sized videos. All our workshops were centered around a restorative approach to ensure we did not perpetuate further harm, understanding that the legacy of systemic harm is real and long-standing surrounding the use and misuse of data in BIPOC communities. Examples of this are abundant in the Black community, from the infamous Tuskegee syphilis study to a more recent example in Pasco County, Florida, where the school district shared information about students with the sheriff's office without the knowledge of the students or their parents. The sheriff's office then used a computer algorithm to predict criminal behavior, ultimately using this "predictive policing" to label children as criminals "for crimes they have not committed and may never commit" as reported by the Institute for Justice.[2]

At the end of the summit, we debriefed with participants in restorative circles led by healing practitioners to unpack all they had learned, felt, seen, and heard over the course of the event. We gave ample time for this activity to ensure folks had an opportunity to process any tensions or discomfort brought up by discussions of historical and contemporary racial injustices. The outcome of the summit was a call to action: we would shift our focus from pausing the JPA process to dissolving the JPA altogether. And we did.

In the fall of 2018, in collaboration with In Equality and the Stop the Cradle to Prison Algorithm Coalition, we published *Improving Outcomes for*

Kids and Families: Beyond Predictive Analytics and Data Sharing, a policy brief we intended to use as a tool to better educate elected officials on the gaps and missteps embedded in the JPA. Below are the core messages from the brief.

- **Data sharing initiatives risk racially profiling children as "future criminals."** Predictive analytical tools that draw from data influenced by systemic racial biases will continue to re-inscribe inequalities and will not be accurate reflections of children's individual strengths or challenges. For example, Minnesota is among the states with the largest racial disparities in suspensions and on-time graduation. Suspensions are correlated with race and law enforcement contact; therefore, BIPOC students would be seen as "higher risk" of becoming criminals.
- **Predicting behavior: "Risk" becomes "threat" when applied to children of color.** When preexisting racial biases that over-associate BIPOC folks with crime shape the data, then the "risk" score becomes a proxy for "threat to safety."
- **Assigning risk scores, especially when there is lack of clarity about data chain of custody, will stigmatize children and families.** When children are flagged by the system for services, those scores are likely to leak throughout school communities, further exacerbating implicit and explicit racial biases.
- **Data sharing agreements may divert resources toward study and surveillance and away from services.** The JPA outlined an expensive, resource-intensive, and myopic study of individual children and family "weakness," ignoring ways to address systemic injustice, bias, and harm.
- **By turning to big data to solve problems, local governments in this case obfuscated their own culpability in generating disparities and their responsibility to correct them.** We must stop jumping to making decisions via computer analysis rather than creating authentic, trusting human relationships.
- **Integrated data may be vulnerable to political agendas of those who want to criminalize segments of the community.**

Our coalition, activities, and report drew heavy media attention and public outcry against the JPA. With that pressure and a report of a data breach in the fall of 2018, the JPA was dissolved.

After the defeat of the JPA, our coalition celebrated with community, but we knew we had only scratched the surface. "What you have taken on here in St. Paul is 10 to 15 years ahead of the majority of places across the country and where you have succeeded provides a roadmap for others and should be replicated," Yeshimabeit Milner, founder of Data for Black Lives, said. We catalogued our lessons. Many in the coalition returned to the primary focus of their advocacy, while the Twin Cities Innovation Alliance decided to go deeper into the work, launching the Data for Public Good campaign. The campaign's goal is to educate and engage youth, parents, educators, administrators, superintendents, county officials, and elected officials on the lessons we have learned and to plot a new path forward—one centered on the public good as defined by the public.

WHY WE NEED DATA FOR PUBLIC GOOD

Data for Public Good (D4PG) is not an event but a milestone of a larger movement, a movement that defines shared leadership, vision, and responsibility for the good outcomes we want our data to drive. It is essential that data scientists and local, regional, national, and international governing bodies include the people who will be most impacted by big data and AI. Until these entities learn better practices of authentic community engagement, community organizers must remain vigilant. It is clear from our experience that governments are having difficulty keeping pace with technological change. Big data, new technologies, and new analytical approaches, if applied responsibly and in co-design with those most impacted, have tremendous potential to be used for the public good. But we need local, state, and federal agencies to work with communities to craft policies that establish a basis and expectation of trust that data and privacy is used for the common good, informed and determined by the people. D4PG uses a mix of research, networking, and public events to generate opportunities for community to be involved in data justice learning, activism, and policy shaping.

The Data for Public Good Campaign led a Community Participatory Co-Research National project in 2019 and 2020 with support from a multiracial group of interns under the guidance of Dr. Catherine Squires and in cooperation with the Dignity in Schools Campaign and the Communities for Just Schools Fund, whose members and grantees represent more than 125 communities with whom we engaged across the United States, resulting in the release of several reports and a toolkit at our national online conference in November 2020.

The toolkit provides information and resources regarding data-centric technology; student data sharing and privacy; infographics on districts; comparisons on district transparency about which agencies have access to their data; types of data collected; explanation of data shared; transparency about third parties; and clarity of the consent process. Also included are a case study with a comparison of the dos and don'ts of data sharing agreements for school districts; a frequently asked questions guide on FERPA and education technologies for families; PowerPoint presentations for sessions on student privacy; a data primer to break down jargon; a reader's guide for the Data for Public Good Book Club; and a compilation of videos and articles on data-centric technology, with current examples of immediate and secondary impacts, to share with communities.

In addition, we have created the No Data About Us Without Us Fellowship and community institutes. The fellowship is a six-to-nine-month cohort-based co-learning experience for parents, youth, educators, and community members. It is designed to build data literacy and data advocacy skills to empower the fellows to disrupt the ways big data, predictive analytics, and engineered consent are currently weaponized against marginalized and BIPOC communities, especially in education. Sessions are designed to be interactive and experiential, and the fellowship is grounded in relationships. We meet fellows where they are at and move at the speed of their understanding and trust. At our No Data About Us community institutes, participants learn about and deepen their understanding of big data, predictive analytics, algorithms, engineered consent and other terms; understand the historic arc of BIPOC communities and the misuse of data; review existing policies and laws meant to protect us and the problems and gaps that have been created; and learn to do research on the use of data in local communities and school districts and the existence or absence of transparency. They are supported in creating site-based campaigns and projects. In 2020, TCIA partnered with the Education Partnership Coalition to pilot the fellowship across six Minnesota communities, including Red Wing, Northfield, Farmington, Saint Cloud, Saint Paul, and Minneapolis. Sessions were simultaneously translated into Spanish, and materials were provided in Spanish as well. No Data About Us Without Us community institutes are delivered in partnership with government systems, nonprofits, and funders to ensure program recipients and grantees have agency over the use of their personal data.

WHAT'S NEXT

TCIA is now co-developing and piloting a national and state policy tracking, analysis, and collaboration tool with Civic Eagle. As we roll out this tool, we will continue to collect and use the Data for Public Good Campaign survey data to design interactive heat maps that will provide a bird's eye view on this emerging policy trend. TCIA will work with local and national partners to introduce and track a constellation of policies that subsequently protect privacy, safeguard data, and ensure community trust.

All this is what authentic community engagement and data justice looks like. Our community-centered work is the opposite of what many governmental bodies, nonprofit agencies, private companies, and technical assistance providers put forth as "community engagement." As the story of the JPA demonstrates, if community partners are not involved when technological solutions are brought into the mix, just data practices will not result. Data fixes generated by systems built on injustice will most likely replicate those injustices. Communities disproportionately injured by bad data practices need to be at the center of discussions and designing any use of technology that purports to address those injuries. We insist on authentic engagements and conversations between communities and data scientists, tech vendors, foundations, and government agencies that want to apply technology to solve inequality. When we say, "No data about us without us," we mean it. Our well-being depends on it.

NOTES

1. Twin Cities Innovation Alliance, *Defeating the JPA: A Story of Community Empowerment through Education & Coalition Building*, 2020, www.tciamn.org/cpa-journey.

2. See Institute for Justice website, https://ij.org/case/pasco-predictive-policing/.

PHILANTHROPY'S RIGHTFUL ROLE IN EVALUATION

FOSTERING LEARNING AND EMPOWERMENT

JOHN BROTHERS

Early in my tenure as a philanthropy professional, I was fortunate to meet with a community leader who was seeking support for a new initiative in her neighborhood. It was my first meeting with a community leader—until then, I'd had only a few interactions with members of the community. So, I prepared for the meeting by researching the leader's background. I learned about her long track record of accomplishments and her impressive personal history. I also learned that she valued her neighborhood and her family's relationship with that neighborhood. Needless to say, I was excited about meeting and learning from her.

As our meeting began, I welcomed her and thanked her for her service to the community. Immediately, she began to present materials about the initiative, including several charts and graphs.

"I know you care about the numbers," she said, sharing what she believed I wanted to see. After a while, I asked, "Why are these metrics important to you? What data will help you tell your story?"

She looked confused and hastily responded, "We'll measure whatever you tell us to measure!" I was taken aback by her response. After all, she was a highly regarded community leader and I was just starting as a philanthropy professional. Yet, she had immediately ceded an important part of her work to someone whom she had just met and who did not know anyone in her community.

I think this happened simply because she assumed that the nature of my job automatically came with the power and authority that she didn't think she had. As we ended our meeting and agreed to meet again, I had a growing knot in my stomach. I wondered: *Is this how philanthropy works?*

I think of that meeting to this day because it illustrates so much of what is wrong with the social good sector. Based on that meeting, I have outlined four ways in which communities—and the social good sector—can achieve the impact we seek.

FINDING EVIDENCE OF COMMUNITY SELF-DETERMINATION

My meeting with the community leader illustrates a dynamic that occurs often in the relationship between philanthropy or government funders and their community partners. The community leader believed the value and impact of her efforts was best showcased through charts and graphs.

But, behind the numbers were even more impactful stories of champions in her neighborhood, which she did not share. It is these stories that have the ability to change our world, either by galvanizing supporters, changing policy, or advancing their narratives to a wider audience. Finally, and most troubling, she was quick to give funders the power to chart her community's direction. I am sure this was not an isolated incident for her, but one she had experienced previously and, ultimately, acquiesced to over a long period of time.

Finally, the data from a community that is included in an evaluation inherently—and rightfully—belongs to that community. It represents the goals and the stories of the people who live in that neighborhood. If a community is defining its future—improving their youth's reading ability, reducing crime, or increasing the number of new trees that line their streets—the goals and objectives should, ultimately, be decided by the members of that community. This includes the metrics and impact they hope to achieve.

When funders or academics look for community data, they must understand that they are guests in that community and cannot own or determine that community's outcomes. For example, if someone decides they must lose ten pounds, ten becomes the metric for how much weight they should lose. By owning that data, the person will take ownership for losing weight. They may seek guidance from experts on how to lose the weight, but they own their data and they own how it is used. This is called self-determination.

Similarly, a community's self-determination is fundamental to the relationship between the community and its partners. The community's data, metrics, and information are part of that self-determination. This must be at the forefront of how community impact is evaluated.

ADVANCING COMMUNITY LEARNING

A majority of nonprofit organizations are small and locally based. These organizations deliver important services to our communities and foster social safety. But, with most having fewer than ten staff members, it is safe to assume these organizations do not have Research and Development departments to help with evaluations. In a study on nonprofit evaluation capacities, Tara Kolar Bryan, Robbie Waters Robichau, and Gabrielle L'Esperance (Wiley 2020) outline the capacity elements that nonprofits need for effective evaluation—the organization's ability *to do* evaluation and the organization's ability *to use* the evaluation.[1]

Acknowledging again that most of the sector lacks the human and technical resources to conduct robust evaluation, we also acknowledge that, since the 1990s, through works such as Peter Senge's *The Fifth Discipline*[2] and many other publications, we have learned that thriving and successful organizations are ones that learn for their *own* development and success. Unfortunately, small, community-based organizations often do not learn for themselves but, instead, learn for others, like funders.

Additionally, while funders often aggregate large amounts of community data, provided through reports from their grantees, funders often are not strong learners. Thomas D. Cook, professor of sociology, psychology, education, and social policy at Northwestern University and a world-renowned expert in education evaluation, observed in 2006, "Evaluation is often something that funders want to be seen doing, but not what they value being done. They're feeling the winds of accountability, and they're passing it on to their programs."[3]

Fifteen years later, not much has changed since Dr. Cook's observation. As a former researcher, I remember meeting with several leading philanthropies in a large American city and asking them what the major developments were in the field of poverty alleviation, an area they specifically funded. Although these funders collected multiple reports from hundreds of their grantees each year, none of them could answer a simple question about an area for which they had a roomful of grantee reports. Sadly, we are not much further along today than we were then.

To achieve impact in our communities, under-resourced community-based organizations and the funders that support them will need to create a joint learning agenda that finds more avenues to collect and share data to advance their practices, propel their strategies, prove their missions, and advance their communities. They will have to partner to reestablish a commitment to learning together.

ORGANIZATIONAL HEALTH OVER PROGRAM OUTCOMES

Over a quarter of a century ago, the strategic philanthropy movement was created to transform philanthropy, making it more business-like and data-driven, and often created grantmaking processes centered on goal setting, strategy development, and measurement. The movement grew in popularity among many philanthropic leaders, but in hindsight, many have learned that strategic philanthropy also may have been damaging to communities. Darren Walker, president of the Ford Foundation, said, "Strategic philanthropy too often minimizes or ignores complexity because it is difficult to understand and predict."[4]

One complexity caused by strategic philanthropy was an over-focus on program outcomes and an under-focus on the organizational health of nonprofit organizations. Although I work in the corporate community, I am not one to believe in the notion that nonprofits should be more like the corporate sector. However, this often is believed in the nonprofit community, often by its board members with business backgrounds. Each sector has unique and valuable attributes, and there is much the nonprofit sector can teach the corporate community, especially around areas of equity and inclusion.

On the other hand, one area where the business community has shown strength, especially from my viewpoint working at a global financial services firm, is that a healthier, stronger organization is a better investment than a structurally weak organization with a potentially strong product or program. In the nonprofit sector—partly because of an over-commitment to strategic philanthropy—we often have taken the opposite approach, valuing programs over organizational health.

A popular example, seen in nonprofits but not in for-profit companies, is the push by funders for low overhead rates and commitment of funding resources to be solely dedicated to programs but not to operations, the idea being that more program funding will mean more community impact. Additionally, this belief has been propelled largely without any data or evaluation to support it.

In 2016, as part of our philanthropic efforts at T. Rowe Price, we started to evaluate the strength of the nonprofit sector in Baltimore, the home of our global headquarters. Our evaluation found that Baltimore's nonprofit sector has a number of glaring challenges, especially in comparison to other Rust Belt cities, and that one of the only ways to see impact in our communities was through building stronger community-based organizations. Since then, we have been growing an ongoing repository of organizational health data that shows the strength and challenges of our organizations and our local sector.

To date, thousands of organizational health data points have revealed information helpful for our nonprofit partners, for us as funders, and for the larger social good sector. For example, our data show an interesting irony: most nonprofit organizations have strong confidence in their ability to deliver quality services, but at the same time, they believe they severely struggle in their ability to evaluate their programs. Considering the discussion that our sector (and this book) is having on impact and evaluation, our current data illustrate a significant challenge: nonprofits *believe* they deliver a good service although they do not have the capacity to prove it.

Finally, as we reimagine a future sector that focuses on the importance of organizational health, our data will, hopefully, illuminate a number of areas, including healthy overhead rates, the greatest differences between board and staff members, or the specific organizational challenges that nonprofit industries suffer from. Imagine being able to understand with pinpoint accuracy our sector's organizational challenges, and imagine what a responsive funding community could do with that data.

UNCOMPLICATING EVALUATION

One of my all-time favorite papers on the nonprofit sector is Tony Proscio's *In Other Words: A Plea for Plain Speaking in Foundations*.[5] The piece outlines the use of jargon and other overcomplicated language among funders. In one passage, Proscio discusses how damaging jargon can be:

Among foundations, the result of so much accumulated jargon can be especially hard to penetrate—a lethal combination of the dense and the tedious, a congregation of the weirdest and most arcane words, crammed unhappily together like awkward guests at an international mixer. Most of the time, this happens naturally and unintentionally. It usually is not a conscious attempt to condescend, to pose, or to exclude. Yet that is understandably how it's taken, and all too often, that is the actual effect. That effect is even more destructive in philanthropy than it is elsewhere.[6]

Nowhere has jargon been more challenging in the nonprofit sector than in evaluation. With new evaluation terms added to this more specialized field every day, we have come to a place where only consultants, researchers, and foundation staff have the supposed expertise to understand the areas of impact. At the same time, this creates more confusion for social good practitioners in the field of evaluation.

With the continued, consistent increase in the number of nonprofit organizations over the twenty-year emergence of strategic philanthropy, there has been significant growth in the number of experts dedicated to nonprofit evaluation. All this has occurred against an economic backdrop of a small group of large nonprofits with the R&D capability to obtain the financial resources of a philanthropic sector impressed by advanced metrics that only a small few can produce.

The challenge is that if evaluation can be achieved and owned only by a small section of the nonprofit sector, where does that leave the others? Evaluation has become less of a sector-wide utility and more of a specialty item, afforded and operationalized by a select few. Until evaluation can be simplified and un-jargoned, and becomes a universal utility for all nonprofits, the outcomes needed in our communities will be out of our grasp.

CONCLUSION: EVALUATION MUST BE ROOTED IN TRUST-BASED PRINCIPLES

In 2010, I wrote an article for the *Stanford Social Innovation Review* titled "Carrot and Stick Philanthropy."[7] I discussed how funders use various levers, or carrots and sticks, to motivate and sometimes control their grantees. One of the carrots funders often use is the tool of evaluation. As stated above, when evaluation is used and understood only by a few, it can be used by a small group to shape and shift others. I experienced this firsthand in my early days as a philanthropy professional during that meeting with the community leader.

Since that time, T. Rowe Price has become recognized by the Trust-Based Philanthropy Project as a national leader of trust-based philanthropy principles.[8] Part of this work is rooted in the belief that if philanthropy is to be an active agent for change in our communities, then philanthropists must regain trust with the communities we aim to serve. This begins with the *how*, or the operations, of our philanthropy. Since evaluation is part of our bedside manner in the community, we must find evaluative approaches that help communities use their own data for their own self-determination while at the same time building the capacity of our under-resourced community-based organizations to measure and grow their impact.

When I think back to the meeting with that impressive community leader, I reminisce on the subsequent conversations that have resulted in a strong relationship today, but I also remember that it took months to move beyond the idea that our foundation valued only the strength in her metrics rather than the strength in her community's experiences. We recently met for coffee, and what I found most rewarding about our discussion was that her discussion of the results in her organization centered around the stories of people gathering together around an issue they cared about and how that combination of energy and passion were helping her fellow neighbors reach new heights. Yes, sometimes there were numbers used to describe this work, but those numbers were supported by the names of her community members and their amazing stories of beauty, grit, and grace.

What was most memorable in this meeting was the leader who was telling me about this work and the difference in our two meetings—one beginning on paper with lines and graphs and the other ending with energy, passion, commitment, and hope. I thought again about philanthropy and its place in these two meetings, and it was very clear where my profession has struggled and where philanthropic support needs to evolve. Most of our funder colleagues continue to struggle with having authentic relationships with the communities they aim to serve. A holistic approach to evaluation—seeking stories from the community punctuated by useful data, could help increase authenticity and build strong ties. Let's hope that we pursue this.

NOTES

1. Tara Kolar Bryan, Robbie Waters Robichau, and Gabrielle E. L'Esperance, "Conducting and Utilizing Evaluation for Multiple Accountabilities: A Study of Nonprofit Evaluation Capacities," *Nonprofit Management and Leadership* 31 (2021): 547–569, https://doi.org/10.1002/nml.21437.

2. Peter M. Senge, *The Fifth Discipline: The Art & Practice of the Learning Organization* (New York: Doubleday, 2006).

3. Alana Conner Snibbe, "Drowning in Data," Stanford Social Innovation Review, Fall 2006, https://ssir.org/articles/entry/drowning_in_data#.

4. John Kania, Mark Kramer, and Patty Russell, "Strategic Philanthropy for a Complex World," Stanford Social Innovation Review, Summer 2014, https://ssir.org/up_for_debate/article/strategic_philanthropy.

5. Tony Proscio, *In Other Words: A Plea for Plain Speaking in Foundations* (New York: Edna McConnell Clark Foundation, 2000), www.comnetwork.org/wp-content/uploads/2010/08/inotherwords.pdf.

6. Proscio, *In Other Words*, pp. 8–9.

7. John Brothers, "Carrot and Stick Philanthropy," *Stanford Social Innovation Review*, March 19, 2010, https://ssir.org/articles/entry/carrot_and_sticks_philanthropy.

8. Trust-Based Philanthropy Project, "How Can Philanthropy Redistribute Power?," accessed May 16, 2021, www.trustbasedphilanthropy.org/.

EARNING COMMUNITY TRUST IN DATA-DRIVEN INTERVENTIONS AT THE DUKE ENDOWMENT

RHETT MABRY

A MISSED STEP

In 2005, Hurricane Katrina hit the New Orleans area, causing more than 1,800 deaths and approximately $160 billion in damages.[1] At its high-water mark, the category 5 storm left nearly 80 percent of the city and many of the surrounding parishes underwater.[2] More than 1 million people were immediately displaced; thirty days later, some 600,000 remained unable to return to their homes.[3]

In the aftermath, the Red Cross used a variety of techniques to help those in crisis locate loved ones living across the United States. The Red Cross's search system led to many successful connections, and leaders who work in the child welfare system began wondering if those same techniques might help locate relatives of children in foster care. Adoption placement agencies also saw potential for locating family members who might be unaware of a child's circumstances.

In North Carolina, leaders from the Department of Social Services, The Duke Endowment, and a prominent private state adoption placement agency met in Raleigh in 2008 to discuss using family search strategies to connect children with kin, whether to establish an ongoing relationship or possibly entrust the child with a caring relative as a more permanent placement solution.

In short order, a plan was hatched. Nine North Carolina counties were selected for a randomized control trial to test if identifying more relatives of children in foster care might lead to increased permanent placements with kin. Early results were promising. Using internet and other search techniques, cases randomized into the treatment group identified, on average, almost ten times more relatives than control cases in which traditional methods were used. Even after excluding relatives who were not interested in or could not commit to a relationship, approximately four times as many family members in the treatment group were willing to commit to the child than in the control.

In the end, however, there was no difference between treatment and control cases in achieving permanent kinship placements.[4] What happened? Truthfully, we cannot be sure, but we think we have identified what went wrong and what we might have done differently.

In our zeal to test this new, promising approach for locating family members, we failed to adequately engage caseworkers to capture their input before subjecting the approach to a rigorous trial. We failed to consider how front-line social workers, who typically are overburdened and carry caseloads above recommended standards, might respond to having additional family members (in the treatment group) to vet for possible placement options. More specifically, we did not account for the additional time the expanded options would require in determining the best placement for the child. Understandably, with nothing taken off their already full plates, demanding schedules limited the number of families with which social workers could work. Not surprisingly, then, the inability to capitalize on the increased family contacts resulted in no difference between the treatment and control groups in achieving a permanent placement.

Had we been more patient, we would have tested the approach on a smaller scale before rolling it out across nine counties. Doing so likely would have identified implementation challenges and surfaced ideas for freeing up necessary time. For instance, one possible solution would have been for staff who conducted the family searches to proceed with finalizing placements—which, in effect, could have served to extend caseworker capacity. Instead of vetting extra placement options on their own, caseworkers could have played a supervisory role.

ADDRESSING CHALLENGE

The unfortunate reality is that our experience is not unique. The United States spends billions of dollars each year to provide social supports for housing, food access, education, medical care, transportation, job training, and more. In many cases, this money is spent without knowing if the intended impact was achieved.[5] For the past twenty years, The Duke Endowment has sought to increase the body of evidence for emerging promising practices, such as using family search techniques in foster care. We also have placed a priority on replicating what works. The list of evidence-based programs we support is long, including Nurse-Family Partnership (NFP), The Incredible Years, Triple P, Strengthening Families Program, Multisystemic Therapy, and Trauma-Focused Cognitive Behavioral Therapy. For us to call these models "evidence-based," each is required to have at least two randomized control trials documenting impact. While admittedly simplistic, this definition helped to guard against the tendency in social services to overuse and dilute the meaning of a catchy term like *evidence based*.

With the assurance that a model has demonstrated that it *can* have impact, our focus turned to effective implementation to ensure replication with fidelity. For instance, NFP has accumulated impressive short-term and longitudinal data supporting its impact across three randomized control trials dating to the 1970s. Spurred by this encouraging data and resulting cost savings, NFP has expanded dramatically during the past two decades, reaching more than 60,000 families across the country in 2019 alone.[6] In most communities, public health departments serve as hosting agencies even though NFP's more targeted and intensive home visiting approach does not always sync with the traditional public health orientation of providing lighter touch interactions and broad-reaching, community-wide strategies. Consequently, we suspect that, as NFP evolves, it will increasingly need to consider how it best integrates within existing community systems as opposed to operating as a standalone or adjunct program within health departments.

NFP leaders appear to agree with this direction and are taking steps to identify families who benefit most from the program and determine when home visiting services are (and are not) the most efficient use of resources. This may lead to accepting more targeted referrals and discharging families sooner based on progress assessments. NFP recently merged with Child

First, a home-based intervention that helps vulnerable children and families heal from the damaging effects of trauma, stress, and adversity, indicating its interest in integrating with complementary programs. Embedding services within a broader community tapestry may require adjustments to the model and will push funders enamored with strict model fidelity to think more adaptively.

In our work to build evidence for promising practices, we have not always engaged in a sufficient formative evaluation to ensure proper systems integration. Systems are complicated and difficult to change. Introducing a new, well-researched or emerging practice for broader adoption requires far more than the naïve "plug and play" mentality we and other funders sometimes assume. Our work with the National Implementation Research Network in Chapel Hill, North Carolina, continues to inform our understanding of the importance of planning for and addressing implementation challenges before, during, and after introducing an innovation.

Relatedly, attempting to bring an evidence-based program to a much greater scale can have significant challenges. For starters, there is the question of whether there are enough high-quality or adequately trained service providers (social workers, nurses, clinicians, etc.) to deliver the intervention as designed. If not, service quality might be compromised. Scaling also requires logistical and technology enhancements. Consistent program delivery must include continuous staff training and recruitment, seamless telecommunications, and sufficient working capital to manage through reimbursement delays. Few challenges are insurmountable, but even smaller obstacles must be managed.

UNINTENDED CONSEQUENCES

At least two unintended consequences have emerged as foundations support replication of evidence-based practices. One is that the insistence on evidence and model fidelity may inadvertently stifle further innovation. Evidence-based solutions are coveted by funders and nonprofits alike, yet interventions that have documented impact from one or more randomized control trials are few. From 2002 and 2013, the U.S. Department of Education conducted some ninety randomized control trials, and nearly 90 percent produced weak or no effects.[7] These results are consistent with those reported in other fields such as psychology[8] and medicine.[9] Once encouraging data are confirmed, the instinct is to lock the program, cease further development, and proceed with high fidelity replication. This works fine for a while. Eventually, however, the world changes: Medicaid expands, telemedicine takes off, women stop smoking during pregnancy, smartphones start tracking vital health and exercise statistics. Interventions must constantly adapt to become more efficient and effective.

To be clear, the blame for this stagnation lies more with the funders, who insist on replicating "proven" models, than with the purveyors of the interventions, who would surely value research funding to continuously improve their approaches. The solution may be for funders to continue to demand and support strong implementation of well-documented and evaluated programs while also funding pockets of innovation in select communities. An example of testing new aspects of evidence-based programs in the context of a broader initiative is underway in Guilford County, North Carolina, where NFP, Family Connects, and Healthy Steps have agreed to meld their programs into a cohesive suite of services for young children and families. This collaboration, undergirded by an integrated data system, should allow NFP to serve highest-risk mothers and for other community providers to receive timely referrals for less intensive interventions. Another example is to capitalize on the popularity of telemedicine during the recent pandemic and use virtual connections to increase efficiencies and the number of contacts with families.

The second challenge with focusing on evidence-based programs is that it likely has funneled more resources to well-funded, established nonprofits at the expense of smaller grassroots organizations.[10] Evidence-based models require not only evaluation expertise but also development staff to raise money for costly enhancements, along with policy advocates to tap into sustainable public funding. Without that capacity, many smaller organizations, which might have deep expertise in the issues faced by the communities they serve, are frequently passed over by foundations.

Fortunately, as philanthropy seeks more equitable solutions, foundations are realizing the importance of grassroots organizations, many of which are operated by leaders of color and located within or near communities that have been marginalized. Investing in these organizations for a sustained period would help them build and deliver interventions with documented results. The combination of earned community trust and data-driven interventions may prove potent for improving outcomes that have been difficult to change.

WHERE DO WE GO FROM HERE?

Given the challenges with building, testing, implementing, modifying, and sustaining evidence-based interventions, what might be a path forward? Abandoning support for those tried-and-true programs does not seem to be a good option. Perhaps there is a compromise.

For the past several years, in North Carolina, select churches in more than fifteen rural communities have agreed to offer The Duke Endowment's summer learning program for elementary-age students and help us test its effectiveness. Challenges along the way have required us to adopt a developmental approach to accumulating evidence. The initial goal is to refine the intervention by better understanding student profiles, pinpointing specific learning obstacles, creating an effective curriculum, developing teacher skill sets, and establishing student selection protocols while also collecting pre- and post-test data to discern directional impact. We hope options for summative evaluation will evolve over time as we amass a preponderance of evidence in support of the model's effectiveness as opposed to an all-or-none designation of the model as "evidence based." Boston-based Project Evident is helping us design a systematic approach for accumulating compelling evidence.

This iterative approach has broader applications beyond rural communities. The spate of place-based interventions emerging across the country will likely also need to adopt similar tactics. Placed-based investments face considerable hurdles, not the least of which is measurement. This is particularly true in the early childhood field, which is the focus of many place-based efforts and for which there is a dearth of administrative data covering outcomes prior to school entry.

This lack of administrative data rules out the most straightforward designs for impact evaluation. Instead, evaluators are exploring multi-pronged designs and strategies for collecting a variety of data that produce a "basket" of converging evidence. For instance, primary or original data collection—family-by-family surveys and interviews—may be used to capture information on children's social and emotional development. A convincing evaluation surely requires a credible counterfactual, so data collection will need to occur in both the treatment community and matched comparison sites. Kindergarten readiness assessments, currently a hodgepodge of tools administered with varying levels of rigor and credibility, will need standardization. Individual program-level data will be tracked to assess contributory effects and directional trends. These data on outcomes will need to be complemented by documenting implementation details—numbers served, satisfaction with services, referrals made, funding redirected, reimbursement revenue opened, and policy changes enacted. The latter output measures should help shed light on the progress of system changes underway.

The difficulty of data collection for place-based investments, especially those targeting early childhood outcomes, applies to other large-scale interventions as well. Implementation costs are high, formative and summative evaluations are expensive, and integrated data systems are complicated to build and frequently fail to achieve sufficient buy-in. It is worth considering a collaborative investment between philanthropy and government to establish systems and administrative processes for routinely capturing data where gaps exist. Such a system would decrease both the expense and risk of evaluating large-scale, community-wide interventions. Accessible data also would be useful in calculating associated cost savings, which are important drivers for changing policies and practices.

An important caution about adopting a "preponderance of evidence" approach to determining likely effectiveness of a program or initiative is that it may give comfort to those who are fundamentally uninterested in measuring outcomes. Abandoning clear-cut definitions for what constitutes sufficient evidence—such as having at least two RCTs with statistically significant findings—may be seen as an invitation to ignore data altogether. Doing so would be a mistake. Assuring impact requires developing a discipline (by funders and practitioners) of capturing qualitative and quantitative data in carefully planned feedback loops and using that data to improve our approaches. Measuring outcomes is an important part of such a discipline. When the most rigorous methods ("the perfect") are not available, the best data available ("the good") will have to suffice. Either way is preferable to ignoring measurement and data altogether.

Philanthropy, too, often mirrors government in this regard. The complexities that accompany rigorous evaluation should not thwart efforts to use data. Rather, they should spur us to continue to seek new solutions and approaches. No outcome data, regardless of evaluation design or rigor, is absolute, and no findings, no matter how weak or strong, are fixed indefinitely.

I am confident that advances in technology and evaluation approaches will continue to drive improvements in the social sector. The first human coronavirus—the cause of the common cold—was identified in 1965. Since then, seven coronaviruses have been known to sicken us, and scientists had little success developing vaccines until the recent breakthroughs in response to SARS and COVID-19. That success built on decades of work studying the mechanisms of viral transmission, messenger RNA transcription, and therapeutics.[11] Just as a combination of meticulous laboratory observation, epidemiological studies, and clinical trials surely drove our vaccine success, research in the social sciences must follow a similar painstaking path, iteratively building on prior findings with an appropriate mix of methods. In addressing society's complex challenges, let us be encouraged and committed to that journey.

NOTES

1. See Britannica website, www.britannica.com/event/Hurricane-Katrina.
2. Ibid.
3. Allison Plyer, "Facts for Features: Katrina Impact," The Data Center, August 26, 2016, www.datacenterresearch.org/data-resources/katrina/facts-for-impact/#:~:text=The%20storm%20displaced%20more%20than,housed%20at%20least%20114%2C000%20households.
4. Child Trends conducted the evaluation.
5. See the Coalition for Evidence Based Policy website, http://coalition4evidence.org/.
6. Annual Report 2019—Nurse-Family Partnership.
7. See the Coalition for Evidence Based Policy website, http://coalition4evidence.org/wp-content/uploads/2013/06/IES-Commissioned-RCTs-positive-vs-weak-or-null-findings-7-2013.pdf.
8. The "replication crisis" in psychology was triggered by a report that just 36 percent of studies published in top psychology journals could be replicated. Andria Woodell, "Leaning into the Replication Crisis: Why you Should Consider Conducting Replication Research," APA, March 2020, www.apa.org/ed/precollege/psn/2020/03/replication-crisis, and Ed Yong, Psychology's Replication Crisis is Real," *The Atlantic*, November 2018, www.theatlantic.com/science/archive/2018/11/psychologys-replication-crisis-real/576223/.
9. Reviews have found that 50 to 80 percent of positive results in initial ("phase II") clinical studies are overturned in subsequent, more definitive RCTs ("phase III"). John P. A. Ioannidis, "Contradicted and Initially Stronger

Effects in Highly Cited Clinical Research," *Journal of the American Medical Association* 294, no. 2 (July 13, 2005), pp. 218–28.

10. Cheryl Dorsey, Jeff Bradach, and Peter Kim, "Racial Equity and Philanthropy: Disparities in Funding for Leaders of Color Leave Impact on the Table," The Bridgespan Group, June 4, 2020, p. 15.

11. See https://www.cdc.gov/coronavirus/2019-ncov/vaccines/different-vaccines/how-they-work.html?CDC_AA_refVal=https%3A%2F%2Fwww.cdc.gov%2Fcoronavirus%2F2019-ncov%2Fvaccines%2Fdifferent-vaccines%2Fmrna.html.

THE DUKE ENDOWMENT SUMMER LITERACY INITIATIVE

QUESTIONS MATTER AND METHODS SHOULD MATCH

HELEN I. CHEN

The Duke Endowment's Summer Literacy Initiative is an unusual multi-year collaboration between a supporting funder, rural churches that operate summer reading programs, local schools and districts, evaluators, and other stakeholders. The Summer Literacy Initiative is designed to help United Methodist congregations improve early childhood literacy in North Carolina's rural communities. The programs combine six weeks of literacy instruction with enrichment activities, family engagement, nutritious meals, and wrap-around services for rising first- through third-graders who read below proficiency for their grade level.

The Duke Endowment (TDE) has supported the initiative's development and implementation—starting with one pilot program in 2013 and expanding to fifteen communities across the state by 2020. Along the way, TDE has invested in formative evaluations to assess student outcomes and to build evidence that can inform its efforts to continuously improve the program model and expand the initiative's reach.

Today, seventeen church-based sites serve about 250 students annually, most of whom are from low-income families and about half of whom are Black, Latinx, or Native American. As a funder, TDE has made a long-term commitment to supporting this initiative by building a learning agenda, making evidence-based decisions around funding, and incorporating program evaluation as a key component in sustaining and scaling it.

TDE played a unique role as both the funder and the programmatic "home" of the Rural Church Summer Literacy Initiative (SLI). The Rural Church team worked with program staff, the evaluation team, and external partners in its efforts to build an evidence-based summer literacy model that leverages the strengths and resources of rural churches. Partners include local United Methodist churches, local school districts and schools, the North Carolina Department of Public Instruction (DPI), United Methodist Conferences in North Carolina, independent researchers and evaluators, and other community partners.

While TDE initially defined the learning agenda and evidence priorities, teachers and site directors were critical in shaping learning questions each year, in particular around assessments and best practices for reading instruction. School districts and DPI provided student data and shared assessments, and partnered in learning questions about which children were served by TDE camps and which were served by district-sponsored camps.

USING EVIDENCE TO INFORM A SCALING STRATEGY

The expansion of the SLI program has been accompanied by the development of a program model, guiding principles, logic model, evidence roadmap, plans for a multiyear research design and accompanying implementation protocols, and a myriad of other considerations.

2013–2015: Laying the Foundation for a Program Model

In 2013, TDE's key questions centered on the feasibility of the church programs to produce positive student outcomes and the supports and learnings needed to do so. TDE asked two sites to use formative reading assessments to track student progress over the summer. After three years, TDE and churches had anecdotal evidence of positive student outcomes as well as informal formative assessments given by teachers. To confirm these gains, in early 2016, TDE engaged an external evaluator, Dr. Helen Chen, in what was to become a multiyear process of building and measuring evidence.

2016: Bringing in Outside Evaluation

Dr. Chen noted that the initial two sites were each committed to student reading outcomes, but functioned independently. If TDE was to launch a "program," it needed a program model and guiding principles. Dr. Chen developed principles and protocols for implementation; surveys for stakeholders; and a plan for measurement and evaluation with the church directors to answer their questions.

2017–2018: Building the Framework for Rigorous Evaluation

In 2017, Dr. Chen conducted an implementation analysis to confirm that the guiding principles accurately represented TDE's vision for the reading camps, and to assess how well sites adhered to these principles. Her recommendations focused on standardizing data collection and instruction (with input from teachers); improving student recruitment; and identifying the landscape of reading camp options available in each district. In addition, feedback from the churches indicated challenges with the use of the pre- and post-assessment tool, which led to a change in the assessment in 2018.

In 2018, the key evaluation question was whether the summer reading programs produced intended outcomes in student reading, as measured by the Gates-MacGinitie Reading Test (GMRT), weekly teacher-administered formative assessments selected by teachers themselves, and expanded qualitative measures. The TDE team elicited input from site pastors, directors, and teachers, which impacted evaluation efforts in a way that was empowering to teachers, effective for measurement, and helped bring consistency and fidelity to implementation.

2019: Additional Evaluation Resources

By 2019, twelve churches were queued up to host summer reading camps, and TDE needed a comprehensive roadmap, given their goal of a large-scale impact study to demonstrate the effectiveness of the Summer Literacy Initiative. Project Evident was brought in to develop a strategic evidence plan that aimed to leverage practitioner and community voices and advanced actionable knowledge needed by TDE and its partners to make decisions about how to scale the SLI to best serve its communities.

The year 2019 also saw a new partnership with North Carolina's Department of Public Instruction (DPI), in which DPI endorsed TDE's reading camps, giving churches credibility as they recruited. The partnership also involved a shift in assessments used, adding to DPI's database so they could track student progress beyond the school year.

In 2019, based on Project Evident's evidence roadmap, evaluation focused on how the program might be scaled and adapted for a broader range of rural communities. There was great variability across the twelve sites in North Carolina, with differing local contexts, community demographics, and needs and agendas of partner districts. Teachers and site directors proved to be sophisticated and thoughtful partners in examining the data, asking how the assessments and classroom instruction were aligned. To better understand differences across sites, Dr. Chen added classroom observations to rate across three domains: emotional support, classroom organization, and instructional support.

2020: Toward a Large-Scale Impact Study

In early 2020, TDE added the American Institutes for Research (AIR) to the evaluation team, to support the work of building evidence and moving toward a large-scale summative study. Over the years, TDE's leadership team had expressed a preference for a randomized control trial to establish evidence of SLI's impact. Pushback from church partners and DPI, concerned about students not receiving an intervention they might need, helped reshape TDE's expectations. Project Evident and AIR proposed several research designs that address stakeholder concerns while offering rigorous, evidence-based studies that can show the effectiveness of the SLI. The evaluation team anticipates an impact study in the next couple of years, with all stakeholders working toward a research design that will, ultimately, serve the greatest number of students.

CHALLENGES AND RESPONSES

In order for the Rural Church Summer Literacy Initiative to successfully build actionable evidence, several aspects of the work required special attention.

Research Design that Balances Rigor with Stakeholder Needs

TDE places a high value on randomized control trials. However, pastors, site directors, and teachers were very concerned about the possibility of excluding any students who needed support. The partners continue to explore designs that will best serve key stakeholders (for example, a quasi-experimental design option in which SLI students would be compared to state DPI administrative data). Interestingly, 2020's COVID-19 restrictions offered opportunities to make variations to programs such that some sites offered a condensed four-week camp and still saw increases in student reading at the end of the summer. This planted the idea of offering back-to-back four-week reading camps, with half of the students assigned to the first session and half to the second. In the first session, AIR would then have a control group of students who had not yet received the intervention.

Fidelity versus Flexibility

As site numbers have increased, what program elements must be adhered to strictly, and where can we allow for differences in local contexts and extenuating circumstances? TDE has approached this with the recognition that practitioners know their communities best. For example, churches were allowed to admit the occasional "extra" student (who might fall outside eligibility criteria) who receives the full range of instruction, enrichment, and wrap-around services but does not participate in evaluation activities. Another site, with a student population that is 100 percent Native American, asked to close reading camp during an annual homecoming week celebrated by their entire community. TDE saw both these cases as acceptable variations that respected the norms of the community, and in both cases, data were collected, evidence built, and practitioners' judgments validated.

Bringing in the Lenses of Trauma-Informed Instruction and Cultural Humility

Church congregations and the populations they serve in the SLI often come from very different racial and socioeconomic backgrounds. The TDE team introduced trauma-based approaches to instruction and cultural humility into the training of site teams and church volunteers. Some sites began to lay the groundwork for forming parent advisory councils to incorporate parent voices in the summer reading program. Other sites responded by making sure their summer camp materials were translated into languages used by their families, and one site intentionally adjusted some of their teaching practices to meet the tactile learning styles they learned might meet the learning preferences of their Native American students.

RESULTS

The Rural Church Summer Literacy Initiative has resulted in stronger student outcomes as well as greater engagement among stakeholders.

Student Outcomes

Evaluation findings indicate statistically significant student reading growth, positive changes in student attitudes and behaviors related to reading, and positive changes to the child's home literacy environment.

- **2016:** Three months of reading comprehension growth as measured by the Iowa Test of Basic Skills (ITBS); increases in reading accuracy and speed as measured by running records; positive effects on student reading behaviors and attitudes.
- **2018:** Statistically significant gains in raw scores and National Percentile Rank across all sites and all grade levels; weekly gains in comprehension, fluency, and decoding; positive effects on student reading behaviors, attitudes, and intrinsic motivation, as well as increases in parental engagement with their children around literacy activities.
- **2019:** Statistically significant gains in Reading Success Probability and its component domains as measured by the Lexia RAPID assessment; gains on the DIBELS formative assessments; positive effects on student reading behaviors, attitudes toward reading, intrinsic motivation, and home literacy environment; increases in parental support of their children around reading activities.

Strengthening Partnerships

Each year, TDE shares aggregate findings in a large-group meeting and a site-specific "data card" with each site specifically. In addition, TDE puts together a public-facing document that each site can share with school and district partners as well as their congregations. Churches are keen to use these findings each year to improve engagement, their instructional practices, and their wrap-around services. This incremental approach to building evidence and making transparent the research goals has reaped big benefits in stakeholder buy-in as well as funder commitment to the SLI. Practitioners believe their contributions are not only welcome but valued, and new learning questions are generated each year in a collaborative way. TDE continues to invest in building actionable evidence, understanding that a long-term commitment is needed to yield results that will lead to a sustainable, scalable program.

REFLECTIONS

The Rural Church Summer Literacy Initiative is an innovative approach that leverages the infrastructure and social capital of United Methodist churches to support families with literacy. Building evidence of student outcomes and using evidence for continuous improvement has been at the core of the initiative's evolution. The Rural Church program area took an R&D approach—a continuous process for testing, learning, and improving—that is often rare in the education and social sector. In addition to investing in regular evaluations, TDE worked with Project Evident to develop robust evidence tools—a theory of change, a learning agenda, and a strategic evidence plan—to drive a more intentional and disciplined approach to building evidence, grounded in strategic priorities for the initiative.

Most importantly, TDE centered its grantmaking strategy and evidence building efforts on improving outcomes for children and families, and it empowered churches and practitioners to equitably participate in the evidence building and learning process. As TDE's Robb Webb said:

All questions ultimately lead to: How do we make this better for the students? How can we make this more impactful? That clarity and that focus is driving this work. The evolution of the Summer Literacy Initiative and how we've strengthened our evidence over the years has been remarkable. And the churches have wanted to come along for the ride because they care deeply about how they are impacting students, and they want to have an impact. It's been an incredible learning journey for us as a department and for the churches we work with.[1]

The Rural Church Summer Literacy Initiative is a prime example of the "Next Generation of Evidence"—one that centers on community needs and voices, embraces continuous improvement, empowers practitioners, and prioritizes collaborative learning and accountability among funders, researchers and practitioners.

NOTE

1. The Duke Endowment and Project Evident, "Mobilizing Rural Churches to Improve Early Childhood Literacy in North Carolina," 2021, https://www.dukeendowment.org/resources/mobilizing-rural-churches-to-improve-early-childhood-literacy-in-north-carolina.

CRIMINAL JUSTICE LAB

EMBEDDING UPSTREAM BEHAVIORAL HEALTH SOLUTIONS INTO THE CRIMINAL JUSTICE PROCESS

KATY BRODSKY FALCO AND CHRISTOPHER LOWENKAMP

Across the country, millions[1] of individuals come into contact with the criminal justice system not because of criminal behavior but, rather, because they struggle with mental illness and substance use disorders.[2] Many of these individuals repeatedly cycle through our jails, often because they are never provided with services that can address their underlying problems. After all, jails and prisons incarcerate and punish but rarely address underlying issues. Because we fail to safely address significant, underlying drivers of crime, we pay an enormous cost—both human and financial.

Until now, we have not given the police—or other first responders—the tools they need to successfully identify individuals who suffer from mental illness and substance use disorders. Without this information, there is no way to know objectively who could be safely diverted to treatment that, according to research, stands as one of the few proven pathways to reduce crime and re-arrest.[3] Today's criminal justice system puts enormous pressure on police officers to follow the traditional law enforcement path of identifying someone who has broken the law and arresting them. When an officer deviates from that model, they are relying on their subjective judgment and assuming personal and professional responsibility and risk in doing so.

For these reasons, we believe a simple, accurate screening tool that can be administered in the field can dramatically move the needle on diversion and future criminal justice involvement by providing officers with an objective basis on which to identify individuals with behavioral health issues.

Many police departments have begun thinking about new ways to address crime's underlying drivers instead of waiting for the next crime to occur, and many are rethinking how they respond to individuals with behavioral health issues. We have been fortunate to cultivate partnerships with Indianapolis, Indiana, and McLean County, Illinois, two jurisdictions that have shown a strong desire for change to their policing practices. These two partnerships have been instrumental in the success of our project. We hope this tool will be viewed within the broader national framework to develop alternative responses by police and provide training and tools for de-escalation.

CREATING THE HEALTHLINK DIVERSION TOOL

Early in her tenure as New Jersey Attorney General overseeing the Camden Police Department, Criminal Justice Lab's founder Anne Milgram realized the large intersection between public safety and public health. A majority of all Camden arrestees—67 percent—made a trip to the hospital emergency room at least once during the study's timeframe, with 54 percent of arrestees making five or more visits during the same timeframe. There was a clear relationship in the data between the high use of hospital emergency rooms and frequent arrests.[4] This research has driven many of the Criminal Justice Lab's research priorities and led us to wonder what type of tool we could build to integrate treatment for underlying upstream drivers of crime into the criminal justice process itself.

To achieve this goal, we first convened experts in mental illness and substance use disorders to see if we could design a short questionnaire with high predictive validity that police officers could use to identify and divert eligible individuals. The tool's requirements at the outset were that it could be short and easy enough to administer in the field without extensive training. While many instruments exist to diagnose mental illness, substance use disorders, and suicidality in a medical setting with high accuracy, they are far too long and detailed for a police officer to use in the field. Moreover, the vast majority of existing tools require administration by a trained healthcare professional, which makes them unsuitable for use by law enforcement. Our team's goal was to design a tool that would require minimal training to administer so the cost and time to train police officers would not be prohibitive to its implementation. Our team of experts successfully developed a ten-question tool that can be rapidly administered by police officers in the field to flag individuals who could safely benefit from diversion out of the criminal justice system and into behavioral health treatment.

One challenge we had to tackle at the outset was being diligent to choose language to solicit honest answers and ensure the questions did not sound accusatory. We wanted to place all the questions in a framework of health rather than criminal behavior. This is challenging when the tool's administrator is not in a health setting. As this relates to questions about drug use, we have observed that certain groups are more reticent to reply honestly about an illegal activity to a police officer based on negative historical and personal interactions with law enforcement. The more distrust between an individual and the police the greater chance they will not be forthcoming about behavioral health issues, resulting in a lower likelihood of being recommended for diversion. This concern is a subject we are actively engaged in with racial bias experts and behavioral economists.

The Criminal Justice Lab engaged in a two-step validation process to test whether this instrument could work. In 2018, we developed an application for tablet computers that was rolled out to a small group of police officers in Indianapolis, Indiana, to determine whether the short questionnaire could be administered in the field and whether officers would want to use it. These officers reported that individuals understood and responded to the questions, that it took less than five minutes to complete, and that they would be willing to adopt and use it. This was critical feedback because, if officers were not amenable to using the tool, it would not succeed regardless of its accuracy level.

The Criminal Justice Lab then proceeded to test whether the short list of questions could accurately identify individuals with mental illness, substance use disorders, and suicidality. We administered the tool to 712 individuals at booking in jails in Indianapolis, Indiana, and McLean County, Illinois. We then concurrently administered a validated and widely accepted diagnostic tool—the MINI (the Mini International Neuropsychiatric Review)—to the same 712 people. We found that the correlations between the MINI and our tool were strong. We created three different scales—one for

FIGURE 3.6.1 **HealthLink Tool AUC-ROC Rates**

Mental Illness	
Score	% Diagnosed
0	9%
1	16%
2	39%
3	59%
4	46%
5	74%
6	84%

Suicidality	
Score	% Diagnosed
0	2%
1	7%
2	18%
3	38%
4	61%
5	65%

Substance Use Disorder	
Score	% Diagnosed
0	10%
1	30%
2	47%
3	68%
4	86%

mental illness, one for substance use disorders, and one for suicidality—with scores generated for each scale ranging from 0–4, 0–5, and 0–6, respectively. Each scale uses a different subset of the ten questions on the tool, based on the strength of the correlation between the answer to the question on our tool and the MINI diagnoses. Figure 8.4-1 shows the area under the receiver operating characteristic curve (AUC-ROC) rate for each component of the tool, which range from 0.78 to 0.88. The AUC-ROC is a measure of predictive accuracy, with a range of 0 to 1. Our range of 0.78 to 0.88 is classified as excellent accuracy.[5] To build support for the tool, we had to show it did a significantly better job than human judgment alone; otherwise, the extra time to administer the tool might not outweigh its benefits to police. It is unique for a tool so short to achieve a similar level of accuracy as an extensive, validated tool used in a medical setting.

As this figure demonstrates, using just ten questions, the HealthLink tool can identify individuals with mental illness, substance use disorders, and suicidality with a high degree of accuracy. The shading shows the cut points established in consultation with our team of experts; based on these, individuals in the shaded zone would be recommended for diversion. When police officers interface with the application of the tool, they will not see the scores for each scale; they will simply see yes or no as an indication of whether the person is eligible for diversion, and based on which of the scales. We wanted to streamline the display for the application so it did not require extra work to understand the tool's results, so as to both encourage use of the tool and reduce human error.

THE PROMISE OF DIVERSION

As noted above, the scoring cut-off points identified 48 percent of all people in the pilot study as eligible for diversion. This means that *almost half of all arrested and screened individuals during the pilot study could be eligible for diversion*. When we consider that 48 percent of the people screened in our validation study would have been eligible for diversion from a behavioral health standpoint, we begin to see the tremendous impact this tool can have on policing and the entire criminal justice system.

To give a sense of scale for this impact, consider that, for Indianapolis, a city of 800,000 that averages about 30,780 arrests and criminal summons annually, an estimated 14,774 people would be eligible for diversion. While the financial savings of diverting up to 14,774 people is calculable, the additional cost savings to communities is immeasurable. As we have seen in recent research findings, averting initial entry into the criminal justice system has the greatest benefits in terms of reducing future criminal justice involvement without increasing local crime rates.[6]

At this time, we are ready to deploy the tool to all law enforcement officers in Indianapolis, Indiana, and McLean County, Illinois. The Criminal Justice Lab would like to scale the tool beyond our first two implementation sites, first to an additional two to four sites and then nationally, with the goal of improving how police departments identify and divert individuals with mental illness, substance use disorders, and suicidality.

To make the adoption and scalability of the use of this tool easy and accessible to all police departments, we are building an application that can be used on Android or Apple systems, that can be run on a phone, tablet, or computer, and that will be free for use. The application also can be used to track outcomes and other metrics, including demographics, numbers eligible for diversion, and those actually diverted, as well as use of the tool. We have secured private funding to build these applications and cover the cost of the first two years of cloud hosting fees to allow for simple and free national scaling.

As our research has revealed, almost *half* of the arrestees in these pilot jurisdictions could be safely diverted from the criminal justice system. The prevalence of underlying behavioral health factors, which remain largely untreated, shows the immense power of a tool like this to change the entire system. When we think about the scale of the problem, we begin to see the extent to which a tool like this can dramatically enhance community safety, improve long-term outcomes for police-involved individuals, and provide a new path forward for the law enforcement community.

NOTES

1. See Jennifer Bronson, Jessica Stroop, Stephanie Zimmer, and Marcus Berzofsky, "Drug Use, Dependence, and Abuse among State Prisoners and Jail Inmates, 2007–2009," Department of Justice, Special Report, August 10, 2020, www.bjs.gov/content/pub/pdf/dudaspji0709.pdf.

2. See "Federal Prisons: Information on Inmates with Serious Mental Illness and Strategies to Reduce Recidivism," GAO, February 2018, www.gao.gov/assets/700/690090.pdf.

3. Samuel R. Bondurant, Jason M. Lindo, and Isaac D. Swenson, "Substance Abuse Treatment Centers and Local Crime," Working Paper, NBER, September 2016, www.nber.org/papers/w22610.

4. Anne Milgram, Jeffrey Brenner, Dawn Wiest, Virginia Bersch, and Aaron Truchil, "Integrated Health Care and Criminal Justice Data—Viewing the Intersection of Public Safety, Public Health, and Public Policy through a New Lens: Lessons from Camden, New Jersey," https://www.ojp.gov/ncjrs/virtual-library/abstracts/integrated-health-care-and-criminal-justice-data-viewing.

5. S. L. Desmarais and J. P. Singh, "Risk Assessment Instruments Validated and Implemented in Correctional Settings in the United States" (New York City: Council of State Governments, 2013). See, also, M. E. Rice and G. T. Harris, "Comparing Effect Sizes in Follow-Up Studies: ROC Area, Cohen's d and r," *Law and Human Behavior* 29, no. 5 (2005), pp. 615–20, which characterizes AUC values over 0.714 as large.

6. Amanda Y. Agan, Jennifer Doleac, and Anna Harvey, "Misdemeanor Prosecution," NBER Working Paper No. 28600, March 2021, https://www.nber.org/papers/w28600.

PACE CENTER FOR GIRLS

ADVANCING EQUITY THROUGH PARTICIPANT-CENTERED RESEARCH AND EVALUATION

MARY MARX AND KATIE SMITH MILWAY

Pace Center for Girls, a Florida-based, multiservice nonprofit serving middle- and high school–age girls with histories of trauma, faced an ethical dilemma several years ago: The organization and the community it serves, as well as funders and policymakers, sought concrete evidence that Pace was effective in helping the girls who participate in its programs. With this goal, the nonprofit launched a randomized control trial to assess whether performance in school was better for girls in the program than for those not enrolled. But that meant withholding services from some girls (the control group) and referring them elsewhere, at odds with Pace's mission. The longitudinal RCT also would be costly and labor-intensive—and take years—while approaches at Pace and in the field naturally evolved.

Despite the downsides, Pace pursued the RCT. Conducted from 2012 to 2018, it found that Pace girls were nearly twice as likely to be on track to graduate from high school as girls not at Pace.[1] But the findings had limited application, as they focused on standard measures such as attendance and grades. The RCT, as designed, could not establish a causal link with Pace's signature individualized services—such as counseling, anger and stress management, and building self-efficacy—that set girls up for success in life.

Flash forward, and Pace has learned that empirical research can be done holistically and equitably. Instead of relying on an RCT for ultimate answers, Pace blends empirical findings with participatory approaches to learn how processes, policies, and social institutions help the girls it serves. Pace's approach to evidence-building has evolved from conducting one arm's-length study at a time, post facto, to sustaining an ongoing process that directly involves girls, community members, and other stakeholders in designing and answering research questions. By building a robust internal research and evaluation function, Pace has identified causal links between feedback and outcomes for girls. It now immediately incorporates participants' insights into program improvements, thereby strengthening Pace's culture and its participants' self-efficacy and self-advocacy in real time.

Founded in 1985, Pace today serves more than 3,000 girls annually in twenty-two locations in Florida and Georgia with its evidence-based model, and it is recognized as one of the nation's leading advocates for girls in need.

PARTNERING TO GROW CAPACITY

Pace's pivot to participant-centered measurement was supported by an overall pivot to developing a feedback culture as an organization. Because Pace's major funder, from the organization's inception, was Florida's Department of Juvenile Justice, Pace placed a significant focus on compliance. A decade ago, it simply was not part of Pace's culture to be highly innovative in seeking to improve its model for helping girls prepare for the future. Pace grew from a program for ten girls in Jacksonville, Florida, in 1985 to seventeen centers across Florida, serving approximately fifty girls per center by 2006, but then growth stalled.

In 2010, with support from Edna McConnell Clark Foundation (EMCF), Pace brought in new leadership, and EMCF encouraged a thorough review of program data to understand the true scope of what Pace needed to become more impact-driven and to embrace learning for continuous improvement. Financial analysis showed that many Pace centers were not cost efficient and that innovation lagged the field. To renew growth and to initiate in-house learning and evaluation, Pace needed to find more cost-effective ways to evolve its model and reach beyond its physical sites. This led to partnering with leadership and culture consultancy Human Synergistics (HS) to define high performing behaviors and evolve Pace's talent and culture to support the growth strategy. It also led to partnering with Fund for Shared Insight, a funder collaborative building the field of feedback, to ensure that the voice of Pace girls and community members informed the evolving the model.

OUR APPROACH

A core difference between RCTs and participatory approaches is at the heart of the equity argument and our blended measures approach: RCTs follow a treatment and control group over time—and then look back, running regressions, to analyze change. They can't adapt the intervention in response to participant feedback, because the participants are seen as subjects. After all, RCTs have roots in scientific experimentation. In contrast, participatory tools—feedback, surveys, focus groups, testimonials, diaries, participant councils—derive from a social science method called "participatory action research" (PAR) that dates back to community surveys initiated by sociologist W. E. B. Dubois in the late 1800s to understand structural racism.

PAR connects immediate learning with continuous improvements to programs and policies, with participants seen as experts in their own experience.[2] It gave rise to participatory evaluation (PE), which gives program participants, staff, and other stakeholders ownership in designing and managing the evaluation process itself. It emerged in the late twentieth century as a subfield of program measurement, particularly outside the United States among international relief and development nonprofits. PE radically shifted how to gauge social programs' effects on participants and their sense of power, always asking, "What answers are we seeking? Why? By whom? For whom?"[3]

With these participatory approaches in hand, Pace set out to reshape its culture. Yet, culture is not something an organization can change overnight. To identify the behaviors Pace wanted to see as an organization, it used a tool from HS and a corporate culture framework from search firm Spencer Stuart. The HS tool allowed Pace to measure its current culture against constructive benchmarks as well as define an ideal culture to advance its strategy, asking: "What are you expected to do here to fit in?" (from a list of 120 behaviors related to constructive, aggressive/defensive, and passive/defensive work styles[4]), rating each on a scale of 1 to 5).

With culture ratings in hand, Pace could analyze gaps between current culture and its ideal, determine strengths, and focus on areas for improvement. The Spencer Stuart framework helped Pace's leadership zero in on words to describe its ideal culture. The team landed on a desire to be caring, learning, purposeful, and results-oriented.

The HS ratings allowed Pace human resources to work with centers and individual departments to identify any subcultures at odds with the ideal culture. Where alignment was off, team members worked together to create goals and performance and development plans to grow the culture constructively. The thinking was that if at least 75 percent of staff moved to the ideal culture, it would become a norm that talent would start to opt into—or out of.

In keeping with cultural aspirations, Pace built an internal measurement team focused on listening, learning, and improving their work with girls and their communities, which meant soliciting feedback from community stakeholders and program participants. Pace implemented a salesforce.org tracking system to analyze participant feedback along with metrics such as school attendance and juvenile justice involvement. And Pace reinforced functions that facilitated communication with participants, greatly expanding IT efforts and investing in technology that improved connectivity among sites, participants, and staff members' homes.

Pace girls played an important role in this cultural realignment. The new measurement team, led by Lymari Benitez, Ph.D., Senior Director, Program Information and Impact, used girls' feedback (qualitative data), captured by Shared Insight's Listen4Good (L4G) survey system, to identify staff behaviors that aligned with Pace's cultural expectations and to develop trainings to support such behaviors. In 2016, Pace received co-funding from EMCF and the Fund for Shared Insight to embed L4G in its measurement approach, which allowed girls' input to influence the design of Pace's culture model. The L4G survey probed how often the girls felt treated with respect and how likely they would be to recommend the program to their friends. The latter likelihood, scored from 0 to 10, is called a Net Promoter Score, or NPS. Using the NPS system, Pace conducted multiple regression analyses and found that positive feedback was predicted by a higher sense of belonging and feeling safe and respected. Pace also aligned positive culture expectations with girls' outcomes. Data analyses indicates that, in Pace Centers with high social cohesion among team members, girls are more likely to improve academically and have longer length of stays (low attrition).

Ultimately, Pace shifted from being a compliance-driven partner of the juvenile justice system to a future-focused agent of change with a practical goal—of developing "socially, emotionally, and physically healthy, educated, and stable girls"—that permeated all of its departments and its refreshed theory of change. The L4G survey now takes a twice-yearly pulse on what Pace is doing well, what it could improve, the degree to which participants feel treated with respect, and how likely participants would be to recommend Pace to their peers.

To further permeate change, Pace developed feedback processes across organizational functions. It created Girls Leadership Councils at every site—the girls help design, execute, and interpret program research and evaluation; conduct focus groups with peers; aid in interviewing new hires; and contribute to program decisions, where Pace "closes the feedback loop" and lays out areas the girls' input has surfaced for improvement.

Any organization embarking on building an ongoing research and evaluation function needs a way to fund the high-quality talent and technology it entails. In Pace's case, then-COO Yessica Cancel and her team found cost efficiencies in changing their approach to health insurance and in reducing turnover. For the former, Pace became self-insured. By paying claims directly versus working with an external healthcare insurance provider, and by educating staff on wellness practices, Pace saved 40 percent of a $2 million healthcare line-item expense while simultaneously expanding coverage. Pace also reduced turnover, which by 2011 was costing $2.3 million a year in recruiting and hiring. Pace did market analysis to ensure it offered competitive salaries and reduced absenteeism through investments in wellness and educating team members on how to become smart consumers of health care. Since embracing a feedback culture, over the past five years Pace team-member turnover has declined by nearly two-thirds, and productivity and engagement have increased more than a quarter.

Pace reinvested dollars saved into active recruiting and in retaining and developing new talent. In the process, Pace and Cancel won a Nucleus Research Award for achieving a human resource breakthrough. Pace also became a "Best Place to Work" in northeast Florida.

CHALLENGES AND RESPONSES

Pace has faced two key challenges to its participant-centered measurement systems since implementing them. The first was to figure out the right blend of participatory measures and empirical data to generate evidence of impact absent an RCT. Pace evaluation lead, Lymari Benitez, uncovered links to outcomes by conducting statistical analysis (correlations, regressions,

ANOVAS, T-tests, and structure equation models) of girls' perceptual responses in the L4G surveys, and empirical data such as school attendance, grades, and interactions with juvenile justice. To date, Pace has found the strongest link between teacher retention and girls' feeling more respected and staying in the program longer, with tenure in program statistically proven to positively influence their results.

A second challenge came with the pandemic and difficulties collecting feedback from girls who were unable to attend programs in person. Here, Pace's investments in technology paid off, and it was able to extend its technology—including internet-enabled tablets, laptops, and Microsoft® Teams accounts—during the pandemic from supporting 527 staff to supporting an additional 2,000 girls—and implemented remote services with feedback channels for the girls. As a result, Pace engaged with more than 90 percent of its girls at least once per week during school closures, and with 75 percent seven or more times per week—delivering food, computers, telecounseling, and other goods and services they needed.

Ultimately, 91 percent of girls completed the program in 2020 (versus 81 percent in 2019); and 88 percent improved academically in the fourth quarter (versus 70 percent the prior year). Meanwhile, EdWeek[5] found that in high-poverty communities across the United States (those with 75 percent or more kids on free or reduced lunch), one of three students had no engagement with schools at the outset of the pandemic.

RESULTS

Pace's foray into participatory measurement has transformed the organization, both in makeup and culture. Pace has new roles that keep it proximate to and learning from the communities it serves. And it has developed an instinct across all team members for responsiveness and resilience. But the greatest payoff in shifting from outsourced RCTs to participant-informed measures has been the way participatory approaches have influenced and empowered the organization to step up advocacy to change the systems—juvenile justice, foster care, and education—that can serve as barriers to young girls' success.

Today, Pace uses direct input from its girls to identify local, state, and federal policies that need reform—and the community members who must be involved. For example, to lobby for misdemeanor and civil-citation legislation so law enforcement could censure girls for petty crimes without arresting them, Pace girls testified before legislative committees and met with individual legislators, with success. In 2011 and every year since, Florida has increased funding for prevention measures to keep girls out of the juvenile justice system. Over the last decade, the number of girls arrested annually in Florida has dropped by about 65 percent.

Pace further evolved into a community catalyst to mitigate and disrupt inequities through a data-driven, collaborative approach that would allow community stakeholders to identify and address specific issues affecting girls. They convened Girls Coordinating Councils (GCC), where these stakeholders, including girls themselves, are given the space to influence favorable conditions for girls and young women's healthy development in their communities. In 2018, a GCC in Broward County, Florida, tackled the county's rate of detaining girls for failing to appear at their court hearings. When the girls interviewed judges, probation officers, and youth, they found that, often, girls who had been arrested forgot their hearing dates or struggled to find transportation to the court. The girls' research team created a video with avatars portraying what arrested girls could expect upon entering the juvenile justice system. They also created cards teenagers could carry with their cell phones in the event of arrest, with hotline numbers for case managers, transit information, and contact information for the court in the event of a delay. The following year, arrests in Broward declined 16 percent, and instances of failing to appear after arrest dropped 27 percent.

REFLECTIONS

Nonprofits serious about building equity and inclusion must ensure constituents are true participants in evaluating program impact to develop more inclusive organizations that empower the voices of the communities they serve. This is the key, too, to building resilient nonprofits and to bringing about complex and lasting social change.

Pace's growth in listening to the girls it serves, gathering high-quality feedback from them, and applying it to advance the organization's goals has empowered girls and changed the nonprofit's culture. Today, Pace connects ongoing learning to continuous improvement of programs, policies, and practices and views its participants as experts in their own experience. As a result, Pace has:

For Girls:

- Increased engagement, measured via girls' attendance, their net promoter scores, and their active participation in feedback loops.
- Increased confidence and self-efficacy.
- Achieved better outcomes.

For the Organization:

- Built high awareness of the value girls' insights bring to program improvement.
- Enhanced the nonprofit's reputation—which increases referrals from girls and their families to Pace's voluntary program.
- Facilitated a shift from work silos to a systemic approach for process improvement that resulted in a more trusting and equitable organization.

For All:

- Developed more equitable relationships between girls and staff.
- Implemented an actionable feedback loop—Pace uses real-time data and participants' insights to inform services and ensure the program addresses girls' needs.
- Expanded the scope of organizational culture to include staff and participants. Aligning all to a common cultural ideal has been the key to process improvement and better outcomes.

It was important throughout Pace's work that cultural transformation remain anchored in Pace's mission and that Pace align investments with aspirations for change. Accordingly, Pace made strategic investments in recruiting, talent management, internal research and evaluation, IT, and learning and development.

It also was important that Pace adapted both its processes and its mindset in interacting with the girls to elicit not only their participation in creating change for themselves and their communities but also their belief in the power of their own voices.

Other practitioners seeking to advance equity through their approach to measurement should bear in mind lessons learned at Pace: that transformation begins with an engaged and competent workforce, ultimately leading to lower attrition of program participants, their greater persistence in the program, and better outcomes.

Meanwhile, funders supporting this work need to ensure that resources are flexible enough to fund the talent and technology needed to gather participant feedback and target their true needs. And they need to ensure that their arc of funding is long enough to sustain change.

NOTES

1. See Pace Final Report, January 2019, www.mdrc.org/sites/default/files/PACE_Final_Report_2019.pdf.

2. Sociologist W.E.B. Du Bois in 1898 used community surveys to understand structural racism. Sociologists Kurt Lewin, Margot Haas Wormser, and Claire Selltiz in the 1940s and 1950s used participatory community self-surveys to understand individual lived experiences.

3. IDS Policy Briefing, "Participatory Monitoring and Evaluation: Learning from Change," Issue 12, November 1998.

4. "The Human Synergistics Circumplex," Human Synergistics International, www.humansynergistics.com/about-us/the-circumplex.

5. Stephen Sawchuck and Christine Samuels, "Where Are They? Students Go Missing in Shift to Remote Classes," *Education Week*, April 10, 2020, www.edweek.org/leadership/where-are-they-students-go-missing-in-shift-to-remote-classes/2020/04?cmp=eml-contshr-shr.

SECTION 4

EMBRACING A CONTINUOUS R&D-LIKE APPROACH

Evidence, for the most part, is an exercise in innovation: how to make processes work better, how to develop better products or combinations of services. At its best it is really about continuous improvement.
–BRIAN SCHOLL, "THE UNFINISHED BUSINESS OF EVIDENCE BUILDING: DIRECTIONS FOR THE NEXT GENERATION."

The best way to improve program implementation, promote innovation and assess impact is to think of evidence building as an R&D function that informs and is informed by strategy. This section acknowledges shortcomings of traditional data and evidence practices and explores advances in the design and execution of more actionable evidence building. The approach combines analytics and data science with conventional evaluation in an intentional and continuous practice. For too long we have focused on research (the R) without the development (the D). A next generation of evidence calls for both.

In the "The Unfinished Business of Evidence Building," Brian Scholl extolls the need for researchers to "work backwards" from practical outcomes to design worthwhile studies. Christopher Spera explores common challenges to conducting evaluations and calls for a learning lens and increasing internal capacity of practitioners to evaluate their own programs.

In the spirit of innovation, authors Kevin Corinth and Bruce Meyer discuss overcoming the limitations of any single data source to measure poverty through the new Comprehensive Income Dataset. Kathy Stack and Gary Glickman speak to improving data analytics at the state and local government levels, and Neal Myrick of pandemic driven innovations in R&D by practitioners. David Yokum and Jake Bowers articulate the value of pre-analysis plans, and Jim Manzi what we've learned about how RCTs can best be used to assess social impact, and where they've fallen short.

Four use cases include criminal justice organization Center for Employment Opportunities, a later stage organization, which shows the importance of establishing and staffing internal R&D capacity. Children's media innovator Noggin (chapter 4.9) uses multiple strategies to quickly iterate on content and ensure it continually improves. First Place for Youth shows how a foster youth transitions organization streamlined its measurement and evaluation to zero in on one "North Star" metric: youth income and living wage. Gemma Services, which offers youth-oriented psychiatric care, describes an approach to building "evidence on demand," inspired by data science and machine learning algorithms that serve Amazon and Netflix.

Questions addressed in this section include:

1. What constitutes R&D capacity for practitioners and how can one best build it?
2. How does internal R&D capacity add to what we learn from conventional evaluation techniques (both benefits and watchouts)?
3. What do leaders need to know?

THE UNFINISHED BUSINESS OF EVIDENCE BUILDING

DIRECTIONS FOR THE NEXT GENERATION

BRIAN SCHOLL

This chapter covers a few odds and ends about evidence. Evidence is a curious business, in some cases too much business and in other cases not enough business. Much is yet to be tapped on the evidence front, and issues of who does the building and how the evidence is used once built are critical.

The content of this chapter comes from a mix of my own experiences and views of evidence, policy, and implementation that I have developed over a period spanning decades in government, the private sector, and academic research settings. I have worked from the very micro or direct-provider level with organizations so small they cannot rightly be called organizations, on up to the most ivory tower levels of policy, and, really, almost every level in between. Those experiences have enabled me to see an enormous amount of variation in the way organizations organize their work and the obstacles they face, and to develop perspectives on why evidence works for them or why it does not.

When we in the evidence community talk about building evidence, so often our conversation goes to the math and the statistics of it all: experiments, treatment effects, causal estimates, randomization protocols, and so on. Those are so important in so many ways, but also so unimportant in so many other ways. In my mind, it is the organizations, the institutions, and the people that really matter. The wrong people at the top (leadership) can dead-end any efforts to generate evidence. Wrong people generating evidence get to all the wrong questions and all the wrong answers using all the wrong methods. Wronged people at the bottom (beneficiaries) bear the consequences of getting policies and programs wrong, and those beneficiaries might rally with torches and pitchforks if they feel it is their right that someone gets a policy or program to work (even if they are not articulating their concerns as stemming from a lack of evidence). Organizations and institutions can be set up by good leaders to carry the torch even when bad leaders come along. Evidence is critical to getting our work to work and keeping our democracy democratic, but it cannot help if folks have their fingers in their ears.

In this chapter, I am not conveying a read of the evidence on evidence but, rather, a view of informed experience, and sketching something of an ideal that I think organizations can work toward. Hopefully, some of the lessons I have learned over the years are helpful, or will at least help you the reader feel you are not alone.

THE PROMISE OF EVIDENCE

These days, "evidence" is something of a buzzword. Evidence, though, is not so much a fad as a set of techniques for advancing knowledge about particular questions.

Evidence generation is an investment. Think about a government agency or a nonprofit that is trying to make something good happen for a number of people: feeding the hungry, sheltering the homeless, teaching students reading and math skills, or reducing unemployment. As shorthand, we can say the outcomes related to this goal are Y. For example, maybe Y is the number of people who have found homes or the number of people who are not hungry or the number of people who have a job.

The agency has a set of policy levers it can pull to accomplish its goal, like providing food, shelter, or education, and, perhaps, it has the ability to invent new levers to try new ideas. Let's call those policy levers "p." The overarching problems are: Which levers does the organization pull? How much should it pull each one? How does it think up new levers to construct?

Since the outcome Y changes when we pull different levers in different amounts, we can think of Y being a function of p; change the mix of levers and we get different outcomes for our beneficiaries. We can denote this relationship as Y(p), with Y being a function of p. (Don't worry, we are not going to do any real math here; we just need a way to communicate that our outcomes Y change when our policy mix p changes.)

There are only so many potential beneficiaries out there, and usually a policy or social sector organization has only limited resources with which to reach them. We can call the best possible value of the outcomes Y*. This is the best we can ever do; so Y* might be the maximum achievable level of employment (or the minimum level of unemployment). Y* is the best outcome the organization can facilitate with any potential combination of activities p.

Achieving Y* is tricky. We will get the best outcome Y* using the ideal combination of policies p*. This is where the evidence comes in. Evidence can help us get closer to figuring out Y* and p*. More important than those optimum values, evidence can help us understand the relationship between Y and p. The evidence does not just land in our lap. To build this evidence, we need to invest effort and time and resources. If we have not been building the evidence, we have no idea what the relationship between Y and p really is. We also do not know what Y* and p* are. Heck, in many cases, we may not even have data that tells us what our current levels of Y are, or what the current mix of activities are that are in p. Without evidence, we are really just stumbling around in the dark without any idea of whether we are helping people or hurting people, whether we are doing our best or less or can improve. The more evidence we build and the better we get at building evidence, the better ideas we will have for trying things that are new.

What we have above is a sketch of what evidence can mean: evidence gives us hope. It is a flashlight when we are wandering in the dark. The further we get from the optimum (Y*), the worse off constituents, clients, and beneficiaries will be, and the more likely we are to be squandering our own resources. For example, if the Fed chair sets the wrong interest rate, or if Congress sets a bad tax policy, they can create distortions in the economy. Those distortions can motivate people to make dumb moves and cause other problems. Even a small social sector organization can inadvertently distort peoples' choices, incentivizing people to put their effort into the wrong thing; for example, by wasting their time in a classroom training that is not working, just because tuition is free. We need evidence of our impact to make sure we are providing the right services for people. Organizational ignorance is a distortion, one that pushes the costs onto beneficiaries (through a lack of an appropriate solution to their problem) and potentially creates other negative consequences that hurt innocent bystanders.

THE PATH TO EVIDENCE BUILDING

Buying into the promise of evidence is the first step in the journey. The next step along the path is figuring out how to succeed in building insightful and impactful evidence.

The Power of Working Backward

Let's discuss a problem that has plagued some evidence-driven organizations: the "dead-end study." Before we even begin: do not panic. There is a solution to the dead-end study: developing a credible theory of change by working backward from desired outcomes in the design process.

What is a dead-end study? Some organizations have built infrastructure, collected data, hired the right folks, engaged consultants, and so forth. In some cases, they have thought their efforts to be fruitless because big, careful, and sometimes expensive studies turned up point estimates of zero or close to it, or repeated tests have yielded inconsistent results, or their tests were simply too scattered to add up to a clear direction.

To be clear, the problem many organizations face is not, in fact, the null result. "Null results" are much maligned, but they can often be informative. Null results are shorthand for: "We tried a couple of different things and either nothing worked or our best ideas didn't show any incremental benefit." Null results feel like a failure because people may have put effort and resources into developing and testing an innovative new idea only to find out that it is a dud. Yet, solid evidence that something does not work is informative. It tells you what not to do (we will return to this point in another way). The problem is really not the null result itself; the problem is when the null result has left the organization without insight on what to do next. This happens when evidence is generated without the conditions to succeed.

I would estimate that at least 80 percent of my time working with organizations to develop evidence is spent on programmatic issues, and less than 20 percent is spent on methodological issues. That is, I spend most of my time asking *Why is the program doing what it is doing?*—a programmatic question—and only a small fraction of time asking *How do I generate evidence to tell if what it is doing actually works?*—an evaluation question. The reason for this is that many organizations—from the small nonprofit to the large government policy institution—do not have a strong, coherent, and *credible* theory of change that links their activities (p) to a set of outcomes (Y) that they care about. *Credible* is the key word here because it is not sufficient to postulate wildly implausible causal links between activities and outcomes that are not justifiable using existing social science theory and evidence.

Without the theory of change and without some soul-searching to identify potential alternative activities, tests typically can comprise only a limited set of questions of the form: *Does our current activity work to effect change in our outcomes of interest?* We can design a test to determine if what the organization is *currently* doing works, but if the answer is no (it does not work), then we have gained little understanding of the factors that can affect Y, and we have little to guide us on what to do next from either an implementation perspective or an evaluation perspective.[1] An organization or program's theory of change is, in my experience, the most overlooked component of successful and meaningful evidence generation.

How do we construct a theory of change? Truthfully, it is not an easy task either methodologically, institutionally, or emotionally. We need to put our pride aside and open ourselves to hard and uncomfortable questions. Then we have to work backward.

Working backward means starting our theory of change at the end: identifying the outcomes of interest (Y), and developing the causal chain by working the pathway backward through outputs and intermediate outcomes and causal mechanisms on through to a set of activities that can credibly produce change in those outcomes.

Working backward not only forces critical thinking about the assumptions needed to connect current or proposed activities to intended outcomes but also can help organizations identify alternative activities that also can lead to the desired proximate and distal outcomes. In contrast, working forward (starting with the current activity mix) often has the potential to push everyone to contrive assumptions and explanations for how and why current activities lead to outputs that cause changes in behaviors that ultimately lead to the outcomes. In a sense, working forward basically *assumes* that the current activities work, whereas, with working backward, one might not ultimately even situate current activities in the set for consideration. In my experience, the assumptions embedded in working forward simply tend to be unreasonable. In working backward from desired outcomes, there tends to be stronger footing, perhaps because ignoring current activities forces everyone to think critically about actions that can result in the stage of the causal chain they are focused on.

For example, suppose our goal is to improve student classroom outcomes and our current activity is to engage students in arts and crafts. In working forward, we might need to make a lot of very tenuous assumptions about the immediate effects of the crafts program to draw a link to better student performance. If we were to work backward from the premise that we are trying to help student performance, we might have to be critical of ourselves and identify the main obstacles to student performance and, in time, come to a recognition that there are better options than an arts and crafts program. To be uber-clear, such an arts and crafts program might be valuable for reasons other than student performance, or it may be an important complement to other student-centric programs, so it is possible that we just chose the wrong rationale for the program, but if student performance is the right outcome, we might need to consider alternative activities.

Working backward can lead to a more critical assessment of whether the activities make sense, since they have to fit within the path models described at more distal stages rather than forcing a path from activity to outcomes. To be fair, some organizations will be limited by cultural, capacity, mandate, funding, and other constraints that will narrow the practical range of activities they can implement; an association of school teachers providing after-school tutoring is unlikely to hop into providing basic income support for local families. Yet, in my experience, even in those organizations, working backward can force a much broader conversation about considering alternative (yet feasible) activity options than working forward can do. Evidence generation developed based on that causal chain also can provide a better basis for understanding the factors that affect the desired outcomes. For larger or policy-oriented organizations, working backward using existing social science theory and evidence is critical to help the organization think outside the box and critically assess activities for which alternatives and alternative methods of implementation are possible. These alternatives, if tested, can provide insights that curtail dead-end research.

I understand some of the reasons theories of change often are absent. Organizations may be sensitive to opening up the theory of change discussion. Theories of change are fundamentally an element of program design rather than evaluation, so even though they are critical to generating meaningful evidence, they can be viewed as outside of the evaluator's domain. Organizations have activities, but they often do not have a credible theory of change for why they are doing what they are doing—or they figured it out so long ago that it may be deep in the recesses of their memory. Evaluators can either work within existing activities to develop tests of efficacy, or they can push the very difficult conversation that involves developing and vetting a credible theory of change. It often may be practical for the evaluator to postpone the difficult theory of change conversation and work on evaluations of existing activities to build trust with the programmatic people, which could facilitate tougher conversations in the future. The problem is that postponing those discussions can lead us to the dead-end study.

The theory of change does not eliminate the dead-end study entirely; it simply lays the foundation for meaningful evaluation. A credible theory of change is a necessary but not sufficient condition for meaningful evidence generation. Evidence still needs to be developed strategically, and in a methodologically sound way, because dead-end studies also can arise with a non-strategic approach or an ill-conceived methodology.[2]

Pragmatic Parameters: Leadership and Resources
Management Integration

One would hope that the push for evidence would come from within the organization and resonate top and bottom through leadership and the rank and file, yet it often comes from external pressure. Management-centric evidence initiatives—ones that feed helpful perspectives directly into the decisions management encounters—will tend to find a warmer reception. Evidence programs should seek to tie into management goals as much as possible so the evidence can help optimize activities along the dimensions management cares about. At the same time, evidence programs can help steer management to a more suitable dashboard if it is not already looking at the right indicators of progress.

Management integration need not be confined to the evidence itself. In some circumstances, it could be creating data tools that support management's direct objectives. For example, consider that the first step in evidence generation often is taking stock of the activities the organization is engaged in, perhaps compiling a dataset of such activities. Instead of conducting this inventory as a one-off research activity, is there a way to create a data collection system that regularly reports out to management? Are there data collection tools that can be delivered to staff that both captures data needed for evidence generation and helps staff perform the tasks they normally do, making their jobs easier and allowing management to capture productivity gains? Evidence needs friendly management to flourish, but it also will find management and the organization more friendly when it makes the organization's work easier. Good evidence programs will seek to align and integrate with an organization's operations and to keep focus on the organization's overall goals.

The Economics of Evidence

Does evidence cost a fortune? In recent years, too many folks have gotten overzealous and proposed large and time consuming studies with costly data collection and so forth. There is definitely a role for that sort of evaluation, and in some settings, such investments may be necessary for big, expensive programs, broad policy issues, difficult to quantify and study issues, and initiatives that will affect a large number of people. These all deserve careful study and attention.

Yet, the economics of evidence is not really about cost control; it is about adapting research to the institutional incentives, constraints, and other realities that each organization and its leadership faces. What can an organization do? What are they required to do? To what measures are management held accountable? What are their major operational problems?

Unfortunately, the reality is that time and money tend to be focal points of those institutional incentives. One thing I have endeavored to do in organizations with which I have worked is to lower the costs and shorten the length of the evidence life cycle. Decision makers who are told that research will take three to five years and cost several millions of dollars will be unenthusiastic when their job tenure may last only a few years. Reducing the costs and time may be essential.

I cannot say I have a one-size-fits-all solution to money and time problems. Each situation should be examined based on the organization's own resources, opportunities, and constraints. In almost every organization, I find that focusing on better recordkeeping is a first step to supporting operations while also providing administrative data that can be organized by researchers to study the work of the organization. In large and complex organizations, I have found it often is important to make investments to build standing capacity for data collection, and to build internal technical expertise, which will lower the marginal cost of evidence generation and make individual projects more easy to approve.[3] Smaller organizations will need to be creative but can do some things to generate good evidence without breaking the bank: partner with graduate students who hunger for interesting problems and unique datasets; start out with small qualitative research programs; leverage outreach networks to conduct data collection. Evidence programs can start out with baby steps.

Valuing Evidence Investments

While evidence building does not need to bust the bank, it does need a reasonable amount of support. That is always an uphill battle, especially when decision makers are not researchers. Many professionals tend to see only the perspective of their own profession and cannot appreciate what it takes to generate informative evidence. Unfortunately, many non-researchers in leadership positions seem to believe that evidence generation can be free and instantaneous. The view may arise from ignorance about the costs and benefits of evidence generation, cynicism about the value of evidence, a lack of resources available for evidence, or a view arising because "research" in the form of a Google search feels so fast and free and easy that all evidence generation must be similarly quick and costless. Whatever the cause, this view is obviously unrealistic. Evidence is an R&D-like investment, which can be viewed and evaluated though the lens of cost per outcome.

For example, if status quo intervention A costs $500,000 and intervention B costs $100,000, then, in principle, there is a large gain from finding out that both interventions are equally effective. There are many ways to look at the value in this setting and how much one should be willing to spend on an evaluation, but a framework with the flavor of Return on Investment (ROI) is often reasonable. In fact, thinking in terms of ROI gives us a different context for the null results discussed above. In this example, the null hypothesis is that intervention A and B are equally effective. Failing to reject the null here feels disappointing from a programmatic and evidence-generation perspective; you did not come up with a better mousetrap. Actually, though, you have achieved an impressive win. We can implement intervention B and save $400,000 for other projects. Even if A and B have similar costs and identical impacts, externalities may differ considerably. One intervention might create distortions or adverse investment incentives or adversely affect local markets in other ways even if intentions are noble.[4]

While it is fair to think in terms of long-run ROI, evidence programs have other benefits beyond their direct focus of study. They can develop as an early warning system to better understand emerging risks and how to respond to them; they can develop internal expertise to be able to identify and address other problems; and they can inform a host of management decisions in a variety of contexts. In the case study by the Camden Coalition in this volume, a Health Information Exchange launched in 2010 provided benefits in the fight against COVID-19.[5]

The point here is that evidence should and can have value, but it will do so only if we are crafting it and evaluating it in the right way. If, for example, the academic value of evidence for researchers or the compliance value of evidence for funders is prioritized over more actionable evidence with practical value to those providing and receiving the services being evaluated, this can contribute to the sense that evidence is not worth what it costs, since it is not answering the questions that matter to these stakeholders and it is taking resources from other priorities.

The question that arises often enough is: Could we be over-investing in evidence? Are we doing too much? In my experience, that concern most frequently arises within organizations that are doing almost nothing, and often comes out of a fear of change. While it is clearly conceptually possible to be over-investing in evidence, I can think of no organization that is actually doing so. In my view, while this issue is often fretted about, we are nowhere close to a world where opportunities for evidence generation have been over-exploited.

Building an Infrastructure for Evidence

Building credible evidence often requires technical skills: economics, statistics, other social sciences, econometrics, experimental design, qualitative research. All these skill areas can come into play when building a program of evidence or particular studies.

Most organizations do not have these capacities lying around (statistician in the cupboard?). Often enough, organizations reach out to consultants to augment their capacities. Consultants can bring expertise and experience and a fresh perspective on the work of the organization. After all, consultants have not been in the trenches trying to deliver the goods and services and policies, and they can ask "smart dumb questions." Smart dumb questions are naïve questions about the organization, its work process, its goals and motivation, and so forth that are asked until there is mutual clarity about activities and rationales. In my mind, smart dumb questions are essential, and organizations need to have the patience for these questions if they are to succeed in building evidence. These probes can be uncomfortable for the organization because they not only will include questions like: *What are you doing?* There also will be questions like: *Why are you doing this? Why do you think this works? What made you draw that connection?*

Those questions can be unsettling because organizations have spent a whole lot of time figuring out how to do what they do and may not remember the original rationale, or they may feel their decisions or motives are being challenged, or that they are being told they do not know anything. I often have needed to ask the same questions over and over again to make sure I have gotten it right and that I have understood the motivations and the details. That is, I ask a lot of smart dumb questions, which I find helps me if the questions are dumb enough.

Relying on consultants alone may not solve the evidence problem. They may leave the organization with an interesting research result or fresh perspectives, but rarely do they build the organization's internal capacity along the way. I have personally spent a lot of time worrying about the gap in knowledge and expertise between an organization and external help and the role of expertise in helping an organization achieve its goals. The issue really is the disparity in knowledge between the consultant and the client, and the propensity for that disparity to end in a whole lot of nothing.

Maybe an example will illustrate better than pontificating. When the term "impact evaluation" was all the rage, there were organizations that were externally bullied or forced to march down Impact Evaluation Highway. In many circumstances, consultants were hired to conduct impact evaluations, which sometimes generated reports that had the words "impact evaluation" written on the cover page. Report complete, external pressure eased, life went on. The problem was that some client organizations did not have the capacity to know what an impact evaluation was. Sure, some people knew a few of the basics, but by-and-large, if a client organization did not know much about impact evaluation, a consultant could pass along a report with "impact evaluation" in the title and with contents consisting of a bunch of nonsense, and no one would be the wiser. That is because the client organization did not have the skills to distinguish an impact evaluation from gobbledygook, or even to know the difference between an expert consultant and a charlatan.

One key protective factor in this realm is how evidence-fluent staff are and how much the organization has bought into evidence. Organizations with a lot of buy-in typically want to have the right people to make sure they get the questions and answers right. Organizations lacking the buy-in typically just want to check a box and move on.

Does every organization need to build its own evidence shop or have a team of PhDs at the top? Clearly no. After all, building an evidence group takes energy and resources that may be inefficient for really small organizations. Yet, in my view, most organizations should at minimum develop enough expertise to be good partners and informed consumers of the products they pay for or depend on. Social sector entities should establish relationships with organizations that have good track records of being honest brokers, particularly ones that have been not only tirelessly working with individual organizations to help them build and manage evidence programs but also trying to change the ecosystem so that evidence is more valued, valuable, and strategic. I expect that, over time, there will be more options for cooperative learning systems where small organizations can work together as an association or consortium to finance collective evidence-generation capacities and share learnings.

The larger the organization and the more complex its work, the more urgent it becomes to build an internal evidence team. In the case of large social sector organizations or government entities that develop wide-ranging policies, it often is critical to build internal expertise with technical specialists who can really understand the organization; integrate into work processes; ask important questions of colleagues; help the organization learn; and liaise with senior academics studying the domain. As the size and costs of programs increase and the need for evidence-generation activities grows, it becomes more important to avoid mismatches between the research objectives and the organization's goals because the consequences of bad programs or policies can be enormous. The fact that so many large policy entities around the world still operate without this internal capacity to generate evidence, and instead pass on the costs and distortions of ignorance to their beneficiaries and stakeholders, is, in my mind, unforgivable; the world is too complex, the policy challenges are too great, and resources are too scarce to be used unwisely.

Internal expertise has another advantage: the organization can potentially in-source aspects of the evidence production process that are costly and time-consuming to outsource. It can put people who care about the research in the driver's seat. Of course, evaluator independence is often important, too, so in-sourcing has to be done in a way to preserve that independence so the research is kept honest.

Creating a Culture of Evidence

In parliamentary debate, Winston Churchill once rebuffed a critic's chiding of a proposal's multitude of changes: "To improve is to change; to be perfect is to change often."[6] This could very well be the motto of the evidence community. Evidence, for the most part, is an exercise in innovation: how to make processes work better, how to develop better products or combinations of services. At its best, it really is about continuous improvement. Yet organizational change is tough unless there is enough buy-in from all the parts of the organization that are involved in the undertaking. A more evidence-friendly culture provides fertile ground for evidence to be developed, to prosper, and to be applied. Building a culture of evidence—a learning organization—is an important ingredient for change.

Learning organizations are organizations that have achieved a heightened state of awareness about improvement. These organizations are "skilled at creating, acquiring, and transferring knowledge, and at modifying its behavior to reflect new knowledge and insights."[7] They have a culture of learning that permeates the organization and provides structures that support the learning agenda. Learning has to have the commitment and space to grow. A culture of evidence means that almost everyone has drunk the Kool-Aid and buys into the importance of evidence. Evidence is part of the ecosystem.

A continuous improvement model, undergirded by a culture of evidence and learning, becomes more essential and pressing when organizations are larger, more complex, and have different activities. In that context, it is hard for one-off studies to cover everything. For example, large, policy-oriented organizations operate in a sea of complexity and often do not have direct levers of control over their outcomes; their actions are moderated by the vagaries of human behavior. In these circumstances, the need for a continuous improvement culture seems essential to keep policy on target. "One and done" studies are tough to consider as an evidence program on their own.

A culture of evidence means getting uncomfortable. Evidence should challenge your assumptions, your closely held beliefs, and all of your opinions. The deeper your beliefs, the more the scrutiny of the process of evidence generation and the evidence itself has the potential to unsettle you. You have to go into it all with an open mind, or you will never learn anything. It is always going to be uncomfortable; you just need to learn to live with it. Relax, let it go, and get uncomfortable.

Leadership is a key ingredient in the culture equation. Leadership can set a tone for, reinforce, dismantle, or circumvent attempts at building evidence, guiding the ship to open sea or into the rocky shallows. If management does not embrace a learning culture, the learning agenda will persistently toss on turbulent waters. Strong, visionary leadership can prioritize evidence and ensure that learnings are used. Myopic leadership can stonewall an evidence program, divert an evidence initiative to inconsequential points of inquiry, or banish the results to a basement repository. If key players in the organization are brainstorming reasons why evidence generation cannot happen instead of brainstorming ways to make it happen more easily, only strong leadership can get everyone on board. With time, good leadership can foster strong culture, and strong culture can keep the organization on the path even if bad leadership comes along.

TROLLS AND TRAPS ON THE PATH TO EVIDENCE BUILDING

Methodological Fundamentalism

In my research, I do a lot of experiments in which participants are randomized into different conditions. This allows me to estimate the effects of the different alternatives I study with a high degree of precision and little concern that alternative explanations might be driving the difference in results for the different groups. Randomized control trials can be highly informative in building evidence in a number of contexts. Regrettably, RCTs still are underutilized, and there often still is considerable sensitivity and resistance to using them in some contexts. In the early days, we spent lots of time explaining to various stakeholders the ethics of withholding treatment to a control group, but I do not personally find myself debating the ethical merits as much these days. I think people better understand that withholding a project from beneficiaries is bad only if it has demonstrably positive benefits, and that most programs do not have enough resources to reach everyone in the first place. As Jim Manzi highlights in his chapter in this volume, RCTs have an important role to play.

At the same time, there has been something of a cult of RCTs emerging in the evidence community: methodological fundamentalism. This religious zeal for RCTs—this rejection of all non-RCT evidence—seems almost as problematic as the apprehension some have had about using RCTs. Good evidence can come in many forms, and methodologies to create it must be faithfully followed, attuned to the circumstances and the research questions. Process evaluations, qualitative research, observational studies, and other approaches all can have a role, *if executed using strong methodological standards*. Moreover, an absolute focus on RCT evaluations can undermine research and learning on topics that are hard to study with these methods, such as studies where treatment cannot be withheld for ethical reasons or studies of issues that cannot be randomized in a practical way (e.g., historical political events). An emphasis on RCT-only research undermines credible evidence that can be generated in a variety of contexts. Much evidence generation can be obtained through other methods leading up to an RCT. Qualitative research can provide important background and contextualize quantitative work, and if you have the data to do an observational study, often it will provide important results that may inform a future RCT. The fields of econometrics and statistics have developed robust tools for dealing with data for which random assignment was not available. At the same time, while causal inference methods often provide a powerful set of tools, the focus on causal identification can at times create a "research bias" analogous to the well-known publication bias (the biased perspective that researchers gain when academic editors are biased against publishing null results).[8] Such research bias deters researchers from taking up the mantle of important research questions for which strong causal identification methods are not available, limiting research in important areas where knowledge generation is desperately needed.

The bottom line is that evidence comes in many forms. We should strive to find the highest quality research design appropriate to the question, circumstances, and problem, and to apply methodologies rigorously, but not shy away from tackling questions that add value even if the research design does not conform to some religious view on what evidence is about.

The Dark Side's Abuse of Evidence

There is a dark side of evidence: it may be co-opted in a way that seeks to deceive rather than inform. For most of my career, I believed that co-opting evidence was not possible because bad evidence could be critiqued openly in public debate in order to debunk bad methodologies or faulty claims.

In recent years, I have had experiences that changed my mind. These were experiences for which the cake was already baked; evidence, rather than being pursued to discover and inform, was curated to justify decisions that had already been made. In these cases, the curation of evidence was intended to mislead or misdirect the public about the decisions being undertaken. If the public raised questions about projected outcomes, projections were generated using contrived assumptions to demonstrate the impact that would occur or that concerns raised were unfounded.

Much of this "evidence" was not evidence at all but, rather, just numerical tricks in the guise of evidence. In some cases, actual programs of evidence generation existed, but in those cases, spokespeople abused the programs.

Why did any of this happen? I am not sure I know all the reasons, but I will provide a few observations. In some cases, decision makers used processes to quell any questioning of the results that were presented. They ruled questions out of order, or ignored them, or asked the critics to discuss the issue in a sidebar conversation that never materialized. This failed to give critics a venue to raise legitimate objections. Where critics voiced concerns, the proponents of policies could easily marginalize or bury queries in a mountain of paper and talk, weaponizing process control.

Crucially, the lack of an independent watchdog to call foul created the conditions for such abuse to take place. This may be a byproduct of modern media, the busy lives of ordinary citizens, and a lack of the requisite head-space to fully understand the implications of thousands of pages of policies and proposals. A lack of local coverage—with all media attention grabbed by sensational national headlines—in this new information order, a deep dive into complex policies or local issues does not gather much attention. The lack of local issue coverage is of extreme concern. Many decisions that can make people's lives better or worse happen at the local level, but the demise of local press over the past few decades means little is monitored or dug into deeply. More generally, the truth often is in the details, but it is hard to communicate deep truths in a 280-character limit Twitterverse.

More evidence sorcery lies in crafting questions to curate proof points for a desired position. This includes survey questions like: "Do you feel this project is: a) a great project; b) the best project ever; or c) all of the above!" Decision makers can control the evidence-generation process to forestall asking any questions that might be meaningful or challenge a course already decided upon.

This is not a full list; other tricks abound. The important thing to note is that all these efforts are evidence in name only. Attempts to deceive rather than inform fly in the face of any acceptable standard. The evidence community should pay attention, as this tarnishes us all, whether or not we personally sully our hands. These approaches use the banner of evidence to deceive or misinform, often waving this banner in front of folks who are not evidence gurus and may not be sufficiently trained to debunk details of sham findings. They may notice something smells wrong with the "evidence" but may not have the snout to ferret out what is rotten.

It is the evidence community that needs to find a way to police such shenanigans and help regular people understand that this is not what evidence is intended for. We may not want to be the beat cop, but it might be a role we cannot forgo. The evidence community may need to develop ethical standards that weed out those that distort evidence and evidence methods to the detriment of constituents.

In my view, and the view of the true professionals in the field, evidence is supposed to be for the people. Our tools are intended to find new ways of helping make lives better. That is true whether or not the folks we are trying to help have the background to understand regression analysis or causal chains. It has become more and more obvious that the evidence community has not been doing enough to democratize the evidence we ourselves generate, by bringing stakeholder voices into each step of the process and ensuring we listen to them as well as communicate our research findings. That is a tall order, for certain. Generating evidence is hard enough, and some of the masters are neither skilled at nor feel they have time to listen to participant input and circle back with findings. But they should.

We need to go even further. We need to hold ourselves and others accountable for the evidence (or "evidence") they generate. We need to fight against the dark side to avoid the tyranny of fake "evidence."

The No-Evidence Trap

Entities that do not have robust evidence programs may have, in some circumstances, fallen into a "no-evidence trap," where it is difficult to build evidence. Legislators, external watchdogs, funders, and others may decry the lack of evidence yet find that, no matter how hard they press the organization, there is little movement toward examining effectiveness.

The trap may be the by-product of incentives within and external to the organization. If generated, evidence could be used to defund rather than reform; people within the organization might feel the need to decrease transparency just to protect the organization. As frustration grows on the part of external observers, their calls for evidence may become sharp and aggressive. Such threats further provoke concerns that the evidence will be used to curtail good work rather than to increase organizational effectiveness, perpetuating a cycle where the organization is less willing to develop solid evidence and ask meaningful questions.

One can interpret the intransigence of the rank-and-file staff of no-evidence organizations in a number of ways. On one hand, it can seem that when entrenched individuals within an organization have interests in maintaining the status quo they will do anything to resist change, even if it means blocking their pro-evidence colleagues (who also are trying to serve the organization's interests). In some entities in which I have observed this dynamic, personnel within the entity seemed to have an entrenched anti-evidence culture that I found hard to sympathize with. They seemed so concerned about protecting the organization from the immediate threat of potentially derogatory evidence that they became blind to the longer-term threat of being an organization that is completely ineffective or causing harm. On the other hand, these often are well-trained professionals who have dedicated their lives to a cause, so they may view preserving the institution and its mission as paramount, regardless of effectiveness: you can't win if you don't play. Any way one interprets a reluctance to generate evidence, it is important for key stakeholders (particularly funders) to understand that the path to successful reform involves both carrots and sticks.

Traps of these kinds have arisen in all sorts of organizations. The trap persists in some government entities whose very existence is contested. This also can be the case for social sector nonprofits that are caught pivoting between participant and donor demands, or feeling the squeeze of unrealistic budgets. I also have seen versions of the no-evidence trap in settings where funders have a track record of changing course and failing to update accountability measures in synch with the change in mission.

Reducing the if-then mentality ("if results are not demonstrated, then resources will be cut") will be key to building better evidence in cautious sectors. It certainly is difficult to engage in honest research when so much is on the line. Evidence needs to be built within a partnership between implementors and their funders. It is an exercise in innovation but, at its best, also a joint exercise with stakeholders in discovery.

BUILDING THE NEXT GENERATION OF EVIDENCE

The past decade or two have reinforced the link between evidence and political democracy. Over that period, citizen belief in the existence of objective truth has wavered, and the nation's ability to agree on basic facts has thrown us all into a tailspin. It seems so much of this effect is based on a lack of appreciation for evidence. If we could just tie down facts and evidence, there might be more consensus about the problems we face and the solutions that are feasible. Of course, it is more complicated than that.

The politics of evidence at the national level—in our current political climate—are problematic to say the least. On one side, you have folks who are unwilling to budge an inch on hard-won programs. On the other side, you have folks who want evidence simply so they can dismantle programs and who are equally willing to filter out any inconvenient truths. While there are definitely some heroes who have tried to chart a middle course, there are too few honest partners. An evidence-based approach to the issues recognizes not only that a given problem exists within society (e.g., people are hungry or children are not doing well in school) but also that existing methods can be improved, restructured, reformed, or reorganized to better meet society's goal (e.g., fewer hungry people). The measure of success of a policy or program should not be whether or not a difficult-to-achieve goal (such as complete eradication of world hunger) has been attained without considering the many millions or billions of people who may have been helped out by imperfect interventions. Unfortunately, a balanced view like this increasingly comes across as contradictory in our modern politics: one part heresy for each tribe.

Society's problems do not just disappear when one ignores or manipulates evidence: people are still hungry or unemployed, or students still are falling behind. The only thing that happens when citizens' needs are not met is that people lose faith in society. Ultimately, government and institutions are a reflection of the preferences of people, so a loss of faith in institutions runs the risk of becoming a loss of faith in the entire economic and political system. After all, institutions create the setting in which an economy can prosper and meet the needs of its people, and where individuals have protected rights.

Evidence may not solve the crisis in democracy on its own, but it certainly can provide some fundamental truths on which to latch. The more decision making can be attuned to actual facts rather than political ideologies, the more likely we are to find common ground. Maybe that is too idealistic. Maybe our ideologies trample all reason and love of country. But I do believe that, if we are able to find the right track, evidence must guide us.

Society faces enormous challenges, from dealing with climate change and its effects to dealing with life-altering consequences of new technologies to health issues; racial and wealth inequities; changes in the structure of work; evolving demands on social programs; persistent challenges of how to best educate the next generation; and so on. As life continually becomes more and more complex, and financial and other resources become more and more pressured, the inefficiencies of ineffective policies and programs become ever more difficult to overlook. Evidence, if thoughtful, strategic, and well executed, can illuminate the path so we can focus on getting from Point A to Point B rather than tripping over ourselves in the dark. Evidence itself seems a necessary though not sufficient condition for democracy to prosper.

The best organizations (and institutions and democracies and economic systems), the ones that are going to be leaders in developing the next generation of evidence, are going to ask real questions and build interesting and stimulating environments for building evidence. Environments that attract people with key skills, environments that those folks relish—not simply because of the paychecks they pocket but because their work—both stimulates their synapses and is valued. These organizations will provide a place where evidence-generation gurus get a voice in how projects and policies are conceived and implemented rather than being told to stay in their lane. These folks will be a key part of the leadership and management of the organization, and their impact on the organization will grow over time as learning leads to deeper and deeper understanding of the ways things work.

Organizations that are next-generation leaders are going to be the ones that embrace continuous evidence approaches; that take a strategic approach to evidence generation; and find ways to keep those in their field honest. These leaders will include evidence experts throughout all stages of *both* project development and process improvement. They will understand the importance of equity in their activities and give those experts a real voice. These organizations will not only offer their ear to advice from evidence producers but also give their own internal experts a real career path within the organization, one that allows them to rise to the highest ranks of leadership. No one wants to hit a glass ceiling, which signals they are not valued. At the same time, evidence experts will not truly be impactful within an organization if they are a bunch of folks with hammers and screwdrivers looking for things that look like nails and screws. Rather, these internal experts need to be attuned to the organization's goals and mission. They will need to serve as effective translators between social science, methods, and practitioners, and they will need to be adaptable.

Scale is important in all of this, and the idealized state alluded to in this chapter is not for every entity. Small organizations can use evidence to get better at their work, but it is not reasonable for them to become evidence-first institutions. Yet, as we have discussed, for larger organizations, particularly those that affect large groups of people or enact policies that have wide-ranging effects on people, the economy, or other aspects of society, marginalizing evidence generation can create distortions that hurt people and society and can undermine trust.

For those organizations, it is time for reform, even if it is painful. It is time for action. It is time to shake up old norms and bad habits and take evidence seriously. The challenges and crises we face are too extreme to ignore. The resources are too scarce, making the inefficiencies too glaring to gloss over. It is time to turn on the lights and stop stumbling in the dark. Society, and lives, may depend on it.

NOTES

1. Even if the answer is yes (it does work), we have probably missed an opportunity to identify something better.

2. Three big-picture methodological misalignments are: omnibus program evaluations, overstating of findings, and failure to take account of heterogeneity. These are three examples of where one could still go wrong even when generating evidence using a credible theory of change. "Omnibus" program evaluations lump all activities under "the program," leading evidence generation to fail to distinguish the effects of different program components (for example, your grants program may be working, but it could be undermined by a counterproductive training program); approaches that do not allow one to assess what parts are working and what parts are not also can lead to a dead end. Researchers and practitioners commonly overstate the generalizability of findings when, in fact, the activities work only in a really specific set of conditions. This can lead to erroneous conclusions that a certain activity

works everywhere, leading to misdirected programmatic effort and inconsistent results. And a failure to account for heterogeneity among beneficiary groups is one of many factors that also can lead to inconsistent results (see, for example: Christopher J. Bryan, Elizabeth Tipton, and David S. Yeager, "Behavioural Science is Unlikely to Change the World without a Heterogeneity Revolution," *Nature Human Behaviour* 5, no. 8 (2021): 980–989.

3. Standing data-collection capacity often provides the opportunity to evaluate not the total cost of research initiatives (which typically have high fixed start-up costs) but, rather, the marginal cost of a new project. With standing capacity, negotiations with contractors for data collection can be on more favorable cost terms. In my own efforts, standing capacity has reduced timeframes considerably (from timescales in years to timescales in months or even in weeks), and the reduction in time and costs can be enough to take time and cost issues off the table. Internal expertise also can be helpful with cost cutting, by allowing internal experts to in-source key parts of the production process, and also can provide other benefits.

4. For example, tax and investment incentives that have been used to encourage, preserve, or create affordable housing might lead to less housing affordability if higher-end units are put into the development at a higher rate than affordable units (thus diluting the prevalence of affordable units in the community, reducing income diversity, and raising average housing expenses in the community overall). In other contexts, unemployed workers may desperately seek to augment their skill set with free job training, but if workers are acquiring antiquated skills in dying industries, they may invest their time poorly and find little or no benefit in the job market. Initiatives that seek to promote environmental preservation may backfire if they increase negative attitudes toward the cause of interest because they are punitive for households that do not have the option to adjust their lifestyle, say, due to a disability.

5. See chapter in this volume. [AU: Why is this highlighted? Which chapter?]

6. House of Commons, June 23, 1925, https://api.parliament.uk/historic-hansard/commons/1925/jun/23/finance-bill-1#S5CV0185P0_19250623_HOC_339.

7. David A. Garvin, "Building a Learning Organization," *Harvard Business Review*, July-August 1993, https://hbr.org/1993/07/building-a-learning-organization.

8. Chalmers Lain, "Underreporting Research Is Scientific Misconduct," *JAMA* 263, no. 10 (1990): 1405–1408; Phillipa J. Easterbrook, Ramana Gopalan, J. A. Berlin, and David R. Matthews, "Publication Bias in Clinical Research," *The Lancet* 337, no. 8746 (1991): 867–872; Annie Franco, Neil Malhotra, and Gabor Simonovits, "Publication Bias in the Social Sciences: Unlocking the File Drawer," *Science* 345, no. 6203 (2014): 1502–1505.

FACING EVIDENCE FEARS

FROM COMPLIANCE TO LEARNING

CHRISTOPHER SPERA

While there are hundreds of programs across the United States to tackle a variety of health and social policy topics, as Jon Baron, formerly of the Arnold Foundation states, "U.S. social programs, set up to address important problems, often fall short by funding specific models/strategies ('interventions') that are not effective."[1] A program is a strategy or intervention that has been provided to a group of people to achieve a desirable consequence; in short, programs are developed and implemented to solve a social problem. Given this, the question becomes how society at large can implement more high-quality program evaluations to ensure the public is receiving interventions that are proven to work or, at a minimum, hold promise to work. More specifically, the question becomes: How can the field use a learning framework (versus a compliance framework) to evaluate programs and continuously build evidence with a research and development mindset? Here, I tackle this by looking at three core issues: 1) common challenges to conducting evaluations; 2) viewing evaluation in a learning versus an accountability lens; and 3) increasing internal capacity of organizations to evaluate their own programs and/or oversee a high-quality external evaluation. I will discuss practical experiences and observations I have had in my time as an evaluator in the field for over twenty years at ICF International (formerly Caliber Associates), AmeriCorps, and Abt Associates.

Common Challenges to Conducting Evaluations. While there are numerous challenges to conducting evaluation, in my experience, there are four core ones that need to be tackled to advance the field. The first common challenge is the notion or feeling from several stakeholders that evaluation is too expensive and will divert resources from program delivery. I have heard this notion repeatedly, especially during my time as the director of research and evaluation at AmeriCorps (formerly the Corporation for National & Community Service). The general feeling is that social programs are underfunded from the get-go, so why would you take additional funds away from program delivery to evaluate the program? My response to folks who ask this is to ask them whether they would invest their personal savings in investments without any data on whether those investments would generate a return. They often then quickly seem to understand. In general, we need to continue to look for ways to generate investments in evaluations—which include everything from congressional set-asides to philanthropic investments to enhanced internal capacity within organizations that require evaluation. In short, there is proportionately very little money for evaluation compared to the amount of money put into program delivery; my estimate is less than a tenth of 1 percent, but that is just a guesstimate. It should be somewhere between 3 to 5 percent to really have a cadre of evidenced-based programs to turn to when we need them.

The second limitation is the belief by stakeholders that evaluation is too complex and too hard to understand to put to practical use. Randomized controlled trials are somewhat easy to understand in the sense that one group gets the intervention/program and the other does not and you compare the differences in outcomes. However, once you enter into the world of quasi-experimental designs that require statistical matching and techniques to generate impact estimates that control for extraneous variables, it becomes hard to understand for policymakers and others not trained in evaluation techniques. In short, we need to find a way to educate folks or develop reports that provide more concise evaluation findings and methods in a way that is easier to understand. There have been pushes by federal agencies, such as the Centers for Medicare and Medicaid Services (CMS), to make some progress here in terms of presentation of findings in infographics. Despite this, researchers are incentivized to publish in academic journals, and sometimes the incentives to push toward actionable evidence in user-friendly language are less than needed.

A third limitation is the belief that evaluation will burden staff. It is true that program staff, especially those doing service delivery, are overburdened, but evaluations often can be done with minimal interference with their daily duties and responsibilities. I have found that evaluations that rely on staff administering measures to participants often can result in too much for staff to handle, which leads to issues; however, sometimes this is necessary to complete an evaluation.

The most common challenge to the use of high-quality evaluations is the fear by program staff and funders of what happens in the event of "negative" or "null" findings. The concern is related to the implications for the program—up to and including elimination—if the evaluation shows it generates no impacts. This creates the feeling that the program is taking a risk by conducting an evaluation. The counterforce to this over the last several years has been program funders simply requiring evaluations to continue their funding, making the risk-reward scenario very different. In addition to this counterforce, viewing the program evaluation from a learning versus an accountability lens becomes very important, which is our next topic.

Viewing Evaluation through a Learning versus an Accountability Lens. When I teach program evaluation each fall semester at Carnegie Mellon University's Heinz School of Public Policy, I tell my students early in the semester that there is an inherent tension between using evaluation for learning and program improvement versus accountability and funding decisions, describing it as a tug of war. This is not a new concept as Michael Quinn Patton (2008) described utilization-focused evaluation as designed to answer specific questions raised by those running the program,[2] and Michael Scriven (1980) described judgment-focused evaluation as the technique evaluators use to make determinations about the value or worth of programs.[3]

I recall when I first started in my role as the head of program evaluation at AmeriCorps, program directors of national programs ran the other way when seeing me in the hallway for fear I would evaluate their program and hold them accountable. I quickly changed this by meeting with them and asking them questions like: What do you want to learn about your program? What keeps you up at night? How can I help? This quickly changed their attitude toward evaluation, and they began to see it as a tool for learning and program improvement rather than an up-or-down vote on their program. This takes trust that can be built only over time, but it worked then.

I also know that several other federal agencies have begun to develop and implement learning agendas designed to have evaluators create a collective roadmap for helping programs better understand their progress. While there always will be an accountability lens for programs due to the funding going into them, using the learning lens is more productive and can be used to generate more evaluations in the long run to determine what works. I would argue that program evaluation should be used for program elimination only when programs have run their course trying to improve fidelity and program delivery and cannot make any progress with impact. Very few are at this point, and viewing evaluation as a learning tool can help make progress.

Increasing Internal Capacity of Organizations to Evaluate their own Programs and/or Oversee a High-Quality External Evaluation. A final challenge to nurturing more high-quality evaluations is the lack of internal capacity within organizations that need to evaluate their program. I see this routinely in my work at Abt Associates and saw it also in my former work at AmeriCorps. For a program to implement an evaluation, it requires an individual(s) within the organization who can design an evaluation solicitation, hire a local or national evaluator, and oversee their work. External evaluators also need help from internal teams and stakeholders to design a strong evaluation and interpret results when they are available. An alternative option to an external evaluation is for a unit within the organization that manages monitoring, evaluation, and learning to be responsible for conducting an internal evaluation. Either way, this requires evaluation capacity within the organization. In my experience, there is a paucity of evaluation professionals within the federal and state governments relative to the need (except for a few exceptions, like the Department of Education) and even more so in the philanthropic sector. The recently completed Social Innovation Fund (SIF) requires internal capacity within organizations, which is a promising shift. Building internal capacity will be key moving forward.

In conclusion, in my twenty years within the program evaluation field, I have seen significant progress. More programs have been evaluated, evaluation demand has spiked and surged, especially during the Obama administration, and new techniques such as rapid cycle evaluation have been woven into the fabric of the field. With that said, when you compare the progress the field of program evaluation has made in twenty years and compare it with the progress made in technology, for example, I feel like progress has been slow. Too many policy and programmatic decisions are made based on personality and politics versus evidence and data. As we move into the next decade of "big data everywhere," I am hopeful the field will begin to gallop ahead and the use of evidence to drive major program and policy decisions will become the norm versus the outlier. As a field, we need to embrace this new wave of data everywhere, regardless of whether the data was initially gathered for research purposes, and more quickly harness it to improve programs that help the lives of our fellow citizens. I am excited and optimistic for the future!

NOTES

1. Prepared Statement of Jon Baron, President of the Coalition for Evidence-Based Policy, at a hearing before the U.S. House of Representatives, July 31, 2013, https://www.govinfo.gov/content/pkg/CHRG-113hhrg81981/html/CHRG-113hhrg81981.htm

2. Michael Quinn Patton, *Utilization-Focused Evaluation* (Newbury Park, CA: Sage Publications, 2008).

3. Michael Scriven, *The Logic of Evaluation* (Iverness, CA: Edgepress, 1980).

HOW TO BETTER MEASURE POVERTY

THE COMPREHENSIVE INCOME DATASET

KEVIN CORINTH AND BRUCE D. MEYER

INTRODUCTION

In 2015, Kathryn Edin and Luke Schaefer released their book, *$2.00 a Day: Living on Almost Nothing in America*, making the extraordinary and widely disseminated claim that over 3 million children in the United States live on less than $2 per day.[1] A report of the Special Rapporteur for the United Nations Human Rights Council in 2018 declared that "5.3 million [Americans] live in Third Word conditions of absolute poverty."[2]

If millions of Americans—many of them children—were truly living in a state of deprivation as bad as that faced by the poorest people in the poorest countries in the world, our entire approach to alleviating poverty in the United States would need to be reevaluated. Not only would policymakers need to quickly mend the scandalous holes in the safety net, but on-the-ground interventions by social service providers would warrant an overhaul. Social service providers would need to shift their focus to ensuring that families could simply subsist before they could even consider the possibility of helping them climb the ladder of opportunity.

Fortunately, it turns out that the shocking claims of extreme poverty in America portrayed by Edin and Schaefer and the Special Rapporteur to the

United Nations were wrong. Thanks to the Comprehensive Income Dataset (CID) Project—an unprecedented effort to link government surveys with dozens of sources of administrative data on taxes and government program receipt—we could, for the first time ever, accurately measure incomes at the very bottom of the distribution. When linking all the data sources together and including all sources of income, we could no longer find a single child in the linked dataset living on less than $2 per day.[3] While deprivation is very real in the United States, it does not, in fact, rival the severe levels of deprivation found in the poorest countries in the world.

This example illustrates that understanding the big picture of deprivation in the United States is imperative, not only for policymakers but also for social service providers. Unless we know who suffers from the highest levels of deprivation and the types of challenges they face, it will be difficult to ensure specific interventions are targeted to those most in need, and that they focus on the biggest problems. Nor will we be able to assess how the successes of practitioners and communities add up to overall societal progress.

Unfortunately, existing evidence on the big picture of disadvantage in the United States is inaccurate and incomplete. It relies on surveys that suffer from growing reporting errors and misses some of the most vulnerable segments of the population.

The CID Project seeks to improve on the existing evidence by creating the most accurate dataset on economic well-being ever created for the United States.[4] The CID combines survey data with an unprecedented set of administrative data from tax filings and government programs. We conduct rigorous research and apply cutting-edge statistical techniques to combine these data sources in a way that maximizes the accuracy of our well-being measures. We also are able to capture populations, such as those experiencing homelessness, who are missed from most major household surveys. As a result, we are able to create a much more accurate and complete picture of disadvantage in the United States than has ever before been available.

The CID benefits from broader ongoing efforts in the federal government. The Evidence-Based Policymaking Commission and the Evidence Act sought to make data more secure, more available, and more widely used for evaluation inside and outside of government, furthering efforts like the CID. Specific commission recommendations that have yet to be implemented, such as those on improved access to state data, would further strengthen research and evaluation efforts.

Ultimately, the evidence generated by the CID will inform the areas of greatest need, so practitioners, communities, and policymakers can tackle the biggest problems. And by providing highly accurate evidence on changes in disadvantage over time, we can measure how individual efforts add up to overall societal progress.

PROBLEMS WITH EXISTING DATA ON DISADVANTAGE

Government surveys are used extensively by federal agencies and researchers to assess the extent of disadvantage and broader measures of well-being in the United States. For example, the U.S. Census Bureau's annual report on the official poverty rate, median income trends, income inequality, and health insurance coverage relies on the Current Population Survey Annual Social and Economic Supplement.

Unfortunately, surveys suffer from several problems that reduce their accuracy and completeness. First, people increasingly fail to respond to surveys, which can lead to difficulty in attaining a sample representative of the U.S. population. Second, those people who continue to respond to surveys may provide inaccurate information. A vast body of research has shown that many categories of income—such as means-tested benefits, social insurance, and private pension income—are greatly underreported in surveys.[5] Third, particularly vulnerable segments of the population (for example, individuals experiencing homelessness and unauthorized immigrants) are either under-surveyed or missing from surveys altogether. Fourth, some surveys completely exclude certain types of income, such as housing assistance and capital gains. These issues are likely to bias any survey-based analyses of income distribution, poverty, inequality, and the effects of government taxes and transfers. As a result, existing evidence on the most disadvantaged Americans is biased and incomplete, limiting the ability of service providers and policymakers to target programs to those most in need.

The problems with survey data have led some researchers to turn to administrative data, instead. But administrative data sources on their own do not capture the full set of resources available to individuals, and they do not contain the rich demographic information available in surveys that enables a focus on vulnerable groups.

THE COMPREHENSIVE INCOME DATASET

We are building the CID to overcome the inaccuracies in our basic understanding of economic well-being in the United States. The fundamental insight of the CID is that no single data source on its own can provide a full or accurate measure of income or well-being. But when multiple data sources are linked, the strengths of each data source can be harnessed while overcoming its limitations. The CID relies on three main types of data sources—household surveys, tax records, and federal and state administrative program data on government benefits. Each data source has unique strengths. Surveys provide rich demographic information that allows for the construction of families and analysis by race, educational attainment, and other characteristics. Tax data contain highly accurate information on certain income sources, such as earnings, and have near universal coverage, including many non-filers whose tax forms are supplied by employers and government agencies. Administrative data from government programs provide income information that is not captured well or at all by surveys or tax data.

We link all data sources at the individual level using anonymized identification codes created by the Census Bureau to ensure the confidentiality of personal data. We conduct rigorous research and apply cutting-edge statistical techniques to impute missing data, and also to inform broader conceptual decisions about how to optimally combine data sources. For example, we are pioneering a new methodology that uses a novel set of dozens of material hardship measures—such as housing quality problems, food insecurity, and mortality patterns—to validate decisions on how to construct a comprehensive measure of income. This evidence-based approach for constructing income measures represents a major step forward for the income measurement field, and it will ensure our comprehensive income and poverty measures are as accurate as possible.

The CID Project has, to date, linked together four household surveys with an extensive set of tax records and twelve sources of federal and state administrative program data—to our knowledge, the most comprehensive set of linked income-related data ever created for the United States.

EVIDENCE GENERATED BY THE COMPREHENSIVE INCOME DATASET PROJECT

As we continue to build and improve the CID, we already are demonstrating its power to transform our understanding of poverty, income, and well-being in the United States. To this end, our two major strands of research to date focus on: a) identifying those who are most disadvantaged and their levels of deprivation, and b) understanding a particularly vulnerable population missed by most surveys—individuals experiencing homelessness.

As previewed in the beginning of this chapter, one of the earliest papers using the CID examined the prevalence of households in the United States living on less than $2 per person, per day (that is, "extreme poverty").[6] We focused on extreme poverty in our early research because the results starkly demonstrate the capacity of the linked data to change our understanding of poverty; in this case, due to the presence of survey outliers. We find that more than 90 percent of households with survey-reported cash incomes below $2/person/day are not in extreme poverty once we include in-kind transfers, replace survey reports of earnings and transfer receipts with administrative records, and account for the ownership of substantial assets. Contrary to widely cited findings in the prior literature that over 3.5 million children live on less than $2 per day in the United States, we find no children in our surveys falling below such an extremely low standard when using the CID and making the aforementioned improvements.

In research in progress, we extend our CID-based analysis of poverty to standards more applicable to the modern United States, a project directly relevant for informing improvements to the widely monitored official poverty measure estimates. In particular, we examine how poverty rates change when using the CID to correct for measurement error in pre-tax money income and when we incorporate tax liabilities and credits, in-kind transfers, and other non-cash income sources. In addition to more accurately estimating the level of poverty (holding the original poverty line constant), we can identify a more accurate picture of the poor population in terms of family composition, demographics, and material well-being.

In related ongoing work, we use the CID to study the best way to compare the economic well-being of families in different geographic regions. While families who live in high-cost areas may need to spend more resources to meet their basic needs, they may, at the same time, have access to higher-quality public services, more job opportunities, and a cleaner environment. We pair the CID with a battery of deprivation measures (for example, material hardships, availability of appliances, home quality problems, long-term income, and mortality) to provide an evidence-based test for whether geographically adjusting poverty thresholds leads to a more deprived poor population compared to what one would see without applying geographic adjustments. This project helps illuminate the geographic distribution of disadvantage in the United States and, thus, informs where new efforts to innovate new solutions may be most needed.

We also have worked to improve our understanding of individuals experiencing homelessness, who—despite being one of the most deprived populations in the United States—are largely omitted from household surveys and, therefore, not reflected in official poverty statistics and the extreme poverty literature. We overcome this limitation by linking a census of the entire (sheltered and unsheltered) homeless population conducted as part of the 2010 Decennial Census with administrative data on tax records and government program benefits. Our research to date has shed new light on this highly vulnerable but poorly understood population.[7] For example, we find that 53 percent of sheltered homeless adults under age 65 in 2010 had formal earnings during the year, and a substantial 40 percent rate for those found on the streets. In addition, the vast majority of individuals who experience homelessness receive government benefits—89 percent of sheltered homeless adults under 65 and 78 percent of unsheltered homeless adults under 65 received benefits from the Supplemental Nutrition Assistance Program, veterans benefits, housing assistance, Medicaid, or Medicare at some point during the year. While homeless individuals have greater connection to the formal labor market and government benefits than sometimes thought, they still face low levels of well-being that improve little over time. Among all non-elderly adults who experienced sheltered (unsheltered) homelessness in 2010, less than half had more than $2,000 ($200) of annual earnings in any year between 2005 and 2015.

FUTURE OF THE COMPREHENSIVE INCOME DATASET PROJECT

While the CID already represents the most comprehensive and accurate income-related dataset ever created for the United States, we are committed to pushing the frontier as far as possible. We are linking new administrative data sources, extending the CID back in time to cover more than two decades, and developing new statistical and conceptual methods for combining survey and administrative data to maximize the accuracy of income measures.

In addition, we will use the CID to produce better evidence on disadvantage in the United States. Examples include measuring poverty over more than two decades, validating survey-based, real-time measures of poverty using the CID, examining the effects of government programs on well-being, identifying holes in the safety net, and exploring new projects on vulnerable groups under-covered by surveys, including individuals experiencing homelessness and unauthorized immigrants.

In each of these areas, the unprecedented accuracy and richness of the CID will transform our understanding of deprivation in the United States and break new ground on overlooked segments of the population. The new evidence we generate will be essential to inform practitioners, communities, and policymakers seeking to improve the well-being of the most disadvantaged members of society. Already, our research reevaluating extreme poverty has shifted narratives on deprivation in the media, and our methodological work combining survey and administrative data sources heavily informed the recommendations of a recently concluded federal interagency technical working group tasked with developing new poverty measures for the United States.[8] As we learn more about poverty, and especially vulnerable groups including people experiencing homelessness, it will be imperative to ensure both policymakers and those on the frontlines serving these groups are able to use our research to inform their decisions. Ultimately, knowing where disadvantage is most prevalent will provide the big picture needed for service providers to lead the next generation of evidence building as they innovate their services to deliver better outcomes to the people they serve.

NOTES

1. Kathryn J. Edin and H. Luke Shaefer, *$2.00 a Day: Living on Almost Nothing in America* (Mariner Books, 2015).

2. Human Rights Council of the United Nations, "Report of the Special Rapporteur on Extreme Poverty and Human Rights on His Mission to the United States of America," 2018, https://digitallibrary.un.org/record/1629536?ln=en.

3. Bruce D. Meyer, Derek Wu, Victoria Mooers, and Carla Medalia, "The Use and Misuse of Income Data and Extreme Poverty in the United States," *Journal of Labor Economics* 39, no. S1 (2001), pp. S5–S58.

4. For further information, see the CID Project website, https://cid.harris.uchicago.edu/.

5. For example, see Bruce D. Meyer, Wallace K. C. Mok, and James X. Sullivan, "Household Surveys in Crisis," *Journal of Economic Perspectives* 29, no. 4 (2015), pp. 199–226.

6. Bruce D. Meyer, Derek Wu, Victoria Mooers, and Carla Medalia, "The Use and Misuse of Income Data and Extreme Poverty in the United States," *Journal of Labor Economics* 39, no. S1 (2021), pp. S5–S58.

7. Bruce D. Meyer, Angela Wyse, Alexa Grunwaldt, Carla Medalia, and Derek Wu, "Learning about Homelessness Using Linked Survey and Administrative Data," Working Paper 28861, NBER, 2021.

8. See *Final Report of the Interagency Technical Working Group on Evaluating Alternative Measures of Poverty*, U.S. Census Bureau, www.census.gov/content/dam/Census/library/publications/2021/demo/EvaluatingAlternativeMeasuresofPoverty_08Jan2021.pdf.

NEW FEDERAL STRATEGIES TO STRENGTHEN DATA ANALYTICS CAPACITY OF STATES, LOCALITIES, AND PROVIDERS

KATHY STACK AND GARY GLICKMAN

During the Obama administration, we worked at the U.S. Office of Management and Budget, in close collaboration with White House and federal agency leaders, to launch Pay for Success and other innovative grant programs designed to encourage states, localities, and nonprofit providers to use and build evidence to achieve better outcomes for vulnerable populations. These initiatives sparked important conversations and demonstrated how federal grants could incentivize the use of data and evidence. But they did not spur broad systems and culture change in other state and locally administered programs that deliver hundreds of billions of federal dollars annually to low-income individuals and families.

To improve outcomes and address systemic inequities in the delivery of government services, the federal government must do more to help states, localities, and their nonprofit partners break down silos, pursue holistic reforms guided by human-centered design, and create data-driven feedback loops about what is working and what could be improved. To do this, strengthening cross-program data infrastructure and analytics capacity at the state and local level will be essential.

State and local program administrators and service providers can use data and evidence to understand the interactions of health, nutrition, income security, housing, childcare, education, training, and related community supports to meet client needs. The kinds of questions they should be able to routinely answer include:

- Which subgroups are in greatest need of benefits and services, and what are the best channels for reaching them?
- What mix of services and benefits is optimal for different subgroups, and how could their delivery be better coordinated?
- What outcomes are program investments achieving, by subgroup and geographic area, and what gaps must be closed to achieve equitable outcomes for underserved populations?
- What interventions have the greatest impact and cost-effectiveness?
- What upstream prevention strategies produce better outcomes and reduce downstream costs in other programs?
- What operational streamlining would improve the user experience and reduce costs?
- What major sources of improper payments are readily discoverable by merging data across programs?
- What procurement models work best to incent practitioners to achieve the best results and, at the same time, limit gaming and cherry-picking?

Unfortunately, bureaucratic processes—many of which emanate from the federal government's fragmented program structures—have severely impeded state and community capacity to focus on these questions. In our conversations with leaders of organizations that led the Pay for Success movement at the state and local levels, there is widespread agreement that one of the most significant barriers to outcome-focused innovation is the lack of cross-program data analytics capacity. Specific impediments are lack of *funding* for technology infrastructure and analytics, lack of *access to data*, and lack of *expertise* on how to use data to manage toward outcomes.

Let's examine what the federal government has done, and what more could be done, to help federal grantees overcome these three hurdles.

MAKING FUNDING AVAILABLE FOR DATA INTEGRATION PLATFORMS AND ANALYSIS

During the G. W. Bush and Obama administrations, several federal initiatives provided funding for states to integrate data across systems. The Education Department gave grants to states to build State Longitudinal Data Systems for student-level data and, under the Obama administration, required state grantees to make progress linking K-12 data to pre-K, postsecondary education, and workforce systems. After passage of the Affordable Care Act, the Obama administration used waivers providing increased federal Medicaid funding to incentivize states to create integrated eligibility and enrollment systems linking client-level data from Medicaid, SCHIP, TANF, and SNAP.[1] HHS provided financial support for modernized Comprehensive Child Welfare Information Systems to encourage interoperability and information exchanges among human services and related agencies, including Medicaid, education, and the courts.

In the first half of 2021, the Biden administration took other steps to encourage states to strengthen capacity to link data across systems serving vulnerable populations, using funds already appropriated by Congress. Treasury Rescue Plan regulations[2] included explicit authority for states and localities to use some of the $350 billion in State and Local Fiscal Recovery (SLFR) funds to "build their internal capacity to successfully implement economic relief programs, with investments in data analysis, targeted outreach, technology infrastructure, and impact evaluations." This created a financing source to enable grantees to meet the bold evidence and evaluation requirements of the regulations. The Office of Management and Budget (OMB) issued updated guidance on Evidence Act implementation[3] that defines "evaluation" to include data analysis. This clarification, combined with OMB's 2020 change to government-wide grant regulations[4] making evaluation a permissible use of program funds, allows state, local, and nonprofit grantees to finance data infrastructure and analytics capacity with existing federal funding streams. OMB also issued financial management guidance[5] on Rescue Plan implementation that encouraged agencies to adopt "innovative administrative approaches to increase efficiency and effectiveness across programs (e.g., braiding and blending)," signaling that federal funds can be pooled for data infrastructure and other program improvement activities that support multiple programs.

The Biden administration could spur further progress through widespread adoption of Treasury's two-part SLFR strategy that creates demand for data use as well as clarification that data capacity can be financed with program funds. Every major federal program could: 1) raise the standard for the quantity and quality of evidence building that grantees are expected to carry out; and 2) provide explicit clarification that program funds may be used for data infrastructure, analytics, evaluation, and targeted outreach to improve the effectiveness, efficiency, and equitable outcomes of federal investments.

The administration also could strongly encourage states, localities, and nonprofits to build efficient, enterprise-wide data analytics capacity that supports coordinated, human-centered program delivery and meets the analytical needs of multiple programs. The administration could provide technical assistance to help grantees learn ways to pool funds for cross-program data infrastructure while satisfying financial management and auditing requirements, similar to this HHS-USDA cost-allocation toolkit[6] for human services IT systems.

IMPROVING ACCESS TO DATA

State, local, and nonprofit grantees often lack access to one or more data sets that, if linked, can answer performance-related questions. This is the result of both real and perceived barriers to sharing data that largely can be overcome through the use of new technology and privacy-protecting methods for linking data held in different systems. Leading jurisdictions have created replicable models for data-linkage to support evidence-based decision making. For example, Washington,[7] South Carolina,[8] Ohio,[9] and Allegheny County, Pennsylvania,[10] have built internal capacity to securely link and analyze data across programs, systems, and sectors. California recently launched CalData,[11] a state data strategy that will integrate early childhood, K-12, financial aid, higher ed, and health and human services data. A number of states have partnered with universities to allow government data to be held in secure environments managed by universities. These include the California Policy Lab,[12] the Colorado Evaluation and Action Lab,[13] and the Coleridge Initiative, which is helping over forty states learn how to use merged cross-state and cross-agency data in the Administrative Data Research Facility[14] to analyze and improve education and workforce development strategies.

Innovative state and local governments also are building capacity to merge government data with data held by community-based organizations to improve child and family services, especially for marginalized populations. North Carolina supports NCCARE360,[15] a shared technology platform to unite healthcare and human services organizations to deliver coordinated, human-centered services and report outcomes. The Camden Coalition's[16] Health Information Exchange links healthcare and other records across southern New Jersey to better identify and serve individuals with unmet needs.

Not surprisingly, even in jurisdictions with the capacity to merge data across their systems, data gaps remain. One of the most significant examples is employment and earnings data, a critical indicator of the effectiveness of education, training, and other programs to improve economic security. Unless client-level data for people living, working, and getting services in different states can be merged, state and local decision makers will lack a comprehensive picture of how their programs are performing.

The federal government has a unique capacity to dramatically improve state and local capacity to access and link data for cross-program analytics. First, it can provide technical assistance and use cases—drawing on the examples above—about privacy-protecting methods that enable states, localities, and providers to share data while complying with an array of confusing federal privacy laws. As federal agencies implement the Foundations for Evidence-Based Policymaking Act, they are learning new ways to share *federally* held data with each other using privacy-protecting tools. Going forward, federal agencies can equip federal *grantees* with the same tools and knowledge.

Second, the federal government can provide access, with privacy protections, to comprehensive, reliable federal datasets, such as employment and earnings data held by IRS (annual income), the Administration on Children and Families' National Directory of New Hires (quarterly earnings), and the Census Bureau. Exploratory conversations have begun between federal agencies, researchers, and state and local governments about creating scorecards, modeled on the Education Department's College Scorecard,[17] which would link state and local government or provider data with federally held data to produce aggregate outcome statistics for education and training programs. The unit of analysis could be a grant program, an intervention, a training provider, a jurisdiction, or a subset of program participants (for example, based on demographic characteristics). This innovation could be a game changer if it leads to efficient, scalable processes for linking federal, state, and locally held data. In addition to generating reliable outcome data, it could significantly reduce grantee reporting burden and enable quicker, lower cost, higher quality evaluations.

BUILDING STAFF CAPACITY TO USE DATA TO IMPROVE OUTCOMES

The federal government's focus on compliance with program-specific requirements—without an equal emphasis on achieving better outcomes through cross-program integration—has perpetuated program silos and administrative inefficiency at the state and local levels. Scarce grantee staff resources are devoted to documenting compliance rather than using data to learn more effective ways to serve populations in need.

Over the past decade, philanthropy has invested in organizations that offer states, localities, and nonprofit providers outcome-focused technical assistance to improve the lives of underserved and marginalized individuals and families. This new technical assistance model helps grantees learn how to harness data and evidence to understand client needs, target outreach and services to improve impact and equity, develop innovative procurement and payment models that incentivize better outcomes, develop and evaluate more effective service delivery models, and create routine feedback loops to measure performance and adjust approaches. Some of the organizations using this model were early pioneers of the Pay for Success movement, such as Harvard's Government Performance Lab,[18] Third Sector,[19] and Project Evident.[20] States and communities that have received this new type of technical assistance have created proof points for systems reforms and strategic partnerships that are achieving measurably better outcomes for at-risk populations. For example, the Government Performance Lab helped the city of Denver design and implement a supportive housing Pay for Success project[21] that resulted in significant improvements in housing stability, reduced police interactions, and reduced emergency room visits according to a rigorous randomized controlled trial by the Urban Institute. The city is now using its own general fund resources to continue this cost-effective intervention.

The federal government can use its policy levers to increase both the demand for and the supply of outcome-focused technical assistance to states, communities, and nonprofits across the country. It can increase demand by giving priority in grant competitions to applicants that will employ outcome-focused, data-driven approaches that can be institutionalized and reused in the future. It can also provide waivers in discretionary and mandatory programs that increase grantee flexibility and reduce compliance reporting if grantees adopt innovative program designs whose results can be reliably measured. (This waiver authority was established in 2014 OMB grant regulations.[22]) To increase supply, federal agencies can work with the General Services Administration to create "schedules" of pre-approved, high quality TA providers and make it easy for federal agencies and grantees to procure their services.

PUTTING THE PIECES TOGETHER

The building blocks are in place for federal, state, and local governments—working with outcome-focused service providers—to create a shared vision and coordinated implementation strategy for using integrated data, analytics, and evidence to improve decisions that lead to better, more equitable outcomes for vulnerable populations. Exemplary state and local practices could be widely replicated in other jurisdictions if the federal government provided the needed leadership, coordination, and incentives.

While new legislation and funding from Congress might be helpful, much of what needs to be done at the federal level could be done under existing law through administrative actions. Stitching these actions together into a coherent, high-impact strategy will require White House, OMB, and federal agency leaders to prioritize, and share responsibility for, collaborating with state, local, and nonprofit grantees to strengthen their capacity to use data and evidence. Because of fragmented congressional committee jurisdictions that are mirrored in the Executive Branch, it is currently no one's job in the federal government to provide coordinated leadership to do this. As two former executives of the Office of Management and Budget who led evidence-based policy initiatives involving states, localities, and nonprofit providers, we are confident that OMB could do the job.

NOTES

1. Guidance on Funding and Cost Allocation for Health and Human Services Systems," Administration for Children & Families, podcast, April 30, 2013, www.acf.hhs.gov/cost-allocation.

2. Treasury Department, "Coronavirus State and Local Fiscal Recovery Funds," Federal Register, May 17, 2021, www.federalregister.gov/documents/2021/05/17/2021-10283/coronavirus-state-and-local-fiscal-recovery-funds.

3. OMB, M-21-27, "Evidence-Based Policymaking: Learning Agendas and Annual Evaluation Plans," White House, June 20, 2021, www.whitehouse.gov/wp-content/uploads/2021/06/M-21-27.pdf.

4. "Guidance for Grants and Agreements," Federal Register, August 13, 2020, www.federalregister.gov/documents/2020/08/13/2020-17468/guidance-for-grants-and-agreements.

5. Ibid.

6. "Cost Allocation Methodologies (CAM) Toolkit," Office of the Administration for Children & Families, January 15, 2015, www.acf.hhs.gov/css/training-technical-assistance/cost-allocation-methodologies-cam-toolkit.

7. Results for America, Data Policies Agreements State Standard of Excellence, Washington, 2020, https://2020state.results4america.org/state-standard-of-excellence/data-policies--agreements.html#washington.

8. Ibid., South Carolina, https://2020state.results4america.org/state-standard-of-excellence/data-use.html#south-carolina.

9. Ibid., Ohio, https://2020state.results4america.org/state-standard-of-excellence/data-policies--agreements.html#leading-example.

10. See Allegheny County DHS Data Warehouse website, https://www.alleghenycounty.us/human-services/news-events/accomplishments/dhs-data-warehouse.aspx.

11. See CalData page in Cal Gove Ops website, www.govops.ca.gov/caldata/.

12. See Featured News and Updates page on California Policy Lab website, govops.ca.gov/caldata/.

13. See Colorado Evaluation and Action Lab page at University of Denver website, https://coloradolab.org/.

14. "Cloud Enables Platform to Help State and Local Governments become More Data-Driven," Technology, December 23, 2020, www.govtech.com/local/cloud-enables-platform-to-help-state-573468231.html.

15. See About page on NCCARE 360 website, https://nccare360.org/about/.

16. See "Using Data to Identify and Engage Patients with Complex Health and Social Needs Is the Foundation of Our Work," Hotspot Blog page at Camden Coalition, https://camdenhealth.org/connecting-data/.

17. See Home page of College Scorecard website, U.S. Department of Education, https://collegescorecard.ed.gov/.

18. See Harvard Kennedy School: Government Performance Lab website at https://govlab.hks.harvard.edu/.

19. See Third Sector website at www.thirdsectorcap.org/.

20. See Project Evident website at https://www.projectevident.org/.

21. "Denver's Supportive Housing Social Impact Bond a 'Remarkable Success,'" Denver: The Mile High City, July 15, 2021, www.denvergov.org/Government/Agencies-Departments-Offices/Agencies-Departments-Offices-Directory/Department-of-Finance/News/2021/Independent-Evaluation-Finds-Denvers-Supportive-Housing-Social-Impact-a-success.

THE POWER OF NONPROFIT SECTOR R&D

CREATIVITY DURING COVID

NEAL MYRICK

2020 was the year of the pivot for nonprofit organizations and funders. While scrambling to deal with all the challenges, many of us in the sector noticed something encouraging: some nonprofit organizations with access to flexible capital and visionary leadership seized the opportunity to upend their business models to not just survive but to thrive through the pandemic. That success was driven, in part, by flexible funding that gave those leaders the freedom they needed to research, develop, and implement their ideas.

Data also played an essential role in tracking how well the newly implemented programs were meeting their goals. Measurement was done in near real-time because the chaos of the pandemic demanded it. Using timely data to measure program success helped foundations and nonprofit leaders navigate the daily chaos, identify priorities, and execute in ways traditional measurement and evaluation practices could not have supported.

The pandemic pivots, driven by the dueling forces of survival and ambition, taught us something. When encouraged and funded well, we saw that research and development and the real-time use of data could help nonprofit leaders solve our world's most complex challenges—even during the toughest of times. It should be more evident than ever that this model for supporting and driving continual learning and change isn't just suitable for a pandemic response—it is a best practice for future social impact work.

I am encouraged by what I see. I have spent the last eighteen years in the nonprofit sector, including six years leading Tableau Foundation, focusing on building data capacity in nonprofits in more than 120 countries. When we started Tableau Foundation in 2014, I pledged to provide unrestricted, multiyear funding and let our nonprofit partners lead, recognizing that they are the experts in solving problems in their communities. Our unrestricted funding, along with software and training grants, allows courageous nonprofit leaders to use data for research, development, and tracking the impact of the innovation they scale. The pandemic was brutal, but I am more encouraged than ever.

One of the courageous leaders who led her organization through a massive pivot was FareStart CEO, Angela Dunleavy. FareStart's mission is to transform lives, disrupt poverty, and nourish communities through food, life skills, and job training. Before the pandemic, they operated restaurants, cafes, and a catering business, serving 950,000 meals each year. The pandemic, however, shut down nearly every line of business. Yet today, FareStart is serving more meals in a more mission-aligned way than ever before. How did they do it?

I remember first hearing about FareStart's challenges in February 2020. We heard that Farestart was pivoting its business to focus almost exclusively on delivering packaged meals to shelters, homelessness organizations, and schools. Shifting to packed-meal delivery would keep FareStart's staff employed while also meeting the needs of thousands of people at risk of going hungry. So, we asked how we could help.

Their ask was simple. FareStart raised funds to provide three days of meal service per week and needed $137,000 to scale to seven days of service. Filling that gap seemed like a great idea, but I had one question: Who was funding the capacity they needed to pivot their entire business model? I could see supporting the seven-day meal service, but I also wanted to understand how they would have the ability to innovate so much in so little time.

Answering my question on March 10, they wrote, "The administrative staff time spent on capacity building and response, as you know, is not something we have funding for." They estimated it would cost $183,350 to shift their business model and plan for closing down some programs while scaling up food delivery—all while managing the risks the pandemic posed to their staff and communities.

One day later, on March 11, Tableau Foundation approved an unrestricted $320,000 grant. We wanted to help them move to a seven-day service, of course. Food insecurity was peaking. But, more importantly, we wanted to help FareStart have the administrative capacity to do the R&D required not just to survive but to thrive through the pandemic. The result was a win for everyone.

FareStart delivered 950,000 meals in 2019 and more than 2 million in 2020. They provided weekly updates using data to show exactly how many meals were delivered to which organizations and where there were still needs. The data told a fantastic story about how FareStart was successfully pivoting and, more importantly, that they understood and could respond to an ever-changing environment. Their ability to communicate their command of the situation built trust and confidence among their stakeholders, positioning them to respond to the crisis while creating a new business for a world that would no longer be the same.

Imagine if we and all their other funders had restricted grants to providing meals, not allowing any funding for FareStart to develop the essential, complex business models. The support would have fed people, but FareStart would have struggled to leverage their staff's ingenuity to turn their emergency response into a long-term, scalable business.

When speaking with Angela about this in early 2021, she said the pivot not only allowed them to respond to the needs of the community; it also allowed FareStart to "align our work more closely with our mission by reaching more vulnerable people in our communities." Their pivot worked. It doubled the number of meals they provided and allowed them to build for the future in a more mission-aligned way.

FareStart and many others that successfully pivoted proved that we as a sector can turn on a dime and have more impact as a result. They confirmed that, when adequately funded, R&D and the use of near real-time data play an essential role in helping nonprofit leaders tune their organization's service delivery to more closely meet the needs of the communities they serve.

Supporting these entrepreneurial efforts does not just benefit the outcomes. It also has a multiplying effect on organizations, including improving staff morale when their innovative ideas are listened to and acted upon.

In the book *Intrapreneurship: Managing Ideas Within Your Organization*, author Kevin C. Desousa writes, "Employees that engage in the innovation process, especially in the generating, advocacy, and experimenting with ideas and emerging concepts often find themselves more connected to the organization. While it may sound paradoxical, the staff actually gain more when they play the role of idea advocates for their peers' ideas and participate in experimentation processes with new concepts than when generating and/or advancing their ideas."[1]

Once acted upon, using data to track the success of newly implemented ideas on a near real-time basis allows everyone—staff and funders alike—to witness the sensation as it happens. Seeing success as it happens helps build morale, trust, and confidence in organizations and their leaders.

When considering the power of R&D, it is necessary to define it, because it is not discussed as broadly in the nonprofit sector as innovation is. And innovation is one of those buzzwords everyone talks about but understands differently. The word is often used interchangeably with R&D, putting at risk the progress we could make if the concepts were more universally understood and appropriately funded.

Stefan Lindegaard, the founder of Growth Mindset Lab, once wrote that R&D "turns money into knowledge."[2] Translated into nonprofit terms, R&D is the process of turning funding into knowledge. When knowledge from R&D gets turned into a new product, program, or process, it becomes an innovation. Thinking of innovation as a noun helps keep it differentiated from R&D.

When we funded FareStart's administrative capacity to pivot, we funded R&D—their ability to adapt, experiment, and use their best thinking to develop new ways of serving new constituents. When they turned their knowledge into a new packaged-food delivery business model that more than doubled the number of meals they delivered, their knowledge became an innovation.

Tableau Foundation has always funded R&D and the active use of data to measure whether innovations achieve expected results. We fund these projects expecting them to be successful. Each time we make funding decisions, we ask ourselves, "What is our plan when this succeeds?"

We think this is important because, far too often, we hear funders say they support innovation but not scale. They actually are supporting R&D that produces knowledge about how to solve a problem. Still, they have no intention of funding the process of turning that knowledge into an innovation that works at scale in the real world. That approach encourages "pilotitis," which is the incessant funding of "innovation" without any plans for scaling innovations that work. It is killing our sector, yet it continues to happen with no end in sight. My friend and colleague Kate Wilson, CEO of the Digital Impact Alliance at the United Nations Foundation, wrote a great article outlining some best practices for avoiding the condition.

One great example is with our partner CEPEI, a global think tank based in Bogota, Columbia. Founded by Philipp Schönrock in 2003, CEPEI has provided nearly twenty years of policy solutions and insights to optimize global leadership engagement on governance, finance, and data for sustainable development. In 2019, Tableau Foundation supported R&D efforts to help CEPEI build new knowledge about improving the data ecosystem to help drive the United Nations Sustainable Development Goals (SDGs) across Latin America and the Caribbean.

When we first discussed a partnership, Philipp had clear goals and a few ideas for building the data ecosystem, including an idea for a new data center to help develop a data culture across Latin America and other countries. He knew continual measurement and learning could help. Still, he needed to research and develop his ideas to determine precisely how data analytics and visualization could support his vision.

We provided a small software grant and unrestricted funding to support CEPEI's R&D efforts in early 2019. Months later, in November 2019, Philipp stopped me at a conference and said our unrestricted support gave him the "liberty to experiment," noting that the freedom was transformative and unlike anything he had experienced before. Little did we know how transformative his R&D efforts would, indeed, be.

At the beginning of 2020, Philipp and I discussed another investment to help CEPEI implement a new program built from the knowledge he gained through his R&D efforts. But, by March 2020, everything had changed. Like Angela at FareStart, Philipp realized that merely adjusting to the new normal was not going to be enough. Instinctually, he knew he needed to leverage the pandemic's dark power to revolutionize CEPEI's work. Doing so would meet his stakeholders' needs and allow CEPEI to come out of the pandemic stronger and positioned for having a more significant impact.

Philipp acted quickly. Through their R&D efforts, CEPEI learned how governments and organizations could use timely data to transform decision making and improve transparency. The CEPEI team turned that new knowledge into an innovation—a COVID-19 data and innovation center that would help the United Nations and other partners use data to monitor progress against COVID-19 response and recovery goals.

CEPEI knew that demonstrating the effectiveness of this approach during the pandemic would have an immediate impact while providing evidence that using timely data to guide decisions was an approach that should work all the time. Therefore, their pandemic response positioned CEPEI to scale an innovation long into the future.

When Philipp pitched us his new idea, we knew three things. First, the R&D efforts we supported in 2019 were about to take root in a powerful way. That was exciting and gave us confidence in the data center idea.

Second, turning knowledge from the R&D effort into a valuable innovation takes time, effort, and money. Therefore, we needed to provide seed funding to turn the concept into something tangible.

And third, the work required to adapt the data center to meet the global community's needs during and after the pandemic would not be fast or easy. Therefore, if it was to be successful, we needed to commit to funding it for more than one year. We understood that we were not investing in a pilot project; we were investing in a leader, an organization, and an idea that would require continual adaptation. The result was fantastic.

CEPEI's vision came to life on December 3, 2020, when CEPEI, Tableau Foundation, the United Nations, and more than twenty other partners launched the COVID-19 Data and Innovation Centre. The Centre's purpose is "to deliver information, evidence, knowledge, innovation strategies, territorial requirements, and policy recommendations to the UN COVID-19 Multi-Partner Trust Fund in their purpose of strengthening response and recovery actions in the Global South."

As UN's deputy-secretary general Amina Mohammed said during her remarks at the Centre's launch, "The Centre's data will help tell the story of what is happening across geographies, including the Global South, economic sectors, and diverse groups of people. It is my hope that the COVID-19 Data & Innovation Centre will thus enhance the openness, transparency, and quality of what we do across analysis, programming, and results monitoring. And in doing so, it will shed light on new paths forward and ways to tailor our assistance to help those falling behind."[3]

Looking back, I realize that, at its simplest, Philipp provided vision, leadership, and courage. We provided support for a robust data infrastructure and unrestricted funding with trust.

Thankfully, we were not the only foundation to provide flexible, trust-based support during the pandemic. Many foundations leveraged the pandemic's urgency to loosen grant requirements, offer more flexibility, and reduce reporting requirements. By March 2020, the Council on Foundations, Ford Foundation, and more than forty other philanthropic organizations led the call for funders to commit to more flexible funding to respond to the pandemic. During the announcement, Kathleen Enright, president, and CEO of the Council on Foundations said, "One of philanthropy's key strengths is its ability to pivot and adapt as circumstances require. That flexibility is needed now more than ever."[4]

Pia Infante, steering committee chair of the Trust-Based Philanthropy Project, explained in the same announcement:

> The reason a trust-based approach works in this short-term, emergency frame is that it works for the long term. When we deeply resource our leaders, organizations, and movements, we enable the adaptivity necessary in the chaotic, complex times we are living. It may have taken a global pandemic for some philanthropies to let go of restrictions, arduous processes, and exacting expectations of already overburdened nonprofits, but let's hope one benefit is that we now jettison all that no longer serves the greater good.[5]

Enabling the "adaptivity necessary" requires support for R&D and robust data infrastructure to collect and use data in near real-time so all stakeholders involved can assess whether an innovation is or is not meeting communities' needs. It requires funding that organizations can turn into knowledge, which becomes the source for innovative tools and programs that, in turn, need data and funding to continuously adapt to a world that has always been, and always will be, chaotic and complex.

So, what does this mean for nonprofit R&D? It means the pandemic taught us that R&D is essential to progress, not just during a pandemic but all the time. Arundhati Roy, an Indian author and activist, recently wrote that pandemics have historically forced humans to imagine a new world. She described pandemics as a portal between the old world and the new world.

Many of us walked through that portal and are never looking back. The pandemic gave us a once-in-a-lifetime opportunity to experience the power of adequately funded R&D at an unprecedented scale. We now have a once-in-our-lifetime chance to fight for it to stay.

As Latanya Mapp Frett, president and CEO of Global Fund for Women, wrote when reflecting on her lessons learned from the pandemic, "Creativity comes out in crisis. Let's embrace that. Now is a once-in-a-lifetime opportunity to reimagine and reshape the future. Our most innovative, ambitious ideas are needed at this moment; we can't afford to waste them."[6]

NOTES

1. Kevin Desousa, *Intrapreneurship: Managing Ideas Within Your Organization* (Toronto: Rotman-UTP Publishing, 2017).

2. Stefan Lindegaard, "What Is the Difference Between R&D and Innovation?," LinkedIn, September 27, 2016, https://www.linkedin.com/pulse/what-difference-between-rd-innovation-stefan-lindegaard/.

3. Amina Mohammed, "COVID-19 Data & Innovation Centre: Visualize the Effects of the Pandemic and the Potential Solutions," December 4, 2020, https://cepei.org/en/novedad/covid-19-data-innovation-centre-visualize-the-effects-of-the-pandemic-and-the-potential-solutions/.

4. Ford Foundation, "Top Foundations Pledge Flexible Funding to Grantees in Wake of COVID-19 Crisis," March 19, 2020, https://www.fordfoundation.org/news-and-stories/news-and-press/news/top-foundations-pledge-flexible-funding-to-grantees-in-wake-of-covid-19-crisis/.

5. Ford Foundation, "Top Foundations Pledge Flexible Funding."

6. Latanya Mapp Free, "Letter From Latanya: 19 Lessons We Are Learning From COVID-19," May 20, 2020, https://www.globalfundforwomen.org/latest/article/19-lessons-covid-19-latanya-letter.

THE VALUE OF PRE-ANALYSIS

DAVID YOKUM AND JAKE BOWERS

INTRODUCTION

We describe the idea of a pre-analysis plan (PAP) and explain why you should use one. We emphasize the potential *political* uses of PAPs and how the PAP is, in this respect, a uniquely powerful tool for advancing the next generation of evidence.[1] We give examples from our experiences with PAPs over the past decade.

WHAT IS A PRE-ANALYSIS PLAN?

A pre-analysis plan (PAP) is a document describing how a research project will be conducted, written *before* data is collected or analyzed. The document explains what questions will be asked and how data will be collected and analyzed to answer those questions. The "registration" of a PAP involves publishing the document, with a timestamp, into a public location where it cannot be further edited.[2] A registered PAP is, therefore, a transparent record of what a researcher believed before conducting a study and how the researcher intended to update their beliefs with data.

There is substantial variation in how PAPs are written.[3] A PAP may contain dozens of pages, or maybe only one page or even a few sentences. The description may (or may not) include literature reviews, hypothesis statements, equations, mock figures and tables, code, or data simulations.

People have offered templates, checklists, and guidelines in an attempt to standardize—or at least set minimal standards for—the content and level of detail within a PAP.[4,5,6,7,8] But, ultimately, the researcher must use their own judgment to decide how much detail to include in a PAP, given the context and aims of the study.

WHY USE A PRE-ANALYSIS PLAN?

Pre-analysis plans help individual research teams and evidence-based policy in general in three main ways:

- PAPs enhance research integrity.
- PAPs prompt project management best practices.
- PAPs can be leveraged to facilitate political decision making.

Depending on which uses researchers pursue and to what degree, more or less detail will be required in the PAP.

PAPs Enhance Research Integrity

The first and foremost benefit—and the most common reason PAPs are becoming a standard practice throughout the academic community—is that PAPs *enhance research integrity*. In particular, the publicly registered PAP is a strategy for hedging against risks of *p*-hacking, HARKing, and publication bias.

P-Hacking

In the course of a study, a researcher will make hundreds of decisions regarding the design of data collection and how those collected data will be analyzed and reported.[9] These decisions can substantially affect what results are uncovered and shared.[10] For example, in considering whether the U.S. economy is affected by whether Republicans or Democrats are in office, decisions need to be made about how to operationalize economic performance (for example, employment, inflation, GDP), which politicians to focus on (for example, presidents, governors, senators), which years to examine, whether to entertain exclusions (for example, ignore), whether models should be linear or nonlinear, and so forth. To *p*-hack would be to try combinations of those decisions until "statistically significant" results surface.[11] This could happen intentionally or, much more commonly, unintentionally.[12]

The website FiveThirtyEight provides an interactive tool to build your p-hacking intuitions. Visit the website at https://fivethirtyeight.com/features/science-isnt-broken/ (or search "Aschwanden, Science Isn't Broken"). Toggle values on the "Hack Your Way to Scientific Glory" applet (it is in the middle of the article) to experience firsthand how, depending on your choices, you can reach literally *any* conclusion about the impact of political party on the U.S. economy.

The PAP hedges against p-hacking by forcing researchers to make these methodological choices in advance, based on criteria such as theory or statistical best practice rather than being lured into jiggling choices until a desired result is achieved.[13]

HARKing

To HARK is to "*H*ypothesize *a*fter the *R*esults are *K*nown."[14] HARKing happens when a researcher presents post hoc hypotheses in a research report as if they were, in fact, a priori hypotheses. In other words, a result gets framed as predicted by theory when, in fact, the result was *not* expected given the beliefs held before the study was conducted; it is only upon seeing the results that the researcher updates their beliefs and develops a *new* theory-driven hypothesis that is consistent with the result.

The updating of beliefs is not the problem—quite to the contrary. If properly done, that is the very essence of scientific progress. The problem is how HARKing conceals and distorts the belief updating process.[15] HARKing is alchemy that presents exploratory results as if confirmatory. This sleight of hand is misleading for a variety of reasons.[16] For example, HARKing violates the principle of disconfirmability: if a hypothesis is handcrafted to match already observed data, then there is no opportunity for a hypothesis to be disconfirmed by the study. And it is disconfirmed hypotheses, not confirmed hypotheses, that most efficiently winnow the field of competing ideas and advance our understanding.[17] Consider also that HARKing disregards information: prior beliefs based on theory are ignored, and the hypothesis is, instead, constructed on the sand of currently observed data and cherry-picked rationales.

The PAP prevents HARKing by keeping clear which hypotheses were predicted in advance versus which hypotheses were generated based on new results.

Publication Bias

Researchers are more likely to write up—and journals are more likely to publish—results that are statistically significant, even holding constant the importance of the question and the quality of methods.[18] One study found that research with statistically significant results had a *40 percentage point* higher probability of being published than if results were nonsignificant.[19] Such selective reporting leads to bias in the academic literature. Positive findings become overrepresented. Null or inconclusive findings, in contrast, become underrepresented, condemned to the researcher's personal file drawer rather than shared with the community. When this happens, any review or meta-analysis of the literature is misleading. Zero or contradictory effect sizes are effectively censored, leaving only the positive and largest effect sizes in print—and, thus, false positives are more likely and effect sizes are overestimated. A job training program with two positive evaluations might seem effective, but less so when it is uncovered that ten other evaluations, never published, failed to find any benefits or perhaps even found negative side effects.

To correct publication bias, all results must be openly available, so researchers can potentially summarize the entire body of findings.

PAPs Prompt Project Management Best Practices

The second benefit is mundane but important all the same. It may be the most immediate benefit you feel by adopting PAP practices. The documentation inherent to a PAP *fosters project management best practices*. To properly write out a methodology, the team must plan for a wide variety of details. To explain how randomization will happen, for example, you must determine and map out a suite of implementation details—how exactly will the intervention be delivered and to whom and by whom and when and for how long? In mocking up a data visualization, you are forced to think clearly about what data is needed to create that figure. And so on. You are forced to conduct a sort of "pre-mortem," considering what implementation or interpretation challenges might derail the project. And that, in turn, empowers you to manage against those challenges from the outset. By documenting all these project management details, you also increase communication across the research team as well as build resiliency against staff turnover. Any new team member can be handed the PAP during onboarding to the project.

Note that the PAP process should not actually create any additional work. A PAP should, instead, alter *when* work happens, namely, sooner rather than later. The only way to avoid the PAP work is a naughty one: to plan (even if implicitly) *not* to write up details if you fail to uncover statistically significant results that advance your theorizing.

The registration of a PAP is uniquely helpful in an additional way. There is a tendency for people—especially when busy, which is essentially always the case for practitioners—to carefully review documents only when absolutely necessary. It is common for drafts of reports to be skimmed but not fully engaged. This can lead to the frustrating situation where a document is shared and everyone thinks they agree on its contents, only to later discover—when it is about to *really be* published publicly and so everyone finally really reads the thing—that disagreements or objections linger. In our experience, the fact that a PAP will be registered—it will be public and uneditable at that point—is an excellent catalyst for engaging a partner's full attention sooner rather than later.

Managing a partner's full attention may feel like an added burden. It can slow down the launch of a project because extra time may be needed to clarify questions or negotiate points of debate. But we submit that the advance time is well spent for two reasons. The time will eventually be spent anyway; if not in advance, then after the fact while clearing up confusions about what was done. Indeed, dealing with the consequences of the misunderstanding is usually *more* complicated that averting the misunderstanding in the first place. At the extreme, a partner may want you to redo the work entirely. The second, and most powerful, reason relates to the political uses of PAPs, so let's turn there now.

PAPs Can be Leveraged to Facilitate Political Decision Making

Despite slogans to "follow the science," facts alone cannot determine *any* decision. The reason is that science inevitably involves value judgments, which are created by processes other than measuring and counting.[20] There are necessary value judgments, for example, in deciding what constitutes a meaningful effect size and how much uncertainty should be tolerated in the estimate of that effect size. Resolving these decisions cannot be done on technical grounds. There is technical skill involved in the calculations—there are correct and incorrect ways to calculate a confidence interval or a *p* value, for instance—but subjective opinions always enter when considering whether an impact is big enough, how to balance the risks of a false positive versus a false negative, whether to focus on mean or distributional effects, how to consider the opportunity costs of spending scarce resources on X rather than Y, and so on.

Scientists often make these value judgments entirely by themselves, either deliberately or by default in following a convention, such as setting $p < 0.05$ as the threshold for "statistical significance." In our experience, this is frequently the source of frustration on the part of stakeholders and the lay public. For example, empirical data can be marshalled to estimate how much mask-wearing reduces the transmission of COVID-19. But to step further into a decision about whether people *should* wear masks is to enter a realm of value trade-offs: the estimated benefits of reducing the risk of transmission must be weighed against the downsides of requiring people to purchase and cover their faces with masks, with added considerations for how to manage the risk of misestimating either side of the ledger.

The PAP is *a vehicle to clearly distinguish technical judgments from value judgments, and then to facilitate discussions on both fronts from the appropriate parties.*[21] For the technical components—for example, peer review of whether the randomization scheme was robust or double-checking statistical code—feedback from other experts is usually most fitting. But for the value components, it is usually the case that feedback is needed from the community affected by the research, either directly or via representatives who are making decisions on their behalf.

Consider the PAP used in an evaluation of the Washington, DC, police department's body-worn camera program.[22] Police officers were randomly assigned to wear a body camera or not (this was a randomized controlled trial), allowing the estimation of how much (if at all) body cameras reduced uses of force by way of comparing the group of officers with cameras against the group of officers without cameras. A key question was how long to run the study. From a technical standpoint, the more months of a treatment and a control group, the more precise the estimate will become. But how many months is enough? That is a political judgment. It requires assessments such as: How big of a reduction in use of force would be meaningful in policy terms? How certain do we need to be about that effect size estimate? How much are you willing to pay (in added research costs) to achieve a given precision of estimate? How much downside is there to a false positive or a false negative? And so on. The research team held over ten public events—at schools, in libraries, and beyond—taking pains to explain concepts such as randomization, effect size coefficients, and confidence intervals, so the community could then have a robust discussion about how big of an effect size would be meaningful to them. The PAP was key to facilitating these discussions.

CONCLUSION

The PAP is a uniquely fit tool for advancing the "next generation of evidence," for it empowers all three components identified by Project Evident:

1. **Practitioner Centric**: The PAP, when properly fleshed out and created collaboratively, is geared toward practical decision making and realistic project management. Drafting the PAP requires a clear articulation of: the question(s); the parameters for what constitutes an acceptable answer(s); and how the data for that answering process can be obtained in the field.
2. **Embraces a Research and Development (R&D) Approach**: Despite being a static document, the registered PAP really is geared toward *changing beliefs*, the key nuance being that PAPs facilitate proper belief-updating by way of fostering transparency in when and why beliefs have changed.
3. **Elevates the Voices of the Community**: The PAP is a concrete document that the community can read, comment on, and, potentially, even help draft. The best PAPs are documents, plus associated events or tutorials, that explain the technical components in plain language so relevant stakeholders can engage, regardless of background.

OTHER FAQS ABOUT PAPS

Q1: Do PAPs restrict exploratory research?
A: No, absolutely not. Although PAPs are commonly applied for null hypothesis testing (where problems of p-hacking fester), there is nothing about the underlying concept—making transparent your beliefs and intentions before data collection—that is inconsistent with exploratory research. A 100 percent exploratory PAP could literally just say, "This study is exploratory; there are no predictions and every permutation of data analytics will be attempted and reported." Notice

how this simple PAP hedges against HARKing (no hypothesis at all!); alerts the reader of the many attempted statistical tests (and, thus, vigilance is needed to calibrate uncertainty estimates based on family-wise error rates, to mitigate false positives from p-hacking); and alleviates publication bias by creating a public record.[23]

Q2: Can I deviate from the PAP?

A: Yes, of course. Just be transparent. Insights surfaced during unanticipated, exploratory analyses are the source of many scientific breakthroughs. Not to mention, deviations are often practically necessary if the intervention was implemented differently than planned. The key is that PAPs empower everyone to keep clear on what was predicted versus what was learned through exploration. Register a new version of the PAP if you update before beginning analyses. If after, simply note in your write-up what was planned versus what was not planned.

Q3: Is the PAP process different from community engagement?

A: Yes. Any PAP that leans into political uses must entail community engagement; but community engagement (broadly defined) need not and usually does not entail a PAP. Even when researchers publicly discuss their work with stakeholders, it is relatively rare to facilitate a discussion of value judgments and then to publicly register those agreements.

Q4: Do PAPs have to be made public while a study is ongoing?

A: No. PAPs can be embargoed to have their contents hidden for a specified amount of time. What matters is that the date of their registration be trustworthy to readers.

NOTES

1. Project Evident, www.projectevident.org, describes the "Next Generation of Evidence" as: 1) practitioner centric; 2) embracing a research and development (R&D) approach; and 3) elevating the voices of the community. This is in contrast to the status quo, which more typically involves a point-in-time evaluation geared toward informing academia and external funders. See, generally, www.projectevident.org/nextgenevidence-campaign.

2. Common repositories of PAPs include https://clinicaltrials.gov/, the American Economic Association's RCT Registry (www.socialscienceregistry.org/), the Evidence in Governance and Politics Registry (https://egap.org/registry/), and the Open Science Framework (https://osf.io). The Center for Open Science. The Center for Open Science is dedicated to promoting open

science best practices, and to that end, their website contains a host of additional readings, events, and resources; www.cos.io/.

3. George K. Ofosu and Daniel N. Posner, "Pre-Analysis Plans: An Early Stocktaking," *Perspectives on Politics* (2021), pp. 1–17, https://doi.org/10.1017/S1537592721000931.

4. Joseph P. Simmons, Leif D. Nelson, and Uri Simonsohn, "False-Positive Psychology Undisclosed Flexibility in Data Collection and Analysis Allows Presenting Anything as Significant," *Psychological Science* 22, no. 11 (2011), pp. 1359–366.

5. Jelte M. Wicherts and others, "Degrees of Freedom in Planning, Running, Analyzing, and Reporting Psychological Studies: A Checklist to Avoid p-Hacking," *Frontiers in Psychology* 7 (2016), https://doi.org/10.3389/fpsyg.2016.01832.

6. Anna Elisabeth van't Veer and Roger Giner-Sorolla, "Pre-Registration in Social Psychology—A Discussion and Suggested Template," *Journal of Experimental Social Psychology*, Special Issue: Confirmatory, 67 (November 1, 2016), pp. 2–12, https://doi.org/10.1016/j.jesp.2016.03.004.

7. Abhijit Banerjee and others, "In Praise of Moderation: Suggestions for the Scope and Use of Pre-Analysis Plans for RCTs in Economics," National Bureau of Economic Research, April 20, 2020, https://doi.org/10.3386/w26993.

8. Nuole Chen and Chris Grady, "10 Things to Know about Pre-Analysis Plans," EGAP Methods Guide (2019), https://egap.org/resource/10-things-to-know-about-pre-analysis-plans/.

9. J. M. Wicherts, C. L. Veldkamp, H. F Augusteijn, M. Bakker, R. Van Aert, and M. A. Van Assen, (2016), "Degrees of Freedom in Planning, Running, Analyzing, and Reporting Psychological Studies: A Checklist to Avoid P-Hacking, *Frontiers in Psychology*, November 2016, https://doi.org/10.3389/fpsyg.2016.01832.

10. Simmons, Nelson, and Simonsohn, "False-Positive Psychology Undisclosed Flexibility in Data Collection and Analysis Allows Presenting Anything as Significant."

11. The p value is a statistical measure of the probability of observing results as big or larger than the result observed in a sample, even if, in reality, there is no true effect in the population. So, for example, how likely would it be to get forty tails when flipping a fair coin fifty times? By convention, many scientists consider $p < 0.05$ to be "statistically significant." *See also* Aschwander, "Not Even Scientists Can Easily Explain P-Values."

12. A. Gelman and E. Loken, "The Garden of Forking Paths: Why Multiple Comparisons Can be a Problem, Even When There Is No "Fishing Expedition" or "P-Hacking" and the Research Hypothesis Was Posited ahead of Time," *Department of Statistics, Columbia University* 348 (2013), pp. 1–17.

13. Simmons and others suggest six simple requirements for authors to avoid *p*-packing: 1) decide the rule for terminating data collection before data collection begins; 2) collect at least twenty observations per cell or else provide a compelling cost-of-data-collection justification; 3) list all collected variables; 4) report all experimental conditions, including failed manipulations; 5) if observations are eliminated, also report the statistical results if those observations are included; and 6) if an analysis includes a covariate, also report the statistical results without the covariate.

14. Norbert L. Kerr, "HARKing: Hypothesizing after the Results Are Known," *Personality and Social Psychology Review* 2, no. 3 (August 1, 1998), pp. 196–217, https://doi.org/10.1207/s15327957pspr0203_4.

15. Brian A. Nosek and others, "The Preregistration Revolution," *Proceedings of the National Academy of Sciences* 115, no. 11 (March 13, 2018), pp. 2600–606, https://doi.org/10.1073/pnas.1708274114.

16. Kerr, "HARKing."

17. Karl Popper, *The Logic of Scientific Discovery*, 2nd edition (London: Routledge, 2002).

18. Annie Franco, Neil Malhotra, and Gabor Simonovits, "Publication Bias in the Social Sciences: Unlocking the File Drawer," *Science* 345, no. 6203 (September 19, 2014), pp. 1502–505, https://doi.org/10.1126/science.1255484.

19. Ibid.

20. Richard Rudner, "The Scientist Qua Scientist Makes Value Judgments," *Philosophy of Science* 20, no. 1 (1953), pp. 1–6.

21. David Yokum, "Psychology, Open Science, and Government: The Opportunity," *APS Observer* 29, no. 4 (March 31, 2016), www.psychologicalscience.org/observer/psychology-open-science-and-government-the-opportunity.

22. You can view this at https://osf.io/hpmrt/.

23. Some authors have expressed a concern that *journals* might miss this nuance and drift into only publishing papers that have pre-specified null hypothesis tests, which in effect would chill exploratory research. Indeed, journals should not do that. But note this is a concern about the potential *misuse* of PAPs, not a critique on the value of PAPs rightly used.

RECOMMENDED RESOURCES

Aschwanden, Christie. "Science Isn't Broken." *FiveThirtyEight* (blog), August 19, 2015. https://fivethirtyeight.com/features/science-isnt-broken/. A journalist explains in plain language—and with interactive visualizations—the problems of *p*-hacking and publication bias.

Franco, Annie, Neil Malhotra, and Gabor Simonovits. "Publication Bias in the Social Sciences: Unlocking the File Drawer." *Science* 345, no. 6203 (September 19, 2014), pp. 1502–505. https://doi.org/10.1126/science.1255484.

An empirical investigation of how severe is the problem of publication bias.

Kerr, Norbert L. "HARKing: Hypothesizing after the Results Are Known." *Personality and Social Psychology Review* 2, no. 3 (August 1, 1998), pp. 196–217. https://doi.org/10.1207/s15327957pspr0203_4. Coined the term "HARKing" and explores how HARKing undermines scientific progress.

Nelson, Leif D., Joseph Simmons, and Uri Simonsohn. "Psychology's Renaissance." *Annual Review of Psychology* 69, no. 1 (January 4, 2018), pp. 511–34. https://doi.org/10.1146/annurev-psych-122216-011836. A review of risks to research integrity and how registered PAPs hedge those risks.

A MOUNTAIN OF PEBBLES

EFFECTIVELY USING RCTS IN THE PUBLIC AND NONPROFIT SECTORS

JAMES MANZI

Randomized clinical trials have gained enormous currency as the most reliable way to measure the impact of social interventions, but their application has not reflected the dual issues of high failure rates and difficulty of generalization. This essay offers a short history of RCTs and suggests ways to make them more effective in predicting success.

Attempts to evaluate the effectiveness of interventions by applying the treatment to one group of patients ("test") and not to another group ("control") appear throughout recorded history. We see them in medicine from the biblical book of Daniel to Islamic and Chinese scholars in the eleventh century to James Lind's determination that citrus fruits prevent scurvy in 1747. An enormous challenge in this technique has always been how to hold all other factors constant between the test and control groups so we can know the difference in treatment must be the cause of the difference in outcomes.

The solution to this problem is to randomly assign participants to the test versus control group. The first randomized clinical trial that achieved modern standards of rigor was likely a 1938 U.S. Public Health Service trial of pertussis vaccine in Norfolk, Virginia.[1] One randomly chosen subset of Norfolk's population was selected for vaccination and another was selected to not receive the vaccine. The researchers could, thereby, conclude that any subsequent differences in pertussis rates between the two groups were caused by the vaccine. This is precisely the method used in late 2020 and early 2021 to evaluate the safety and efficacy of COVID vaccines.

Social science researchers quickly observed that this approach could be applied to evaluate the effectiveness of programmatic social interventions, and the RCT has, appropriately, become the gold standard of evidence for the causal effects of social programs across fields including criminology, education, and social welfare.

Evaluation of several decades of these RCTs executed in the developed world leads to two important observations. First, the vast majority of tested social interventions do not produce measurable improvement in targeted outcomes. The second is the problem of generalization; programs that demonstrate gains in experiments often create these benefits only in specific contexts, such as types of recipients, environmental situations, or provider capabilities.

Consider the first observation—that most social interventions fail when tested. Criminologists at the University of Cambridge have done the yeoman's work of cataloging all known criminology RCTs between 1957 and 2004 with at least one hundred test subjects.[2] Twelve of the programs were tested in "multisite" RCTs: experiments in several cities, prisons, or court systems. Eleven of the twelve failed to produce positive results, and the small gains produced by the one successful program (which cost an immense $16,000 per participant) faded away within a few years. This is a 92 percent failure rate. The U.S. Department of Education's Institute of Education Sciences (IES) sponsored a series of RCTs that tested fourteen well-known preschool curricula and found only one curriculum that demonstrated some causal gains in performance that persisted only through kindergarten.[3] This is a 94 percent failure rate. And none of that considers whether either of the two successful programs is remotely cost-effective. We see this same pattern time and again for social interventions.

This high failure rate is not unique to social programs. A National Institutes of Health (NIH) evaluation of 798 drug development programs found that only approximately 6 percent of pre-clinical therapies complete a Phase III RCT successfully and are approved for use.[4] Google has reported that only about 10 percent of on-line changes tested in RCTs create business improvement.[5]

But unlike most medical interventions, even when we find a social intervention that proves impact in an RCT, the problem of generalization rears its head.

We can run a clinical trial in Norfolk, Virginia, and conclude with tolerable reliability that *"Vaccine X prevents disease Y."* We cannot conclude that if literacy program X works in Norfolk then it will work everywhere. The real predictive rule is usually closer to something like *"Literacy program X is effective for children in urban areas, and who have the following range of incomes and prior test scores, when the following alternatives are not available in the school district, and the teachers have the following qualifications, and overall economic conditions in the district are within the following range."*

In 1981–1982, Lawrence Sherman, a respected criminology professor at the University of Cambridge, led an extremely influential experiment that randomly assigned one of three responses to Minneapolis cops responding to misdemeanor domestic-violence incidents: they were required to either arrest the assailant, provide advice to both parties, or send the assailant away for eight hours.[6] The experiment showed a statistically significant lower rate of repeat calls for domestic violence for the mandatory-arrest group. The media and many politicians seized upon what seemed like a triumph for scientific knowledge, and mandatory arrest for domestic violence rapidly became a widespread practice in many large jurisdictions in the United States. But sophisticated experimentalists understood that, because of the issue's complexity, there would be hidden conditionals to the simple rule "mandatory-arrest policies reduce domestic violence." The only way to unearth these conditionals was to replicate the original experiment under a variety of conditions. Sherman's own analysis of the Minneapolis study called for such replications. So, researchers replicated the RCT six times in cities across the country. In three of those studies, the test groups exposed to the mandatory-arrest policy again experienced a lower rate of re-arrest than the control groups did. But in the other three, the test groups had a higher re-arrest rate.

The danger of drawing conclusions based on a single RCT on a social policy topic is obvious in this example. Suppose Sherman had happened to run the original experiment in Memphis (one of the cities where the replication failed). Would we then have been justified in concluding that mandatory arrest doesn't work? Based on this set of replications, whether it works in any given city is roughly equivalent to a coin flip. It is important to keep this in mind when presented with the gold-standard evidence of any one well-designed RCT. The obvious question is whether anything about the situations in which mandatory arrest worked distinguishes them from situations where it did not. If we knew this, we could apply the program only where it is effective.

In 1992, Sherman surveyed the replications and concluded that in stable communities with high rates of employment, arrest shamed the perpetrators, who then became less likely to reoffend, while in less stable communities with low rates of employment, arrest tended to anger the perpetrators, who would, therefore, be likely to become more violent.[7] The problem with this kind of conclusion, though, is that because it is not itself the outcome of an experiment, it is subject to the same uncertainty as any other pattern-finding exercise. How do we know whether it is right? We do so by running an experiment to test it—that is, by conducting still more RCTs in both kinds of communities and seeing whether they bear out this conclusion.

Confronting this difficult reality directly can help us be much more effective in evaluating interventions. RCTs have gained enormous currency as the most reliable way to measure the impact of social interventions, but their application has not reflected the dual issues of high failure rates and difficulty of generalization. I believe public agencies and nonprofit organizations attempting to use RCTs should embrace three simple principles:

> 1. *Kiss a lot of frogs to find a prince.* Based on experience, we should expect to try at least ten very promising intervention ideas before we find one that actually will improve any targeted outcome. This means building the capacity to run many tests at low cost per test. This, in turn, requires using administrative data, semi-automated test design and analysis, and organization and procedures that lower the hard dollar and organization friction costs of running a test.
>
> 2. *Build a mountain of pebbles.* There are no silver bullets for social problems out there waiting to be found through RCTs. Agencies and nonprofits should be looking for lots of small wins through testing, not transformational moonshots. Foundations that fund nonprofits would be better off requesting "Show me the number of tactical RCTs you have done and the results," than "Show me the impact of your overall program according to an RCT."

(continued)

3. *Bottom-up not top-down.* Use of RCTs in social program evaluation often proceeds from observations of their successful use in medicine, but this analogy is far from perfect because the problem of generalization is so much more severe for social interventions. Rather than an image of experts who develop theory-dependent program ideas that are then rigorously tested to find "what works," we should, instead, think of a continuing flow of localized, tactical ideas that emerge from practitioners who then have the capacity (expertise and resource) embedded in their organization to rapidly test these potential innovations and implement the small fraction that create improvement.

NOTES

1. Iain Chalmers, "Joseph Asbury Bell and the Birth of Randomized Trials," *Journal of the Royal Society of Medicine* 100, No. 6 (June 2007): 287–93, https://journals.sagepub.com/doi/pdf/10.1177/014107680710000616.

2. David Farrington and Brandon Welsh, "Randomized Experiments in Criminology: What Have We Learned in the Last Two Decades?," *Journal of Experimental Criminology* 1, (April 2005): 9–38, https://doi.org/10.1007/s11292-004-6460-0.

3. Preschool Curriculum Evaluation Research Consortium, "Effects of Preschool Curriculum Programs on School Readiness (NCER 2008–2009)," National Center for Education Research, Institute of Education Sciences, U.S. Department of Education, http://ies.ed.gov/ncer/pubs/20082009/pdf/20082009_rev.pdf.

4. Tohru Takebe, Ryoka Imai, and Shunsuke Ono, "The Current Status of Drug Discovery and Development as Originated in United States Academia: The Influence of Industrial and Academic Collaboration on Drug Discovery and Development," *Clinical and Translational Science* 11, no. 6 (July 2018): 597-606, doi: 10.1111/cts.12577. Epub 2018 Jul 30. PMID: 29940695; PMCID: PMC6226120.

5. Stefan Thomke, "Building a Culture of Experimentation," *Harvard Business Review*, March–April 2020, https://hbr.org/2020/03/building-a-culture-of-experimentation.

6. Lawrence Sherman and Ellen Cohn, "The Impact of Research on Legal Policy: The Minneapolis Domestic Violence Experiment," *Law & Society Review* 23, no. 1 (1989): 117–44, https://doi.org/10.2307/3053883.

7. Lawrence W. Sherman, Janell D. Schmidt, Dennis P. Rogan, Douglas A. Smith, "The Variable Effects of Arrest on Criminal Careers: The Milwaukee Domestic Violence Experiment," *Journal of Criminal Law & Criminology* 83, no. 137 (1992–1993).

CEO

CREDIBLE MESSENGERS IN RE-ENTRY SERVICES

AHMED WHITT

For more than thirty years, the Center for Employment Opportunities (CEO) has offered immediate, effective, and comprehensive employment services exclusively to individuals with criminal convictions. CEO's programs help participants gain the workplace skills and confidence needed for successful transitions to stable, productive lives. Through our proven model, CEO has made over 25,000 unsubsidized job placements with more than 4,000 employers throughout the country. CEO targets adults of all ages at the highest risk of recidivism and those confronting significant barriers to employment.

The organization's commitment to continuous evaluation is rooted in its 2004–2008 randomized control trial. The study found CEO to be effective in reducing recidivism, particularly for recently released individuals with the highest risk profiles, but no long-term effects were shown on employment. The findings satisfied an immediate need to confirm the core program model but, more significantly, marked an organizational shift from *pursuing proof* to *generating knowledge*. In addition to performing multiple replication studies, CEO initiated a series of projects and strategic hires to consistently test which specific program activities worked best for which profile of participant and what improvements could be scaled to have a more sustaining impact overall on long-term economic mobility. Most recently, CEO has invested heavily in testing program innovations to abate the unique barriers faced by young adults, age eighteen to twenty-five, who comprise roughly 40 percent of our annual enrollments.

BECOMING CREDIBLE MESSENGERS

The Credible Messenger Initiative (CMI) was developed in 2017 for CEO's New York City office as part of broader efforts to improve services to young adults. Credible messengers are individuals with lived experience similar to the people they are seeking to serve. At CEO, this lived experience ranges from being justice-involved, growing up in similar neighborhoods, or being faced with situations similar to those our participants have faced. The CMI model is intended to supplement, not replace, the core program model, specifically the support provided by job coaches (JC) and job developers (JD). CMI was designed to serve participants age eighteen to twenty-five who are identified as struggling or likely struggling through standard CEO programming and need some extra support from staff and peers. Participants are typically referred by their JC but may be referred by other CMI participants or may independently request to join.

CMI was heavily influenced by the Arches Transformative Mentoring program (Arches), which was launched in 2012 and managed by the NYC Department of Probation. Like Arches, CMI combines a group mentoring model with individual case management, consisting of sixteen workshops designed to develop both professional and life skills in young adult. All workshops facilitated by credible messengers are structured as a talking circle and cover topics including networking, goal setting, and time management. These workshops are enhanced by one-to-one support that CEO credible messenger staff offer CEO participants to support their success. The experience and insights of the credible messenger staff were crucial in the implementation of CMI as well as each phase of its evaluation.

CHALLENGES AND RESPONSES

The process and outcome evaluations were designed and executed by CEO's internal evaluation unit. In the project's initial months of CMI, process evaluation data were collected via weekly joint team meetings and a CMI staff focus group. Generally, staff reported strong commitment to the initiative, especially the enhanced case management component. Staff believed the group mentoring sessions were meeting a previously unmet need for many of the "harder to reach" participants to discuss life experiences beyond direct barriers to employment. During planning for the pending outcome evaluation, the staff of CMI and the evaluation team most vigorously debated the possible negative impacts of CMI enrollment on transitional work attendance and how the increased flexibility of the recruitment and discipline standards affected CMI participants as compared to those of standard CEO program model enrollees. The largest hurdle was how the group differences (CMI participants versus other CEO young adults) would inhibit the construction of a valid comparison group.

With these concerns in mind, two key decisions were made prior to the outcome evaluation: 1) enrollment for the evaluation cohort began after an agreed upon maturation point at which the procedures and norms for participation were mirrored across all forthcoming CMI groups, and 2) a quasi-experimental design would be used to allow CMI staff discretion for program recruitment and enrollment. The agreed upon primary research question was "Compared to a matched group of young adults who enrolled in CEO's NYC office and were not offered CMI, were CMI participants more likely to achieve unsubsidized job placement?" Additionally, we tested if CMI participants were significantly more likely to achieve core program milestones and engage with transitional employment as compared to non-CMI young adult participants.

REDUCING BARRIERS TO EMPLOYMENT SUCCESS

The treatment group was comprised of participants who enrolled into CMI between April 2018 and May 2020 in the CEO New York City site (Manhattan and Bronx-satellite locations) (N = 259). Approximately 51 percent of the CMI enrollees received services through the Bronx office. The average age of CMI participants was 22.84 years, and roughly half of participants had less than a high school education. As compared to the overall CEO NYC young adult enrollees during the same time period (April 2018– May 2020; N = 1,991), CMI enrollees were overall more likely to be African-American, female, and referred from a non-parole source. On average, CMI participants were younger and had completed fewer years of formal education than other CEO NYC young adults. All cited differences between the groups were statistically significant ($p < 0.05$). Using historic

CEO program data as a point of reference, the demographic factors to which the CMI group were more likely to identify overlapped with factors of lower comparative engagement and success overall. At this stage of the outcome evaluation, it appeared the CMI staff's recruitment process, which mixed professional intuition with prior interaction with the enrolled participants, adequately identified the "harder to reach" within the NYC young adult cohort.

Between April 2018 and June 2020, CMI participants engaged with CEO program components and staff members at a significantly higher level than other young adults on average. CMI participants completed over 60 percent more transitional employment work hours and engaged in over 60 percent more job coaching sessions than other CEO young adults; both differences were statistically significant ($p < 0.01$). Overall, CMI participants were more successful in achieving program milestones. CMI participants were both more likely to achieve Job Start Ready status (that is, prepared for unsubsidized employment) and more often placed in full-time work positions ($p < 0.01$). The difference between the groups in achieving the 180-day job retention milestone for unsubsidized work was not statistically significant. The structure of rigor of these comparative analyses mirrored those used in Arches evaluation but were, however, alone insufficient for our overall learning goal to inform various scalable strategies to improve young adult performance. In addition to lacking causal inference, the analyses were limited in identifying what specific components would be most worthwhile to test in other offices. We knew that the probability of initially securing funding to scale the entire intervention across all offices would be low.

Given the significant demographic differences between the CMI sample and the overall NYC young adult enrollment population, a propensity score matching technique was performed to extract a one-to-one matched comparison subset. Potential confounding variables included in the analysis were age, date/location of enrollment, level of education, race/ethnicity, gender, and referral source. Eight participants from the CMI sample were removed due to insufficient matches within the non-CMI population; ultimately, 251 CMI enrollees were linked with non-CMI young adult enrollees. In contrast to the differences between the CMI and the full young adult population, the matched groups did not differ significantly on age, gender, race, or educational attainment. The percent difference in referral source (that is, parole versus non-parole) between the CMI and comparison groups remained significant. CEO engagement variables were excluded from the matching procedure to compare the program performance of participant groups with similar demographic characteristics. On average, CMI participants worked 43.4 more hours of transitional work than participants in the comparison group, a significant difference at the $p < 0.01$ level. CMI participants were twice as likely to both achieve Job Start Ready status ($p < 0.001$) and, ultimately, obtain an unsubsidized job placement ($p < 0.001$). Within the matched sample, job retention status at the 365-day milestone was significantly better for the CMI group ($p < 0.01$); the difference at 180 days was not statistically significant. When controlling for participant characteristics and enrollment conditions, engagement with staff and participant race were significantly associated with likelihood of young adults achieving placement in an unsubsidized job. Each additional staff interaction impacted the odds of success positively by a factor 5 ($p < 0.001$). Participants who identified as Black were 65 percent less likely to achieve unsubsidized job placement within the matched sample.

CMI yielded a successful pilot for CEO, particularly when highlighting the comparative engagement and employment outcomes for young adults with the greatest barriers with those of our typical young adult population. However, when we regrouped with our internal and external stakeholders to unpack the results and devise next steps for the program, we focused on the inconclusive results on the full CMI model differentiating outcomes for the "hardest to reach" group. The key driver within that reduced subset was increased interaction with program staff. We facilitated a series of group learning sessions centered on the pilot results with different combinations of our current funders, government partners, organization-wide staff, and current and former participants. As we continue replication studies of the original CMI model, some of the questions we continue to explore include: Should we mandate a certain number of interactions between staff and high-risk participants? Would available methods for identification be accurate and valid? What are the ethical implications of stratifying our young adult cohorts?

REFLECTIONS

The CMI pilot was made possible by years of insights collected from job coaches, job developers, and managers on the unique risks faced by our program's youngest enrollees. The experiences of our young adults were captured in case notes and program feedback, and via individual conversations, but needed to be synthesized to inform a testable hypothesis of how CEO could improve its services. Our practitioners' collaboration with program support staff on data quality yielded a system that easily could store and elevate participant voice alongside individual outcome data. For staff, CMI helped clarify the *why* of how those ongoing investments in evaluation infrastructure, thus closing the misperceived gap between research and practice.

CEO's shift to generating actionable evidence has required buy-in from staff across the entire organization. Internally, the CMI pilot is used as an exemplar for maximizing our data system and optimizing our research and development capability. Similarly, the story of the process has resonated with our regional partner organizations and government stakeholders more than the results of the pilot. CEO is playing an active role as evangelist and collaborator in support of accelerated evidence generation within the social sector.

NOGGIN

LEARNING IMPACT EVIDENCE IN A MULTIMEDIA CHILDREN'S PLATFORM

KEVIN MIKLASZ, MAKEDA MAYS GREEN, AND MICHAEL H. LEVINE

Driving evidence-based outcomes in early childhood education is an urgent national priority: strong scientific evidence about the long-term value of preschool learning and the critical period of early brain development is now broadly understood. The new federal administration has made evidence-based, quality early learning program expansion a large part of its agenda.[1] However, a needed focus on outcomes is a relatively recent phenomenon, tracked back to the first National Education Goal for "readiness," which followed decades of debates about closing performance gaps and many related waves in the K-12 standards-based reform movement.[2]

As a popular media organization, Noggin faces significant challenges in developing evidence-based offerings that will not offend the tastes of our choosy audience of preschoolers!

These days, young children have a sea of choices in the digital kids landscape—from Pokémon to Minecraft to Toca Boca to Scratch—that engage their minds and bodies about three hours a day.[3] Creators must be deft in blending fun and engagement with intentional, outcomes oriented content. One silver lining in this digital wild west[4] is the demand from parents—going back several decades, with the emergence of Sesame Street,

Mr. Rogers, Noggin, Nick Jr., and the Public Broadcasting Service, for educational brands that can help children get ready for school and life.[5] And the present needs of young children, emerging from over a year languishing at home during the COVID crisis, have added urgency to concerns that media time be purposeful.

That is why, in retooling our work at Noggin, the early learning platform first developed by Nickelodeon and Sesame Workshop two decades ago and now a part of Paramount, research has become a key component of the content production pipeline. We use research not only to determine if content resonates with or engages children but also to learn if it helps them acquire key concepts and skills. The latter research, which we call "learning impact research," has a modest but established tradition among scholars who study the potency of informal media, including professional journals[6] devoted to the impact of the changing media landscape, landmark studies of Sesame Street's long-term impact on learning trajectories,[7] and meta-analyses of the educational promise of long-form digital games.[8]

THE CURRENT STATE OF LEARNING IMPACT EVIDENCE

At this time, it is well established that learning products used with children should have proven impact or evidence that those products incite learning. Yet, how we establish what counts as appropriate evidence is still evolving. Starting with the implementation of the No Child Left Behind Act (NCLB) first enacted in 2001, there was an increasing focus on ensuring that educational technology content and products would produce learning. The 2015 Every Student Succeeds Act (ESSA) took NCLB a step further by tying federal funding explicitly to a set of standards for learning impact. These are commonly known at the ESSA Evidence Tiers,[9] a set of four levels of evidence that define what counts as rigorous. As research moved from Tier 4 (Demonstrates a Rationale) toward Tier 1 (Strong Evidence), the level of rigor and quality of the evidence increases.

As much as the ESSA standards are a huge step forward in thinking about learning impact evidence, there have been (a few) criticisms of the standards. First, the government standards themselves were not written with enough detail to be clear on how specific research meets each tier. This has resulted in other agencies offering their own interpretations of how to translate the ESSA standards into practical guidance for researchers, and their interpretations are not in complete agreement (for example, see SIIA,[10] WWC,[11] and Evidence for ESSA[12]).

Second, and most significant for our purposes, the ESSA tiers apply to fully developed products and there is no guidance for rigorous in-development protocols. For a product like Noggin that is a platform continually releasing new content, any point-in-time ESSA Tier 1 study would take one to three years to complete: an estimated 300 to 500 new pieces of content would be added to the Noggin platform during that time, and the study's results would become obsolete by the time they were finally released.

As another angle, the U.S. Office of Education Technology has laid out protocols for how to use rigorous development practices that involve testing and iteration throughout development (see The EdTech Developer's Guide[13] and Expanding Evidence Approaches for Learning in a Digital World[14]). Although the standards provide guidance on when and how to do such work and lay out best practices, there is no guidance on what counts as rigorous, nor any way to prove that a particular product's development process was rigorous.

Another approach to address this problem comes from the organization Digital Promise, with their research-based certification.[15] For this technique, the focus is less on the ultimate product and more on the organization. The organization undergoes a process to certify they have development processes that follow best practices found in the research; once the organization is successful, they are awarded an open badge and acknowledged on the Digital Promise website. The drawback to this approach is that the certification does not note whether an organization itself conducts good formative research on that content. This would be much harder to certify at scale.

In the academic world, Daniel Hickey and James Pellegrino have laid out three general approaches to thinking about assessment of learning impact.[16] They first describe an empiricist approach, which is about measuring facts and the associations between them, and, second, a rationalist approach, which is about measuring the mental models students build up. They note that large-scale, long-timescale approaches have to rely on one of these two models, and the more traditionally rigorous the approach, the more the assessment itself tends to rely on understanding facts and relationships between them (the bread-and-butter of classic multiple-choice achievement

FIGURE 4.9.1 A Summary of the Three Noggin Learning Impact Tiers

Tier	Name	Short definition	Criteria
Tier 1	Directional Evidence	Evidence is trending in the direction that impact exists.	*Must show evidence that is consistent with the idea that learning growth is happening. The evidence is necessary but probably not sufficient.*
Tier 2	Correlational Evidence	Usage of the content is correlated with learning gain.	*Must show learning growth is correlated to usage. That can be either through 1) showing that higher usage corresponds to better results, or 2) pre-post gains occur when the content is used.*
Tier 3	Causal Evidence	Usage of the content causes learning gains.	*Must show learning growth as a result of usage, as compared to a well-defined control group.*

tests). For making in-the-moment measurements of learning, neither of these approaches is sufficient. Thus, Hickey and Pellegrino offer a third perspective, sociocultural, which is about seeing evidence of authentic dialogue and participation in a community of practice. Sociocultural assessments work better in shorter time scales and nearer-transfer assessments, which offers a particularly relevant model for rigor in formative research.

BRIDGING THE GAP: USING IMPACT EVIDENCE IN FORMATIVE RESEARCH WITH MEDIA

To solve those gaps, the Noggin team, which consists of an unusual blend of content developers, instructional design experts, and research scientists, has developed a framework that tests for learning impact throughout the life cycle of a piece of content. This allows us to find learning evidence well before we have the time or resources set aside to run an intensive randomized control trial that produces Tier 1 ESSA evidence. Accordingly, we have developed the following three evidence tiers, described in figure 5.8-1. Lower tiers are considered less rigorous, but moving down, one tier is typically an order of magnitude less costly and time-intensive. Our general approach to impact research is to start gathering lower tiers of evidence first and, once those are proven, spend the time and resources looking for higher tiers of evidence. This avoids having to spend large amounts of resources only to find out something does not work. Additionally, the lower tier research works well with rapid cycle content iteration needs, and ensures the content continues to improve as it is developed.

Let's go through each level individually.

Tier 1: Directional Evidence

This level of evidence indicates there is evidence that is directionally consistent with the idea that learning is happening. Directional evidence can come from: 1) alignment between usage and best practices; 2) observations that learning is happening in the moment; 3) informal measurements that learning has happened over repeat play; 4) ability to transfer learning from the activity to a related task; or 5) a positive but insignificant correlational or causal evidence. We choose one of these five approaches for directional evidence based on whatever makes the most sense given the nature of the content.

Directional evidence typically is found in our formative research process or during the content development process on alpha or beta versions of content, but we also can look for this evidence post-launch. All the techniques are meant to be light and quick forms of evidence gathering that still have elements of quality and rigor to them.

In ESSA terms, directional evidence is most similar to ESSA Tier 4 (called Demonstrates a Rationale), but, really, ESSA does not fully acknowledge this kind of in-process design research as a valid form of evidence. The standard as we have written it is more rigorous than the ESSA Tier 4, as some form of actual evidence is required. This is in the spirit and intent of this fourth level of ESSA, which is to acknowledge products that have not directly measured impact but for which there is good reason to believe they are effective.

Accordingly, our Tier 1 evidence is most strongly influenced by the sociocultural approach advocated by Hickey and Pelligrino, and derives its rigor from that viewpoint. All the levels of evidence typically involve looking for some form of authentic dialogue that represents genuine engagement with the learning content being featured.

Tier 2: Correlational Evidence

This evidence is attempting to make a correlational claim, that some kind of usage is correlated with some kind of learning. There are two general categories that qualify for this level. First is one that directly proves a statistically significant correlation between some kind of usage metric and some kind of learning metric. Second is one where learning gains are seen from a pre-post measure, with use of the learning tool interjected between. This can be thought of as an intervention without a control group.

In either of these cases, the lack of a well-defined control group is the defining feature that results in correlation but not causation. "Well-defined" is the key phrase, and we mainly look to the ESSA standards for the definition of this phrase. Correlational evidence is most similar to ESSA's Promising Evidence. We pretty much follow the ESSA definitions, with the exception that we do not require "statistical controls for selection bias," since that requirement feels overly stringent for a correlational study and, arguably, makes ESSA's Tier 3 evidence no different from ESSA's Tier 2 evidence, as those statistical controls are what makes a control group "well-defined." We follow the SIIA interpretation of ESSA where the ESSA standards lack detail.

Tier 3: Causal Evidence

This evidence is attempting to make a causal claim. The goal is to say that the use of a learning tool causes learning gains, typically in comparison to a control group. The classic form of this study is a randomized control trial, but many newer machine learning techniques are now considered to also make causal claims with various degrees of comparable rigor. One particular category of studies (often bundled as quasi-experimental studies) are ones that define control groups after the fact but do so in a way that ensures there is no selection bias in how the control group is defined, so it is, thus, a "well-defined" control group.

Our causal evidence category combines ESSA's Tier 2 and Tier 1 evidence into one level, which comprises both quasi-experimental and "true" experimental (aka randomized control trial) approaches. Both are combined because both are forms of causal studies, and because several innovations in big-data-driven quasi-experiments (notably those using propensity score matching) are arguably more robust than limited-sample-size RCTs, making this distinction in methodology antiquated. We follow the SIIA interpretation of ESSA where the ESSA standards lack detail.

PRACTICAL APPLICATION OF THE IMPACT EVIDENCE STANDARDS

Below is a brief description of Noggin's general content production pipeline. We describe each of the steps in general terms, as each type of content we make goes through a slightly different form of this process.

Background Research

Our learning and content teams do background research on the topic, looking at best practices found in the field and research literature.

Adviser Feedback

After we have an idea of what we want to produce, we check our designs with an outside expert. We have a robust advisory panel, composed of researchers and experts in the early childhood education space, representing other professionals in academia and other media organizations.

Formative Research

Now we are in production. Usability tests are conducted throughout the various stages of content development, typically at key "alpha" and "beta" stage milestones. The early stage usability tests may or may not test for impact evidence, but at the late-stage test, we make every attempt to incorporate a Tier 1 impact study.

Launch Engagement Analytics

For the first few weeks after launch, we monitor basic engagement analytics. Although not testing for impact, this does indicate if the content is resonating, or is unexpectedly unpopular, and may point to some issues to address.

Post-Launch Learning Analysis

Several months after launch, we will use the performance data to conduct a learning analytics analysis, or do a deeper qualitative research test. This can produce either Tier 1 or Tier 2 evidence of impact, depending on the format of the content and what data are available.

Summative Research Study

Considering the high investment needed for summative research, we selectively employ summative research studies to test our content at large, either groups of content that are meant to be sequenced and done together or our platform as a whole. This gives a zoomed-out view of our content that can produce Tier 2 or Tier 3 evidence.

As a practical example, we have mapped out the research life cycle for a recent piece of content: a vocabulary video series called Word Play (figures 5.8-2 and 5.8-3).

FIGURE 4.9.2 **Word Play Research Life Cycle**

❶ Background Research
The Word Play series was derived from our "Vocabulary" Skill in our Noggin Learning Framework, which intended to teach kids the meaning of specific words using simple engaging visuals and repetition.

❷ Advisor Feedback
Our advisors for vocabulary content include Dr. Susan Neuman of New York University and Dr. Glenda Revelle of the University of Arkansas.

❸ Formative Research
As a short form piece of video content lasting about 1 min in length, the production cycle was too rapid to do in-development testing. Instead we opted for a post launch testing described below.

❹ Launch Engagement Analytics
The Word Play had average launch statistics by both video starts and completion rates.

❺ Post-Launch Learning Analysis
Children were given a PPVT style vocabulary test as a pre-post measure and asked to watch a series of vocab videos at least once a day for two days. Scores showed a statistically significant increase from the pre-test to the post-test.

❻ Summative Research
We are planning to involve Word Play as one component in a larger intervention being planned now for later in the year, which aims to produce Tier 2 evidence for the entire set of content.

REFLECTIONS

The children's media field has had modest but notable success in designing content with measurable impact: industry best practices have been established by leaders such as the Public Broadcasting System, Nickelodeon, and Sesame Workshop. However, the earlier work was done prior to the emergence of learning standards and practices associated with evidence-based outcomes. As children's media leaders, we believe the next round of educational progress, in a post-COVID environment, will require a convergence in the expectations set by educators and content producers. It is our mission to help ensure this new approach is driven by digital teachers and role models that children truly love!

NOTES

1. Cody Uhing, "President Biden Seeks Important Funding Increases for Early Learning & Care Programs in FY2022," First Five Years Fund, April 9, 2021, www.ffyf.org/president-biden-seeks-important-funding-increases-for-early-learning-care-programs-in-fy2022/.

2. See The National Education Goals Panel website, https://govinfo.library.unt.edu/negp/page3.htm.

3. See Common Sense Media, https://www.commonsensemedia.org/sites/default/files/research/report/2020_zero_to_eight_census_final_web.pdf.

4. See Lisa Guernsey, Michael H. Levine, Cynthia Chiong, and Maggie Severns, "Pioneering Literacy in the Digital Wild West: Empowering Parents and Educators," Sesame Workshop, Joan Ganz Cooney Center, August 10, 2014, https://joanganzcooneycenter.org/publication/pioneering-literacy/.

5. See Our Impact, Sesame Workshop website, www.sesameworkshop.org/who-we-are/our-impact.

6. See Taylor & Francis Online, October 1, 2020, www.tandfonline.com/toc/rchm20/10/1.

7. Alia Wong, "The *Sesame Street* Effect," *The Atlantic*, June 2015, www.theatlantic.com/education/archive/2015/06/sesame-street-preschool-education/396056/.

8. See Douglas B. Clark, Emily E. Tanner-Smith, and Stephen S. Killingsworth, "Digital Games, Design, and Learning: A Systematic Review and Meta-Analysis," *Review of Educational Research*, March 1, 2016, Sage Journals, https://journals.sagepub.com/doi/full/10.3102/0034654315582065.

9. See www2.ed.gov/policy/elsec/leg/essa/guidanceuseseinvestment.pdf.

10. Denis Newman, Andrew P. Jaciw, and Valeriy Lazarev, "Guidelines for Conducting and Reporting EdTech Impact Research in U.S. K-12 Schools,"

Empirical Education, April 15, 2018, www.empiricaleducation.com/pdfs/guidelines.pdf.

11. See "Using the WWC to Find ESSA Tiers of Evidence," IES WWC website, https://ies.ed.gov/ncee/wwc/essa.

12. See Evidence for ESSA, Find Evidence-Based PK-12 Programs, n.d., www.evidenceforessa.org/.

13. See "Ed Tech Developer's Guide," U.S. Department of Education, April 2015, https://tech.ed.gov/files/2015/04/Developer-Toolkit.pdf.

14. See https://tech.ed.gov/wp-content/uploads/2014/11/Expanding-Evidence.pdf.

15. See Certified Products page at the Digital Promise website, https://productcertifications.digitalpromise.org/certified-products-2/.

16. Daniel Hickey and James Pelligrino, "Theory, Level, and Function: Three Dimensions for Understanding Transfer and Student Assessment," Research Gate, January 2005, www.researchgate.net/publication/201381886_Theory_level_and_function_Three_dimensions_for_understanding_transfer_and_student_assessment.

FIRST PLACE FOR YOUTH

ALIGNING STRATEGY, DATA, AND CULTURE TO DRIVE IMPACT

ERIKA VAN BUREN, MATT LEVY, AND JANE SCHROEDER

Embedded within First Place for Youth's DNA is a commitment to building evidence from within and leveraging learnings to continuously improve services, raise the bar for programmatic impact, and drive systems change.[1] This case study highlights First Place's continuing journey to generate knowledge and impact that catalyzes programmatic and system-level impact on one "north star" outcome: life sustaining, living wage employment for youth aging out of foster care.

LOCATING THE NORTH STAR

In 2018, First Place partnered with an external evaluator to complete an evaluability and implementation assessment of its core My First Place (MFP) program to inform program improvement strategies and determine its readiness for a more rigorous impact evaluation. The yearlong assessment confirmed where the model was succeeding: outcomes were strong in the areas of employment placements, housing stability, and unplanned pregnancies. At the same time, the evaluation revealed how far the model still needed to go to impart meaningful impact. Youth were enrolling in postsecondary programs but were not necessarily sticking with them; and while youth were getting placed in jobs, those jobs were well below living wage. These findings were echoed by the results of concurrent policy research undertaken at a statewide level.[2]

It had become clear that without developing more specific measures and improvement methods that supported a line of sight toward education and career success, First Place would fall short of advancing long-term economic self-sufficiency for its youth participants. First Place took these learnings as a call to action and, in 2019, formed the Cross-Departmental Strategy Workgroup (CDW) to develop and implement robust organizational and system-level strategies to target disparities uncovered through this internal and state-level research. Members of the CDW included the leaders responsible for strategic oversight of the Evaluation and Learning (E&L), Policy, and Program departments broadly, as well as the specialized employment and education program components implemented through MFP.

Using the Annie E. Casey Foundation's *Accountability for Equitable Results Framework* (Annie E. Casey Foundation, 2019), these leaders first set out to identify a single, measurable result to anchor their strategy development and collaboration, and to provide an ultimate barometer of shared accountability and success. As its north star metric, the CDW selected weekly income from employment and set a target to increase the median weekly earnings for all youth to within 90 percent of the county living wage, using the MIT living wage standard.[3] The metric and associated target allowed for benchmarking that accounted for the disparate cost of living across counties, and for comparison against state-level administrative data from the Bureau of Labor Statistics.

In addition to analyzing organization- and population-level data on youth earnings over time, the team conducted a factor analysis of positive and negative influences on earnings to illuminate the "story behind the data," and identify the belief systems, legal and regulatory structures, and practice norms contributing to inequities for foster youth in obtaining living wage employment. This process resulted in the identification of several factors that became the focus of the CDW strategies at the organizational and systems levels.

Prior to focusing on this north star, First Place had become mired in conflicting priorities—set internally by different departments and externally by public and private funding sources—complicated data collection practices, and competing targets. The CDW recognized that increasing living wage employment would require an organizational reset and the focused alignment of effort across leaders, colleagues, stakeholders, and partners. This focus would help the organization pursue public partnerships and private funding with intentionality, streamline and simplify its data collection and performance monitoring, and reduce program burnout.

ALIGNING TOWARD A NORTH STAR OUTCOME: EMBRACING A "LESS IS MORE" APPROACH

The CDW developed and implemented a set of process strategies to help redirect and align the organization's attention at all levels toward the living wage result.[4] This required taking a "less is more" approach to shape and implement strategies that drive impact. The following examples showcase how a razor-sharp focus on the north star outcomes strengthened team, measurement, and collaborative functions.

Re-Envisioning the Continuous Quality Improvement Team

An unspoken rule of thumb at First Place had dictated that more is always better: more meetings, more metrics, and a constant push to deliver and scale new program innovations. This resulted in a lack of clarity around priorities, roles, and decision making. Without a clear point-of-view on how to facilitate change, the organization vacillated between two extremes in implementation—premature scale *or* abandonment of innovations.

To address these adaptive challenges, First Place re-imagined and reconstituted its existing continuous quality improvement forum into the Practice Innovation Group (PIG). The PIG was rebranded as an implementation off-shoot of the CDW. It brought together key decision makers and leaders from the Program and E&L departments charged with the design and implementation of strategies engineered to impact living wage employment. PIG leaders engaged critical stakeholders, including young people and direct line staff, to shape the design and ownership of CDW strategies. Meeting cadence was shifted from bimonthly to weekly, allowing dedicated time for planning, problem solving, and constant alignment toward the north star result.

Bringing Policy to the Learning Table

As part of its new "less is more" approach, First Place also revisited the relationship between its program learnings and its policy change agenda. The organization had long operated from the perspective that the data generated from service delivery should drive its policy agenda. Over time, however, departmental functions had become siloed, diminishing that focus. Data remained integral to the organization's policy activity, but increasingly, policy priorities were set independently, and data was pulled subsequently to support policy arguments.

Focusing on the living wage result across all departments involved three major shifts to organizational practice, starting with ensuring that the Policy department was "at the table" with E&L as it shaped research questions and metrics for collection and evaluation. This ensured that First Place was asking questions that were relevant not just to the organization but to external stakeholders, systems, and policy audiences. Next, Policy began partnering intentionally with E&L to analyze findings on the impact of living wage strategies to develop recommendations for system reform. Finally, advocacy efforts were targeted on increasing system-wide access to employment and education-related data.

Eliminating the Noise

Prior to the CDW initiative, the E&L department was overburdened, tracking more than 120 internal performance and outcome metrics as well as countless contract metrics reported regularly to its public and private funders. This caused stagnation within the department due to the constant need to focus staff time and effort on producing metrics or fixing calculation errors. These conditions had begun to erode programmatic trust in the accuracy and relevance of the data produced. E&L needed to restore trust and free up staff time to measure impact, analyze trends, and uncover pathways to improved living wage employment outcomes.

After soliciting feedback from front-line staff and organizational leaders, the department reduced the number of internal outcome metrics from thirty down to just five—focused on stable housing, post-secondary persistence and completion, employment, and attainment of living wage. The streamlined metrics aligned directly with the CDW strategies and reflected a more sophisticated, nuanced, and specific assessment of incremental progress toward a living wage. By focusing on youth outcomes at exit and setting specific targets for the next fiscal year, the metrics better aligned staff focus with the mission of the organization.

E&L provided access to all five metrics in a well-designed dashboard updated each morning and available to all staff. Research jargon was replaced with language familiar to social workers, and the development of reports started with staff needs and ideas solicited throughout the creation process. Staff were able to look closely at individual youth progress, drilling down on the data to support outcome attainment. Simultaneously, E&L eliminated countless forms and reports in First Place's case management system that had amplified the data burden felt by staff without providing sufficient value in supporting outcomes. With increased data literacy across the organization, greater focus on living wage employment, and accurate, real-time data, the organization had more capacity to develop and launch new innovations.

CREATING SUSTAINABLE CHANGE STRATEGIES

After creating the enabling conditions to pursue a living wage strategy, the CDW turned its attention to sustainable implementation. First, the organization contracted with BCT Partners to develop a decision support data tool for practitioners using precision analytics and quasi-experimental methodologies. The Youth Roadmap Tool (YRT) helps staff and managers identify and target effective services to optimize each youth's likelihood of success. BCT Partners leveraged five years of First Place data to identify the largest predictors of living wage attainment by program exit. The data modeling was used to create a web-based dashboard for front-line staff and managers, with custom snapshots showing where youth are in programs and which high-impact services and goal areas will most impact the north star outcome.

The PIG successfully led the development and launch of: 1) a post-secondary education coaching innovation that provides youth with the pre-enrollment information, experiences, and networks necessary to increase persistence, attainment, completion, and earning potential; and 2) a pre-apprenticeship and apprenticeship programming model designed to help young people "earn while they learn," receive incremental pay increases, and secure living wage jobs in high-demand markets. For each of these innovations, a measurement line of sight and specific progress indicators were designed to evaluate incremental success toward the weekly earnings target.

Finally, First Place's charge to elevate system-level attention and expectations toward the north star outcome manifested in a data and research partnership with Mark Courtney and the team at the University of Chicago responsible for implementation of the CalYOUTH Study to create greater access to employment and education data among providers and system stakeholders.[5]

RESULTS AND REFLECTIONS ON THE WAY AHEAD

Two years into its execution, promising results of the CDW work are emergent and will unfold over years to come. First Place has released a research and policy brief titled *Raising the Bar: Building System- and Provider-Level Evidence to Drive Equitable Education and Employment Outcomes for Youth in Extended Foster Care* (Van Buren, Schroeder, and York, 2021). The brief shared key findings and emergent learnings from the CalYOUTH and BCT partnerships, and provided targeted practice and policy recommendations aimed at increasing living wage employment system-wide. In the state legislative cycle corresponding with publication of the brief, First Place collaborated on two state bills that would target investments accordingly, and uplifted key findings in conversations with advocates and public policymakers.

The PSE coaching intervention is being actively evaluated, and is demonstrating early evidence of success in supporting young people to persist in educational placements. Likewise, the pre-apprenticeship and apprenticeship program is moving through a generative learning and development phase to determine uptake, retention, and need for refinement. The Youth Roadmap Tool has been adopted *by* practitioners *for* practitioners, and Program and E&L are collaborating on the development of an evaluation dashboard to monitor uptake and success of the tool. These new interventions have taken root as the way First Place works, and the process underpinning their development has generated more ownership of the data, a more individualized and evidence-informed approach to service delivery, and a culture shift around expectation for impact. In fact, an initial round of staff surveys revealed that the data literacy across the organization increased significantly, and practitioners reported they were more likely to apply data to practice.

Perhaps the most powerful result is that First Place is aligned toward equitable results in ways it never was before; the entire organization is hyper-focused on living wage attainment for *all youth* during a time when this outcome could not be more important. Despite the challenges and setbacks from the COVID-19 pandemic, feedback from youth and staff has been positive, and the new tools are expected to help youth rebound more quickly. As strategies build evidence of impact on the core result, First Place will scale these interventions with the same approach used to develop them: an unwavering focus on the north star.

NOTES

1. The story of First Place for Youth is a tale of the relentless, often rocky pursuit of meaningful results and systems change in service to young people aging out of foster care into the world of independence. The organization's guiding vision holds that involvement in the foster care system should not limit the opportunity to thrive in adulthood; in contribution to that vision, First Place's mission is to support foster youth to build the comprehensive skills needed to make a successful transition to self-sufficiency and responsible adulthood. Since its founding in 1998 in Oakland, California, First Place's core program model—My First Place (MFP)—has evolved into one of the largest providers of housing, care management, employment, and education services for youth who were in California's foster care system on their eighteenth birthday. The MFP program provides youth in extended foster care with a stable foundation of housing, along with employment and education services, intensive care management, and a focus on youth-driven skill development in key self-sufficiency areas to promote a successful transition to independence. Its influence has expanded to other states as one of the only evidence-informed placement models nationwide for older youth in care.

2. Dr. Mark Courtney and his research team at the University of Chicago conducted seminal policy research on the impact of California's extended foster care policy on key outcomes for youth (Courtney, Okpych, and Park 2018). Findings from this research mirrored those uncovered by First Place: participation in extended foster care yielded a myriad of positive and preventative outcomes for youth, but the system was not making meaningful strides in post-secondary persistence, degree attainment, or, perhaps most notably, in significant change in income.

3. Massachusetts Institute of Technology, Living Wage calculator, accessed 1/4/23, https://livingwage.mit.edu/.

4. The CDW initiated a radical shift in the way organizational leaders thought about working vertically and laterally across departments. This included clear interdependencies, distributive leadership roles, and shared accountability for leveraging data, building evidence, and holding the living wage result at the center. To optimize the collaborative role each department would play in supporting the result, the CDW established clear aspirations and a container for strategy execution: 1) program teams would implement program strategy innovations using Plan-Do-Study-Act (PDSA) continuous quality

improvement (CQI) cycles; 2) E&L staff would develop mechanisms to provide access to *better* data on employment and education and support CQI design, execution and learning habits, and coaching on data practices; and 3) policy staff would partner with E&L to shape the questions to be addressed through research and data, analyzing learnings to determine an advocacy agenda, and disseminate knowledge to drive systems change.

5. The double bottom line for the CalYouth partnership was: 1) to expand the knowledge base on the impact of diverse foster care placements on post-secondary and employment outcomes throughout California; and 2) to create broader provider-level access to administrative data on longitudinal PSE and employment outcomes in the interest of refocusing their attention and interest on improving living wage employment outcomes among the youth they serve. Policy and E&L collaborated closely to form a stakeholder workgroup of extended foster care providers across the state who helped design an "outcomes snapshot" that would provide multiyear data on post-secondary and employment outcomes for each individual provider, leveraging reliable administrative data from California's Employment Development Department and National Student Clearinghouse. Larger providers also would receive quasi-experimental analysis comparing their risk-adjusted outcomes against state-level averages and other extended foster care providers in similar counties. The ultimate goal in pursuing these dual aims was to support a systems-change (policy) agenda: strengthening the evidence base to push for more targeted investments in the extended foster care system in the areas of employment and education, and democratizing access to data to drive change at the provider level.

REFERENCES

Annie E. Casey Foundation. *Introduction to the Results Count Path to Equity: A Guide to the Accountability for Equitable Results Framework*. Baltimore, MD: Annie E. Casey Foundation, 2019. www.aecf.org/resources/introduction-to-the-results-count-path-to-equity/.

Courtney, M. E., Okpych, N. J., and Park, S. *Report from CalYOUTH: Findings on the Relationships between Extended Foster Care and Youths' Outcomes at Age 21*. Chicago, IL: Chapin Hall at the University of Chicago. 2018.

Massachusetts Institute of Technology. *Living Wage Calculator*, n.d., https://livingwage.mit.edu/.

Van Buren, E., Schroeder, J., and York, P. *Raising the Bar: Building System- and Provider-Level Evidence to Drive Equitable Education and Employment Outcomes for Youth in Extended Foster Care*. Oakland, CA: First Place for Youth, 2021. www.firstplaceforyouth.org/our-work/publications/raising-the-bar/.

GEMMA SERVICES

GENERATING ACTIONABLE EVIDENCE FOR PRACTITIONERS

PETER YORK

Gemma Services—a youth-oriented social services agency that operates a long-term residential psychiatric care program for youth—found that their administrative data system, while extensive, generated little data that could be used by front-line practitioners as they worked directly with youth and families.

First Place for Youth (FPFY) helps youth who have aged out of the child welfare system build the skills they need to make a successful transition to self-sufficiency and responsible adulthood. The leaders of both organizations believed strongly in the need to rigorously evaluate their programs so they could produce the kind of evidence that would advance their programs and practices as well as hold themselves accountable to achieving positive client outcomes.

Like FPFY, Gemma Services considered conducting evaluations using more traditional randomized controlled trials and quasi-experimental evaluation designs. Instead, they chose, as so many providers do, to invest in a program administration data system that would serve to assess, monitor, and evaluate the outcomes of every client throughout their program experience. Both organizations reached the conclusion that, while these data systems served an important program administration purpose, including being able to report to their funders on their amount of service outputs and costs, they were not meeting their evidence generation needs. This was especially the case when it came to their practitioners and clients. Practitioners were the principal data collectors, spending hours every week gathering information and assessments from clients and inputting this data into the system. However, they received no evidence in return on which they could act to learn about and strengthen their program planning and engagement.

ON-DEMAND EVIDENCE

In 2018, the leaders of both Gemma Services and FPFY sought to build a technological solution that used predictive and prescriptive models to put on-demand actionable evidence in the hands of their practitioners on a daily basis. New analytic studies in justice and child welfare showed it was now possible to do so using the program administration data they already were collecting.

With support from BCT Partners, an evaluation and data science firm with a mission to provide insights about diverse people that lead to equity, the precision analytics (PA) approach Gemma Services adopted is designed to meet different needs than traditional summative evaluation designs, including RCTs and quasi-experimental group comparisons. While such designs are useful in examining the overall effectiveness of a program, practitioners making choices about how to treat or serve the client they are meeting right now need more contextualized and precise information. The PA approach is different in that it applies causal modeling to historical program data by finding similar subgroups of cases and determining the ideal set of tailored program recommendations that maximized their success. Put another way, PA finds naturally occurring experiments where some cases within the same "like" group received a set of services, while others did not, and learns which combination of services maximized success.

Practitioner Buy-In

Gemma Services' leaders knew the success of their adoption of precision analytics would require the buy-in and support of their practitioners. As a first step, they scheduled a series of practitioner meetings to present on the concepts of data science and machine learning and to share how these tools, combined with their data, could produce case-specific predictive and evidence-based recommendation insights. BCT worked with the practitioners to co-develop ground rules for the precision analytics approach. One key driver for practitioner buy-in was the explicit support of the program leaders and managers that was garnered through the establishment of these ground rules.

Data Readiness

The precision modeling process began with a data readiness audit, which took approximately ten weeks to complete. Both Gemma Services' and FP-FY's data met the minimum requirements identified by BCT: a minimum of 250 cases that do not have too much missing data; at least two years of reliable longitudinal data; case background, situation, history, and needs; program delivery and transactions; ongoing goals, milestones, and accomplishments; and outcomes. Additionally, both organizations requested conducting additional preliminary analyses to test the feasibility of generating results during the full precision modeling process.

The Precision Modeling Process

Once the audit was completed, there were five steps to the precision modeling process:

1. Develop the Analytic Framework

The first step was to develop and refine the program logic model, review and assess the administrative data, and align the logic model with high-level data constructs that were well represented across the administrative data variables.

2. Get the Data Ready

This step, which consumed the largest proportion of the total project time, began with the process of extracting the data, based on the analytics framework, from each organization's data system.

Then, the data had to be transformed into the final set of variables that would go into the precision modeling step. Additionally, different types of statistical and machine learning scaling techniques were used to construct metrics made up of sets of variables that represented measures of all of the components of the program logic model. The transformation process even included creating "predictive" scales, a technique that builds predictively using machine learning algorithms. This step culminated with selecting the final transformed and logic model–aligned variables for the modeling process and creating prefix tags associated with the different logic model components for each of the variable names to help guide the modeling.

3. Conduct Precision Modeling

The process of building the precision model entailed the evaluator closely collaborating, through multiple screen share sessions, with the data scientists to conduct a series of modeling steps that found matched comparison groups to reduce selection bias, discover what works for each group, and evaluate the effect of what works.

The first step in the precision-modeling process was to *build an "ideal program model."* This model was a predictive model that used all the program dosage, strategy, and goal attainment data to predict the desired outcome.

For example, Gemma Services used their goal attainment data, which represented the tracking of longitudinal changes in thought, behavior, psychiatric, and trauma assessment scores, to predict a child's acuity level at the time of discharge. As noted earlier, Gemma lacked good intervention dosage data; instead, they used assessment score changes as their proxies for what happened to a child during their residential treatment. This first "ideal program model" was able to identify the goal accomplishments most important to reducing a child's behavioral acuity to a level associated with a discharge that was much less likely to return to inpatient hospitalization within the subsequent year. This "ideal model" was approximately 75 percent accurate. Gemma Services' practice expert knew that one size would not fit all children. To reduce selection bias, this goal attainment model produced a probability for every child as to the likelihood they would have achieved a low enough acuity to be considered a success, based on their accomplishment of assessment-based goals.

The next step in the quasi-experimental precision analytics process was to *identify matched comparison groups* based on contextual and baseline intake characteristics that predicted how likely a child was to engage and/or be engaged in the ideal program model.[1] By training machine learning classification algorithms to cluster children into groups based on sharing characteristics that make them equally likely to receive the ideal program model, this step identified matched comparison groups that could be studied and evaluated during the next steps, thereby minimizing selection bias.[2]

The next step in the precision modeling process is to *train machine learning algorithms* to determine which combination of program interventions, dosages, and/or goal achievements predict the highest likelihood of success for children within each matched comparison group. The evaluator guides the data scientist through an algorithmic training process that identifies the program elements that predict the best outcomes for each matched group. These algorithms are able to produce a ranked and weighted set of program elements that uniquely and in aggregate contribute to achieving the desired outcome. The findings are a group-specific combination of program elements that, when combined, increase the likelihood of a matched cluster of children achieving a positive outcome.

The final step in the precision modeling process is to inferentially *evaluate the effect* the group-specific program model had—in the past—on a matched group of children when some received what works and some, counterfactually, did not. The analytic process included conducting inferential statistical tests (for example, t-tests, ANOVAs, etc.) to determine if those within a specific group who received what works achieved a significantly better outcome score than those in the same group who did not. Effect sizes were also calculated.

4. Automate the Analytic Process

The fourth step in the overall process is to engage the data scientists in automating the analytics workflow such that data extraction, transformation, and loading, precision scoring, and results generation all happen at least once a day without requiring any human to initiate the run. This step also requires determining how to keep the data secure, confidential, and HIPAA compliant throughout the entire data transfer process.

5. Produce Daily Actionable Evidence

This final step serves to design and implement a suite of dashboards for an organization's practitioners, program managers, and leaders to receive on-demand insights—actionable evidence—to be used for case-specific, programmatic, and organizational decision making. This step required data scientists with design and visualization training and experience. Most organizations now begin with templates created by BCT that use software programs like Tableau and PowerBI to generate dashboards, reports, and visualizations for practitioners.

CHALLENGES AND RESPONSES

Evaluators and practitioners need to be educated about the use of machine learning algorithms and big data analytics for social science research and evaluation. They need this knowledge to develop an understanding of how to collaborate with data scientists to build these tools. BCT worked closely with practitioners from the very start to co-develop "ground rules" for the use of algorithms and to help them understand concepts such as "selection bias."

Separate individual practitioner performance from the dashboards. Most practitioners embraced the opportunity to have data-driven real-time feedback to guide their work and make course corrections. At the same time, practice experts and program managers quickly realized the dashboards could be seen as a performance assessment tool for practitioners and that this could be problematic. The managers addressed this philosophically by sharing that the focus of the tools was on each child or youth and ensuring their progress, not on job performance. More practically, practice experts worked with managers to develop a set of resources specifically addressing each recommendation, so practitioners could easily access and review how to implement what was being suggested.

Another key challenge is *having enough data to scale this type of work*. Many nonprofit providers do not have the 250+ cases of longitudinal case-level data to get started. However, as the cost of program administration systems and the number of vendors grow, there are many more organizations that have and/or are in the process of setting up and implementing robust program administration data systems. So, there are many more organizations that are getting ready or already there. The learning networks described above also may provide on-ramps for organizations not yet ready to create their own modeling and tools.

A fourth challenge is *ensuring that all identifying data are protected and secure*. Technologies are now in place that, whether within organizations or in the cloud, protect the identity of cases in datasets. Organizations that could be a part of a learning network would not have to share data, but could keep it secure behind their own firewall or behind the secure firewalls of HIPAA-compliant cloud platforms like Amazon AWS. The learning hubs could leverage aggregate findings. If there is a desire to share data, there are efforts of organizations like Brighthive[3] that create data trusts, which, both technologically and through policy agreements, create shared data architectures and processes that protect information.

PROVIDING INSIGHTS IN REAL TIME

Gemma Services has begun implementing their dashboards, and practitioners now are using actionable evidence in real time. The dashboards and tools resulting from precision modeling are designed to be actionable not just for practitioners but also for program developers and managers, organizational leaders, and system change agents, including policymakers and funders. Additionally, the goal of these tools is to improve program delivery in a tailored way that will improve outcomes. The evaluation tools will provide the insights, and the resource guides and the learning and professional support of peers and managers will help practitioners take the actions to increase the proportion of those who achieve the desired outcomes.

The precision analytics process and associated tools led to a culture shift in how practitioners at Gemma Services do their work and how performance is evaluated. Prior to the development of the learning and evaluation dashboards, progress and results of practitioner-led case-specific decisions were not available to practitioners (and were certainly not available in real time, updated on a daily basis). It was now possible to measurably view each child's or youth's progress, performance, and outcomes in a practitioner's caseload as well as the performance of their whole caseload, with updates on a daily basis.

Some of the early results of having immediate and up-to-date actionable evidence include:

- **Practitioners use case-specific dashboard evidence for case engagement.** For example, clinicians working with children in the residential milieu at Gemma Services are using the individual dashboard to understand more about the needs of the children they are working with on a daily basis.
- **Program directors use cluster evaluation information, including the names of children at different current outcome levels, for case planning.** This allows directors to meet with front-line staff to investigate why different youth are in different

places, examine what the data are beginning to indicate, and, most importantly, to more qualitatively examine the root causes for a child's outcome status. Program leaders find these inquiries, made possible by having up-to-date evidence, are strengthening case plans. Real-time feedback also allows for rapid plan changes.

- **Precision tools are motivating program directors and front-line practitioners to gather more data.** In the past, both organizations faced challenges with getting assessments completed on a regular basis. Now, with dashboards that update in real time, including indicating the date when the last assessment was conducted and/or data was entered, practitioners and program managers are visually cued to gather data more frequently.

- **The precision tools are engaging practitioners to help improve the data.** As more program directors, managers, and practitioners make deeper qualitative inquiries into their cases and what is going on, they are beginning to realize they need additional data points to test hypotheses that cannot be answered by the current data. For example, the Gemma Services practice expert learned that program directors felt strongly that parent engagement was a critically important variable that was not currently being captured in the data. They hypothesized that the quantity and quality of parent engagement, both with the child and the clinical staff, were key determinants of achieving many of the goals needed to reduce acuity. So, the practice expert worked with the program directors to design a set of questions that will be tested and modeled in the near future for use on an ongoing basis.

- **Precision modeling findings are being leveraged for policy change.** First Place for Youth's director of public policy, vice president of learning, evaluation, and strategic impact, and academic research partners have used the findings from their precision modeling process to write a policy brief, Raising the Bar: Building System and Provider-Level Evidence to Drive Equitable Education and Employment Outcomes for Youth in Extended Foster Care.[4] The purpose of this paper is to encourage the state of California and federal policymakers to scale First Place for Youth's extended foster care model, in conjunction with their Youth Roadmap Tool learning system.

REFLECTIONS

Gemma Services is raising funds to scale and create a learning network of similar organizations everywhere, all learning together by adopting and/or building their own learning systems. In fact, Scattergood Foundation plans to fund Gemma Services to scale their models to other residential mental health providers throughout the region, state, and country.

Gemma Services' vision is to *become a collective learning hub for similar programs* in communities throughout the United States. First Place for Youth is beginning to plug new affiliate programs throughout the United States into their Youth Roadmap Tool. If a similar program doesn't have enough data, they could, for example, begin by adopting First Place for Youth's extended foster care question sets, algorithms, and dashboards until they have enough longitudinal data to build their own context-specific precision models and tools. If they have enough of their own program administration data, they could build their own precision models. At this point, the build cost is affordable to most larger organizations, and the ongoing support and maintenance cost is sustainable for medium to large organizations.

It is important to note that both organizations voiced a goal of *eventually engaging beneficiaries* in their precision modeling process. However, they wanted to focus on the practitioner for their first precision projects to better understand what the process and engagement actually entailed, to better inform and plan for engaging beneficiaries and their families. Both organizations still are primarily focused on practitioner utilization. That said, Gemma Services has begun developing data gathering instruments to gather information from parents as to their engagement with their child's residential treatment.

NOTES

1. This step is analogous to propensity score matching (PSM) statistical procedures use in tens of thousands of health, education, political science, economic, etc. peer-reviewed observational studies to minimize selection bias. However, training machine learning algorithms to identify matched comparison groups mitigates a significant problem that leading social science researchers and statisticians Gary King from Harvard and Richard Nielsen from MIT, identified in their 2019 paper "Why Propensity Scores Should Not Be Used for Matching," *Political Analysis* 27, no. 4 (2019), pp. 435–54, proving that PSM creates experimental imbalance.

2. This machine learning process, using simple decision tree algorithms, adheres to King and Nielsen's recommendation of using fully blocked matching instead of PSM.

3. See Brighthive's details, https://brighthive.io/.

4. "Raising the Bar: Building System and Provider-Level Evidence to Drive Equitable Education and Employment Outcomes for Youth in Extended Foster Care," https://firstplaceforyouth.org/research-brief-raising-the-bar/.

SECTION 5

REIMAGINING EVIDENCE TO BROADEN ITS DEFINITION AND USE

We must demand that governments, businesses, nonprofits and philanthropies do more to shift the massive amount of dollars to solutions that have measurable evidence of impact. But, we have to also expand our understanding of what constitutes evidence, grow our tent so more diverse voices and perspectives are included and evolve our concept of what classifies as an evidence-based solution from solely programs that meet immediate needs to policy reform that dismantles, disrupts and reimagines the broken systems that have failed far too many.

—MICHAEL SMITH, "SYSTEMS MUST CHANGE: DISMANTLING, DISRUPTING AND REIMAGINING EVIDENCE"

Empirical data can offer proof points but constitute just one element of the evidence equation. To build evidence that is more relevant, timely, and cost-effective, we must broaden its definition to include not only statistical but also practical significance, collecting a broader range of data, encompassing participant feedback, practitioner experience, community signs of change, and more. We must reimagine evidence to allow consideration of context, confidence level, size of impact, speed to insight, and cost of implementation. This is especially critical for state, local, and federal agencies as an influx of federal dollars flows to rebuild U.S. infrastructure.

In this section, author Michael Smith speaks to expanding our classifications of evidence to include a greater range of thought and diversity of contributors, and a broader definition of evidence-based solutions. The latter should include policy reform that disrupts and reinvents failed systems. Coauthors Jennifer Brooks, Jason Saul, and Heather King describe the next phase of evidence-based practice as designed with end users in mind to ensure application, a variation on Scholl's call to work backward. Coauthors Veronica Olazabal and Jane Reisman note increasing use of evidence in policy debates to misinform or disinform, and the need for contextualization that relies on more than mimicking scientific methods. Meanwhile, Brian Komar speaks to building evidence for environmental, social impact, and governance (ESG) efforts.

Focusing on government, Diana Epstein underscores the Evidence act's call to federal agencies to better connect evidence with strategy. Ryan Martin speaks to the need for more small sample studies to find dependent variables—"needles in haystacks"—in the spirit of fostering "a climate in Congress and elsewhere where failure is acceptable, evidence building is prioritized, and those running programs adapt based on what has been learned." Next, Michele Jolin and Zachary Markovits describe a quiet revolution in cities across the United States as they have embraced data-driven transformation to solve intractable problems like the opioid crisis. Vivian Tseng closes out this collection of essays with calls for a movement to democratize evidence-building away from elite powers that shape it today.

In use cases, the Stanford RegLab-Santa Clara County and Camden Coalition both demonstrate the power of cross-sector collaboration in evidence building in response to the COVID-19 pandemic. As a funder, United Health Care shares its strategy of partnering with community organizations to address social determinants of health by identifying those committed to outcomes, building evaluation plans with them and providing the funds to execute plans.

Questions raised and addressed in this section include:

1. How can we expand the definition of what counts as evidence?
2. How can we broaden who is included in evidence building to solve problems collaboratively?
3. What intractable problems can we tackle with a broader definition of evidence?

SYSTEMS MUST CHANGE

DISMANTLING, DISRUPTING, AND REIMAGINING EVIDENCE

MICHAEL D. SMITH

As I was preparing to write this essay, I sat down to watch a talk I gave at TEDxMidAtlantic in 2014. At the time I delivered the talk, I was director of the Social Innovation Fund, a program of the Obama White House and the Corporation for National and Community Service. At the Social Innovation Fund, we sought to combine public and private resources to prove, improve, and scale promising interventions in low-income communities. It is never easy to watch yourself talk, especially after so much time has passed. But watching this one, I found myself wincing even more than usual. Because while my passion about investing in what works has stayed the same, my feelings about how we got here and what we need to do about it have evolved greatly since that talk. In the intervening years, I have spent more time working closely with organizations doing the hard work of building safety nets and springboards on top of what can feel like a bottomless cavern of neglect, institutional racism, and lack of investment where it is needed the most. What I have come to realize is that we need to radically reimagine our approach to evidence in the social sector.

During those remarks almost ten years ago, I discussed why we need to invest in evidence-based solutions if we want to transform the nonprofit sector, find a bigger impact, and get more results. Americans are obsessed with data, rating, and reviewing, but for some reason, that obsession does not apply to the nonprofit sector. Back then, we were spending $300 billion a year on more than 1.5 million nonprofit organizations, but one in eight nonprofits that year spent zero dollars on evaluation, and more than half did not have a theory of change or a logic model. And while these problems seemed to be getting bigger and more complex, we were making decisions on which nonprofits to invest in based on anecdotal stories of success and numbers served. I concluded that I believed the best way to ensure that our limited dollars find their way to the most deserving nonprofits was to follow the evidence of impact, and I cited the Social Innovation Fund as an example. At that time, the fund had invested more than half a billion state and federal dollars to more than 200 organizations that were testing about eighty-six different models with some evidence of impact.

This work matters deeply to me. I grew up in a poor Black neighborhood in a small city in New England. My mom and dad were single teenage parents; with what we knew from data on kids like me, I should have been an unfortunate statistic. My saving grace? The neighborhood Boys and Girls Club. But as I got older and spent my career working in the nonprofit sector and philanthropy, I was forced to wonder why, if that youth center was so transformative for me, did so many of my childhood friends end up struggling in so many different ways? In fact, in a less-than-five-mile radius in my little neighborhood, at least a half dozen organizations, all separate 501(c)(3)s, were doing very similar work, but with diminishing results. In spite of these well-intentioned organizations, my city struggled with poverty, teen pregnancy, and low graduation rates. Too many youths who looked like me were victims of homicide—including my younger brother, who was killed at the age of twenty-seven.

In my talk, I pondered aloud why, despite the fact that my family's story is shared by countless others in this country, we keep doing the same things over and over again, expecting different results. Then I shared some thoughts on what we should do. First, I suggested that individuals, who are responsible for 80 percent of giving to all nonprofits, stop giving dollars to any organization that cannot articulate impact. I defined impact not just as how many people were served, or as isolated success stories but, rather, as how many kids went to college, and stayed in college; how many got jobs, and how many kept them. Second, I shared that we've got to know when to walk away. The nonprofit sector does not face the same market forces that drive dollars away from ineffective solutions in the business world. Sometimes when social sector approaches have been poked and prodded and they no longer work, we are going to have to say "No more money." Finally, I shared that the philanthropic sector would have to push for more mergers or acquisitions to build stronger organizations instead of the "behemoths and masses of ineffective organizations that are out there." (Cringe, I know.) And, we should then bet our money on the winners. Some nonprofits might shut their doors, I thought at the time. Some might go away, and many need to. That might sound harsh. But, perhaps, this focus could put an end to the "Hunger Games" that we've created where nonprofits that aren't growing are fighting each other for scarce dollars.

Now, you can probably tell why I am cringing. So arrogant—so pompous. Somewhat out of touch. I put all the burden on the organizations doing the hardest work; trying to keep their doors open in and out of recessions and near-depressions. And organizations that so many families will turn to in times of greatest need, as we have just seen during the pandemic. Here is what changed my mind.

WHAT INFLUENCED ME?

At the Social Innovation Fund, I worked with hundreds of nonprofits of all shapes and sizes that were struggling to meet basic needs of their constituencies while also building evidence that would pass the scrutiny of funders. Imagine if the organizations could focus on their core missions without constantly trying to prove themselves? In the aftermath of the tragic killing of Trayvon Martin and the shocking trial where his murderer was acquitted, I helped design President Obama's My Brother's Keeper initiative (MBK), and in 2014, I left SIF to lead MBK. We aimed to address the persistent opportunity gaps facing boys and young men of color and ensure all youth could reach their full potential. And that is what the president talked about when he launched the program; he addressed the nation about the urgency of making sure every kid in this nation, no matter their background or neighborhood, knows that their country cares about them, values them, and is willing to invest in them. He also spoke about the urgent need to focus on evidence, data, and results, or, more simply, investing in what works and building on what works. "We don't have enough money or time or resources to invest in things that don't work, so we've got to be pretty hard-headed about saying if something is not working, let's stop doing it. Let's do things that work. And we shouldn't care whether it was a Democratic program or a Republican program, or a faith-based program or—if it works, we should support it. If it doesn't, we shouldn't."[1]

In 2018, the MBK Alliance announced the winners of our inaugural national competition to identify and invest in communities making steady progress to substantially improve the lives of boys and young men of color. The critical importance of the work of these extraordinary organizations and countless nonprofits like them became even more clear in 2021. In the wake of the disproportionate effect that COVID-19 and ongoing racial injustice was having on under-resourced Black and brown communities, these high-performing organizations continued to meet their core operational goals to reduce barriers and expand opportunity for boys and young men of color and their families—but they did not stop there. In the face of so much uncertainty and overwhelming obstacles, they took on even more. They began serving meals; delivering food; handing out personal protective equipment and literature; creating mutual aid networks; helping organize and support calls to action against police violence; and responding to increases in street violence. When I sat back to think about the kind of hoops organizations like these—and many we invested in through SIF—have to go through to prove their work without the resources needed, it convinced me even more of the need to rethink how we approach gathering and applying evidence.

WHAT HAVE I LEARNED IN THESE PAST EIGHT YEARS?

First, I would challenge us to tackle the system, not the nonprofits struggling to hold together a society that was never built to support its most vulnerable citizens. When I gave that TEDx talk, I did not spend any time talking about the massive gap in funding evidence-based programs versus the need to invest in evidence-based policy reforms that seek to dismantle the inequitable systems that created the conditions we need to address in the first place. When we put all the pressure on nonprofits trying to address the base of the hierarchy of need, we give policymakers, business leaders, and everyday citizens a pass on investing in change at society's roots. No amount of randomized control trials and evidence-based interventions will combat the legacy of redlining; Jim Crow; redistricting; under-funded, inequitable schools; and the prison industrial complex. But even when approaching systemic change, there is an opportunity to invest in evidence-based policy reforms that prove they work, such as eliminating external school suspension; increasing support for restorative justice, diversion, and other violence-prevention initiatives; increasing access to public spaces for young people; and facilitating opportunities for trained adults to mentor underserved youth. We also cannot forget that the road to macro reform is paved with lots of micro changes that aren't splashy but are pivotal to transformation, such as government budgeting processes; procurement processes; community engagement; capacity building; and data systems. One example of this micro shift is equity budgeting, which suggests radical intentionality about the inclusion of vendors, contractors, and businesses that are led by people of color and organizations led by residents.

Second, I downplayed the importance of balancing statistics and storytelling, as well as what we consider to be acceptable evidence from the start. Now I see it is not only about better science. It also is about being proximate to the need. It is about letting people closest to the pain be closest to the power because they hold both the causes and solutions in their daily, lived experience. It is about providing the time for rumination and reflection. It is about mirroring the data with the emotion. It is about spending time in communities that don't have the resources to build complex evaluation models but, for some reason, are outperforming the rest. To paraphrase Edgar Villanueva, author of *Decolonizing Wealth*, we have to resist our colonized mindset. We have to resist wanting our solutions tied up in neat, polished packages with the perfect prose from elite institutions and, instead, set our sights precisely on where change is happening.

Here is an example. Early in my career, I was part of philanthropic efforts to bridge the digital divide, distributing computers and internet access across the country. I had one grantee that was not as responsive as others. He submitted reports late, and they were incomplete. I decided to make a site visit to ensure our funds weren't being swindled. I landed at the airport with my MapQuest directions printed out, picked up my rental car, and headed to Ferriday, Louisiana (population 3,312). I drove by shotgun shacks, abandoned homes, and kids playing with homemade toys. I pulled up to our community technology center, which was housed in one of those homes that looked like it was on its last legs. Kids were running in and out, talking, laughing, and learning, and using the computers in a room heated by a wood stove. I met the director, who shared with me that his late mother, who had cared for children in the neighborhood, had left him the house; he could think of no better way to honor her than to create an informal after-school program. It was one of the only safe places in the area kids could go after school, get some homework help and mentorship, use a computer, and get a snack. And the leader was keeping it together out of his own pocket with some occasional grants and some help from the neighbors. No quasi-experimental design could have shown me what I saw with my own eyes and heard in the stories from that servant leader and the children and families I met that day.

We need data. We need evidence-based approaches. We need to trust but verify. But we also need to listen, look closely, create avenues for storytelling, and clear on-ramps to creating social impact for individuals without access to the resources that come with privilege. We need to recognize that the organizations serving the soup and handing out warm coats also may be the best advocates and engineers of revolutionary reforms to address hunger and homelessness. And, we need to reimagine our definition of evidence-based approaches so each of these iterations and innovations at every stage is part of the solution.

Third, if we are asking nonprofits to save lives and stay on top of the science that guides their service delivery and advocacy, we have to do more than talk about it. We cannot tell nonprofits to invest in what works without changing the way government and philanthropy fund operations, administrative costs, evaluation, research and development, data collection, and analysis. And, we need to help build the capacity of organizations of all sizes so they can own their data collection and evaluation instead of having to rely on outside firms with less cultural competency, less understanding of the community, and approaches that turn participants into subjects of a study, which can feel punitive and remote.

We also cannot ignore the fact that smaller organizations, and organizations led by people of color, find themselves constantly facing closed doors when it comes to the kind of resources they need to invest in infrastructure and growth. A recent study by the Bridgespan Group and Echoing Green showed that, in 2019, the revenues of Black-led organizations were 24 percent smaller than those of their white-led counterparts, and the unrestricted net assets of the Black-led organizations were 76 percent smaller than those led by whites. We also know that bias shows up in evidence and evaluation processes on all points on the spectrum. White researchers receive National Institutes of Health (NIH) grants at nearly twice the rate Black researchers do. Changing this paradigm starts with funding small organizations; investing in building a pipeline of more researchers of color from the communities undergoing evaluation; expanding the evidence toolkit to be hyper-inclusive; and leveraging practices such as government budgeting, procurement, and pay-for-performance (paying based on outcomes)—all of which can be tools for creating meaningful community engagement and more equitable structures.

I have long believed in the proverb, "If you don't know where you're going, you'll end up somewhere else." None of us can afford to spend our days tilting at windmills, hoping our work will transform lives. We have to demand that governments, businesses, nonprofits, and philanthropies do more to shift the massive amount of dollars to solutions that have measurable evidence of impact. But, we also have to expand our understanding of what constitutes evidence. We have to grow our tent so more diverse voices and perspectives fit under it and have a seat at the table. And we must evolve our concept of an evidence-based solution from a program that meets an immediate need to include policy reform that dismantles, disrupts, and reimagines the broken systems that have failed far too many. If I could give my talk over today, that is what I would say.

NOTE

1. Barack Obama, address at the launch of My Brother's Keeper, February 27, 2014.

THE NEXT GENERATION OF EVIDENCE-BASED POLICY

JENNIFER L. BROOKS, JASON SAUL, AND HEATHER KING

INTRODUCTION

The current approach to evidence-based policy and its cousin evidence-based practice (referred to collectively as EBP here) primarily focuses on documenting extant evidence while minimally addressing the use of that evidence. This model is static—sharing information about pre-packaged programs with little information on key elements of effective practice. Moreover, rarely are these approaches designed with the end user in mind.

The future calls for more innovative approaches to synthesizing evidence, updating information, and sharing information with end users, including funders, policymakers, and practitioners. This will require a way to systematically review research evidence to identify core components of interventions, a process for keeping evidence reviews updated in real time, and tools to make the resulting information actionable for practitioners, funders, and policymakers.

THE STATUS QUO: A STATIC APPROACH TO EVIDENCE

The rationale behind EBP is strong. Yet, for all the effort put into EBP, it is not clear what it has brought us as a nation. The evidence for improved outcomes is scarce. Evaluations of key evidence-based policies have not shown the desired effects.

The Tyranny of the RCT

One of the strengths of the current EBP movement—its focus on methodological rigor—has, in some ways, stunted this movement's usefulness. Current EBP approaches have privileged causal inference above all other elements of study quality.

RCTs are the best approach for drawing causal inference, but the problem is that their near-total domination in the fields of evidence-based practice limits the actionability of information available to end users. RCTs are most often used to study *packages* of practices—what we will call packaged program models. So, for instance, an education researcher might develop a packaged curriculum model that includes lesson plans, exercises, student assessment tools, and a professional development system for teachers. An evaluation of this packaged model will tell us how those elements—the curriculum, the assessment, and the teacher training—TOGETHER improve outcomes for students. But because the causal inference applies only to the full package of elements, this study would provide little information about the "active ingredients" in the model—those that are necessary versus those that might be nice to have. The study is essentially a "black box," telling us little about why the model worked or did not work.

THE NEXT PHASE OF EBP: MAKING EVIDENCE ACTIONABLE

The reason EBP has not been successful, either in take-up or in achieving outcomes, is because it was never designed with end users in mind. Current EBP models have led with the evidence base—what do studies with strong causal inference tell us and how can we share that information with people? Little consideration was given to what information practitioners want or need.

Two shifts are required to make evidence more useful to practitioners: 1) we must break open the black box of program evaluation to better understand the effectiveness of individual components of practice; and 2) we must share evidence in a way that meets the needs of the end users and reflects the wide variability in those needs.

Step 1: Breaking Open the Black Box by Focusing on Core Components

There is a movement afoot in the world of EBP to better attend to elements of practice, what are commonly called "core components." This movement is not new,[1] but it is gaining traction in the face of the limited success of EBP. Indeed, W. T. Riley and D. E. Rivera note that an emphasis on understanding components of effective practice is critical for intervention research to become a cumulative science.[2]

The ad hoc analysis of core components, however, is not sufficient to drive use and move the EBP field forward. We must find a way to standardize core components so they can be studied systematically. Standardizing core components has significant implications for how we catalogue studies and how information is shared.

Second, we must rethink how we source core components and what evidence is used to analyze them. Restricting core components studies to those meeting strong causal inference standards is likely to lead to a nearly empty database. That is because there are far fewer RCTs that examine elements of practice than there are RCTs of packaged program models. An empty database will not only be of no use to practitioners and funders but also will increase resistance to and frustration with the ideals of EBP.

We also must develop new methods to evaluate the efficacy of core components. Traditional meta-analytic techniques fall short for several reasons. First, traditional meta-analyses are static—large databases built from research studies are developed in silos by academic researchers, with each new field or area of study generating a new meta-analysis. The data are proprietary, nonstandardized, and rarely updated as the field progresses. The information gleaned from the meta-analysis is presented in a set of papers—locked inside PDFs—giving practitioners, funders, and even other researchers no ability to query the data or analyze it to address other questions.

To advance EBP, we, instead, need to find a way to build a common language to taxonomize studies—breaking them down into parts that can be standardized across studies and fields. Information about interventions, samples, contexts, and study design can be coded using a common dictionary. This will prevent the need for a "clearinghouse of clearinghouses," to address the siloed nature of the analyses. Standardization—combined with public access to the data and standardized coding—also will make it easier to update the evidence base over time and better understand variability of effects.

Step 2: Giving Practitioners and Other End Users the Ability to Access Information in a Way that Addresses Their Questions

The wealth of data available through a standardized core component approach will benefit end users only if it is accompanied by tools that allow them to make best use of it. To date, evidence registries have been the primary approach to sharing information about EBP. Those registries are minimally interactive. The information is contained in written reports or syntheses, with little opportunity to find information tailored to one's needs. In effect, these registries are like stagnant pdfs—with the information locked inside whatever format the registry developer deems best.

To make the core components of information most useful, we need to move from this static approach to sharing information to a more dynamic one. We need approaches to EBP that look less like a pdf and more like an app. End users should be able to query the data and tailor the information they receive to their own questions. To get a better picture of this, imagine a shift from static lists of mortgage interest rates to tools that allow buyers to tailor the information based on their context, the amount of money they want to put down, etc. In effect, we need to democratize the evidence, giving access to a broader range of stakeholders to use it however they need.

MAKING CORE COMPONENTS ACTIONABLE FOR PRACTITIONERS: THE IMPACT GENOME PROJECT®

The ideas presented above are what motivated the founders of the Impact Genome Project (IGP). Inspired by the standardization used in the Human Genome Project, combined with the use of algorithms to tailor information for clients on apps such as Pandora, the IGP standardizes information about practices, populations, contexts, and outcomes from research papers and other sources. The IGP mines the core components of practice found across thousands of studies—those small, bite-size, implementable pieces of information that are more easily translated for practitioners, funders, and policymakers.

To avoid the siloing of evidence we have seen to date, the IGP model aims to isolate and identify the "genetic code" (so to speak) that makes interventions effective. This allows the IGP to break down that finite list of practices or approaches common across fields from both each other and the *content* addressed in those approaches. For instance, it allows us to learn about features of more and less effective cash incentive systems, separate from whether the cash incentive is used to promote weight loss or school attendance. By using this approach, the IGP can pull evidence and data from a wide array of sources, ensuring cross-disciplinary learning.

The IGP relies on taxonomic meta-analysis, which uses the component as the unit of analysis rather than packages of components or interventions. Taxonomic meta-analysis is empirically driven, meaning that the taxonomy itself is derived from the literature base rather than established *a priori* from theoretical frameworks. Because the taxonomy is not dependent on discipline-specific theoretical frameworks, it can provide a common language for components that can cut across fields of research.

Some examples of how the IGP has been used include:

- A common genome for childhood obesity intervention research developed by a panel of experts and Mission Measurement with funding from the National Institutes of Health. The genome was then used to conduct a meta-analysis of core components within the field of childhood obesity prevention and intervention.[3] Similar reviews have been conducted in other fields, such as early childhood education, K-12 education, and financial health.
- A component explorer funded by the Chan Zuckerberg Initiative, which delves deeper into the data from the What Works Clearinghouse, allowing users to discover core components of interventions relevant to their work and compare their own program elements to the evidence.

The separate coding of practices, contexts, outcomes, and target populations across fields also allows the IGP to dig more deeply into the nuanced question: *"What works best **for whom, under which conditions, and why?**"* For example, analyses can focus on behavior change, attitude change, culture change, or all the above. They can examine how each of those strategies—or a combination of strategies—work with different populations in different contexts. They also can look at practices based on the type of change they aim for, whether targeting individuals, organizations, or geographically defined populations.

This latter point is critical if we want to address historic inequities both within the evidence base and through using EBPs. To date, most evidence registries have focused on interventions targeted at changing the behavior of individuals—teachers, parents, students, clients. Yet, many of the social problems in the United States reflect long-standing systems-level issues that individual-focused interventions alone cannot overcome. The design of the IGP will allow analysis of the interaction between systems-level levers of change (policy, public-private partnerships, advocacy, community organizing) and individual-level levers of change (individual-focused therapy, training, behavioral interventions). By coding and standardizing all these elements, we can begin to understand not just the components that drive outcomes but also which combinations of components can magnify our impact.

BRINGING END USERS INTO THE MIX

As noted, advancing evidence use will not be solved solely by standardizing the evidence base; we also must shift our focus to sharing evidence with users in a more dynamic and interactive way. By breaking evidence into components, the IGP has set the foundation for a more dynamic approach to interfacing with the evidence base. It does this in two ways: 1) by publicly sharing their coding infrastructure for others to use and add to; and 2) by supporting that infrastructure with user-friendly tools to interact with the evidence base.

In standardizing information, we must make both the data and the coding schemes available to researchers, practitioners, and funders to allow practitioners to compare their own programs to the evidence. This is quite different from how meta-analyses are typically developed, where coding structures are fragmented, hidden behind paywalls and may change over time. It allows practitioners to benchmark their programs and generate scenarios to strengthen their impact. Funders also can use the data to estimate the likelihood of positive outcomes from proposed strategies or compare strategies to one another.

By developing a user interface like the one in the IGP, researchers can provide a simple tool for practitioners to ask tailored questions of the evidence base and get reports that relate to their unique circumstances. In doing so, they can democratize the evidence base, putting it in the hands of those we want to use it. Moreover, interfaces that work with information provided by funders and practitioners also will provide more insight into the needs of the field, such as which questions are of greatest interest or which components of practice are most common.

THE FUTURE OF EBP

As we look to the future of EBP, we must remember that the primary goal of EBP is to improve outcomes through the greater use of evidence. While it is important to enhance the number of rigorous studies and share the evidence from those studies, this approach is not sufficient to promote evidence use. The future calls for more innovative approaches to synthesizing evidence, updating information, and sharing information with end users. The field is on the right track with its emphasis on core components analyses. But, alone, that shift will not meaningfully enhance evidence use. Rather, we will need to revisit how we systematically review research evidence, how we keep it updated over time, and how we make the information accessible and useful to practitioners, funders, and policymakers.

Luckily, researchers do not have to do this alone. We are experts in building and implementing research studies, but we are not necessarily experts at making information available and usable to different audiences. We should leverage the expertise of app developers, data scientists, and others to tailor our evidence systems to those we aim to reach. We should investigate how machine learning can help reduce the cost and delay inherent in our current labor-intensive approaches to analyzing the evidence base. If we can agree on a common language for coding evidence, new articles could be coded by the authors as they are published, so they can be included in the evidence base in real-time.

Most importantly, we must engage with end users to find out how to make evidence more actionable. There must be deep engagement with funders, policymakers, and practitioners—anyone we anticipate using the evidence—to ensure they can easily query the data and get answers to *their* questions. For it is only by addressing the needs of the end users that we will truly reach our primary goal, improving outcomes through evidence use.

NOTES

1. For example, see Embry and Biglan's notion of "evidence-based kernels." Dennis D. Embry and Anthony Biglan, "Evidence-Based Kernels: Fundamental

Units of Behavioral Influence," *Clinical Child and Family Psychology Review* 11, no. 3 (2008), pp. 75–113.

2. W. T. Riley and D. E. Rivera, "Methodologies for Optimizing Behavioral Interventions: Introduction to Special Section," *Translational Behavioral Medicine* 4, no. 3 (2014), pp. 234–37, https://doi.org/10.1007/s13142-014-0281-0.

3. National Collaborative on Childhood Obesity Research, "Childhood Obesity Evidence Base (COEB): Test of a Novel Taxonomic Meta-Analytic Method," accessed June 1, 2021, www.nccor.org/projects/childhood-obesity-evidence-base-test-of-a-novel-taxonomic-meta-analytic-method/.

EVIDENCE AND IMPACT IN A POST-COVID WORLD

VERONICA OLAZABAL AND JANE REISMAN

I want you to act as you would in a crisis. I want you to act as if our house is on fire. Because it is.
—GRETA THUNBERG, 2019 ADDRESS TO THE
WORLD ECONOMIC FORUM IN DAVOS[1]

Greta Thunberg's urgent plea to global business and political leaders in 2019 reverberated around the world. While most notably describing our changing climate, she foreshadowed what was to come—the deadliest pandemic seen in 100 years and a resounding call for racial reckoning—all driving a discourse on building back better (BBB).[2] At the heart of the BBB movement, a reference originally used for natural disasters, is a demand for a more just, equitable, and sustainable world.

This broader view has been burning for some time but was accelerated by the global humanitarian social and economic crisis spurred by the pandemic. As the Organization for Economic Cooperation and Development (OECD) reports, "poverty will rise for the first time since 1998 with 70–100 million people estimated to be pushed into extreme poverty, at least twice as many into poverty, with hundreds of millions of jobs lost and livelihoods affected."[3] Building back better is, thus, more than a catchy slogan. We are at a defining moment in post-industrial society about the purpose of capital in this critical moment, especially as it relates to humanity and our planet, and the use of evidence is at the center of this moment.

A NEW MODEL FOR EVIDENCE USE

We have seen the BBB discourse around evidence play out in many ways recently—from the climate change debates to the disinformation/misinformation around COVID-19, to the United States' story on social and racial justice. The use of evidence across these broader debates shows that evidence-based decision making is about more than generating proof through credible research efforts to inform policy. It is about diverse perspectives and mindsets, uptake, use, and management. The guiding principles of the American Evaluation Association (AEA),[4] the leading industry organization for evaluation professionals in the United States, addresses substantially more issues than the technical aspects of measurement and data needed for an evaluator to generate credible evidence.[5] Half of the AEA principles address matters of integrity, respect, and, importantly, contribution to the common good and advancement of an equitable and just society.

The extension beyond technical matters articulated by the AEA principles is a marked departure from the values-free, neutral observer origins of evidence generation that set itself up on a platform based on scientific methods during its first sixty years and paved the way for next generation evidence. Numerous branches of evaluation, including participatory methods, stakeholder-based methods, developmental evaluation, transformative evaluation, and equitable evaluation,[6] have been advancing the notion that data and evidence generated by evaluators should not be limited to mimicking scientific methods. The current AEA principles reflect this expanded notion of what evidence-informed decision making is, which prompts a rethinking of what it should be to meet the moment we are in (the focus of AEA's conference theme in 2021).[7]

As evaluators, we take these principles seriously, and reflect them into our practice. These principles play out in evaluative practices and become quite significant when evaluators move outside traditional provinces of public and social sector initiatives and engage with the growing body of work aimed at achieving a more sustainable and equitable future. We discuss one such experience below as a use case for next generation evidence.

IMPACT INVESTING AS A USE CASE FOR NEXT GENERATION EVIDENCE

Over the last ten years, as evaluators, we have been paying close attention to the significant growth of private sector engagement and dollars invested in addressing global threats, being both actors and reactors to the world of impact and responsible investing. Initially viewed with a critical and skeptical lens, we have learned that this is a different type of investing that challenges the mainstream model of modern capitalism and expands the previously held notion that the public sector is the only actor influencing the welfare state. The private sector has taken on a new commitment to environment, social, and governance (ESG), otherwise known as "impact" on people and the planet. Widespread adoption of this zeitgeist is apparent in the private sector's increasing alignment with the UN's Sustainable Development Goals (SDG), the growth of sustainability reporting (GRI), and the preponderance of corporate social responsibility (CSR).

Bringing our evaluative lens to this work, there are a few things we have learned through this journey that are highly relevant to next generation evidence, particularly the point that evidence and use are not simply a values-free and objective endeavor. For instance:

- "Impact" is not homogenous; rather, it is "multidimensional" and, thus, hard to easily boil down into one measurable construct or lean set of quantitative metrics.
- Prioritization of dimensions and definitions of what is "impact" are relative to the perspectives of different actors; for example, communities of people with lived experiences, investors responsible for managing capital, policymakers responsible for the public good.
- Not all actors are compelled by the same points or types of proof and statistical confidence levels on what "impact" is, which is meaningful when it comes to evidence-based decision making and action.

Each of these lessons acknowledge varying perspectives on how evidence is generated, what "good enough" evidence looks like for a "data-driven decision," and the trade-offs that result from generating data and evidence that are not rooted in a universal understanding and acceptance of rigorous methodologies or the very meaning of impact itself. For instance, standards of evidence for public policy decisions demand more scientific rigor than is the case for agile private sector business decisions that are likely to be guided by signals and rapid experimentation to influence their evidence-based decision making and actions.[8] Similarly, breadth and depth of intended impact are variables that can be valued differently by people with lived experiences and those who manage capital. The Impact Management Project set out to create a norm for defining dimensions of impact to include five dimensions: What, Who, How Much, Contribution, and Risk.[9]

These lessons matter when we consider the large, and possibly immense, scale of this new investing space, which had an estimated market size that in 2020 ranged between $715 billion[10] and $37.8 trillion.[11] The lower estimated market size restricts the label of impact investing to investments in private sector businesses that intentionally aim to generate positive, measurable social and environmental impact alongside a financial return.[12] The larger estimate, in contrast, is more internally focused and based on ratings of companies' internal (ESG) practices and policies.[13]

CONNECTING ACTORS ACROSS THE EVIDENCE SPECTRUM

Recognizing the growth and scale of this space and the synergies between "impact" measurement and evaluation, in 2016, the American Evaluation Association (AEA) and its members launched several initiatives to bring these discussions together, from bringing together leaders focused on both evaluation and impact measurement to explore these topics, called "Impact Convergence," to launching a new group within AEA focused on what the convergence could look like.[14] Numerous connections and publications emerged between AEA, its members, and intermediary organizations responsible for developing and/or providing measurement principles, standards, tools, and verification, for example, Impact Management Project, GIIN, Toniic, OECD, World Economic Forum, Salesforce, and UNDP. Notably, the forum section of the *American Journal of Evaluation*'s[15] 2018 fall volume focused on "Where Impact Measurement Meets Evaluation" while Salesforce's[16] e-publication about impacting responsibly prompted deeper learning that reflected principles of AEA and, in particular, bringing an equitable evaluation lens to influencing the reimagining of next generation evidence.[17]

The combination of evaluators' (and other technical assistant providers) growing influence across these platforms combined with the growth and scale of private sector investment in social good and sustainability signal what is on the horizon for next generation evidence. Importantly, and given the BBB narrative, an essential part of the formula for achieving sizable and durable impact must focus on advancing inclusiveness and sustainability in how evidence is defined. Along this theme, two areas remain to be developed and are the topics of multiple discussions, particularly as it relates to equitable and inclusive next generation evidence: 1) the relative value of stakeholders, people and communities affected by policies and investments, in defining success and affecting management decisions, and 2) the inclusion of externalities in determining positive and negative impacts.[18]

The Imperative for Stakeholder Engagement

Stakeholder engagement is a necessary part of evidence generation. How can we understand and be accountable for the effects of an intervention, investment, or enterprise's activities without understanding the experiences of those affected by the activities?

Today, there is a clear sense that stakeholder voice matters in the design and provision of products and services[19] and the planning and implementing of activities.[20] Data collection methods for stakeholder engagement are even readily available.[21] Yet, these practices are not regularly instituted, and this must change. Addressing this shortcoming is the focus of a peer learning partnership of evaluators and impact measurement and management professionals supported by an OECD global action initiative[22] funded by the European Union. In particular, the peer learning partnership aims to surface the drivers and barriers to the effective leveraging of capacity building efforts and public policy to foster stakeholder-engaged practices about evidence generation.

Externalities and Unintended Consequences

We next turn to consideration of "externalities" in determining impact. Let's use this working definition excerpted from Oxford Languages[23] to describe externalities as viewed from an economic perspective:

> a side effect or consequence of an industrial or commercial activity that affects other parties without this being reflected in the cost of the goods or services involved, such as the pollination of surrounding crops by bees kept for honey.

In this example, the illustration of an externality is a positive example that creates positive impact, but let's also recognize the potential negative impacts that can occur through exploiting externalities associated with direct operations, supply and value chains, and business partnerships resulting in negative effects on climate and people. These range from pollution to raising the earth's temperature to poor or unethical work conditions, suppressed wages, health risks, and a widening race and gender wealth gap.

As new forms of investing grow as a disrupter to traditional investing (based on the principle of profit first and foremost), so must our understanding about what is material to value creation of an enterprise. The director of the UNDP Standards, Fabienne Michaux, suggested in an April 8, 2021, Impact Entrepreneur webinar, "profit generated from enterprise activity is privatized yet some of the costs of production of these profits on stakeholders outside the company are socialized." The stakeholders Michaux is referring to are society, the environment, and future generations. Along these lines, Social Value International (SVI) has long advocated that the data and evidence required in accounting for business value needs to redefine what is material. Data requirements and the evidence used to inform decisions are based on narrowly focused financial data at the expense of valuing impact, either positive or negative, that have global consequences. This field-level shift would be game-changing for the type of data and evidence that would be tracked and used as a basis for decisions across all private sector business activity. This possibility is not out of reach, particularly among standard-setting organizations for international accounting standards and reporting that are actively engaged in developing sustainability disclosure standards, recognizing the complexity of such an undertaking and the significance of global standards that are interoperable across systems.[24]

CONCLUSION

Throughout this narrative, we have asserted that we are at a pivotal moment in time, one where, as a global community, we need to collaboratively navigate through a post-pandemic era while faced with the dire state of affairs of people and planet.

The AEA has had, itself, to contend critically over the last few years with evaluation's own history in promoting inequitable and exclusive practices, which has led to the explosion of repositioning and reimagining what a responsible evaluation practice could and should look like, from electing its first woman president of color to reviewing its intellectual property rights to enable more equitable membership services to reviewing a more inclusive and equitable approach to who receives its highest awards to explicitly releasing public statements regarding recent hate crimes to African, Asian, and Latinx Americans.[25] All this with the goal of questioning the role of evidence and signaling that, as evidence generators, we must educate, and at times re-educate, ourselves on our own intersectional understanding of race in American and the historical power that the scientific academic method, as the ultimate truth, has had on further marginalizing communities.

As an association, we are finding a new way forward that considers how data and evidence can be designed and then actioned to advance a more equitable and just society. This way forward must include more voices of lived experience (stakeholder voices) and a prioritization on a lens of equity and sustainability (externalities). Critically, more equitable and responsible evaluation practices must permeate not only the traditional places where evaluation works, such as public sector finance, but also be used to ensure that private capital grows to be a force for good and held to account for impacts on people and planet to prevent what is known as impact washing, the practice of overstating or falsely claiming benefits of a product/service to sell more of it.[26]

What will this take? Coming from the world of social sector financing, primarily government and philanthropy, and grounded in the knowledge that using evidence to scale impact is not novel in the context of contemporary business practices, we believe building back better will require a reimagination of how we use evidence to adapt.

And so it goes. We are at a critical juncture in our own human story, where adaptation is a necessity. As discussed throughout this paper, paramount to this next scene is adapting how we use evidence to scale impact to fight against impact washing. This call to action is applicable across all forms of capital, private and public. As we collectively continue to push forward, we can expect the next rational step to bringing the worlds of public and private sector financing in service to people and planet even closer together will be to regulate it. To quench the flames that Greta Thunberg speaks to, this is clearly needed to ensure good intentions translate to scalable impact that advance equity, resilience, and sustainability.

NOTES

1. See "Our House is on Fire: Greta Thunberg, 16, Urges Leaders to Act on Climate," The Guardian, January 25, 2019, www.theguardian.com/environment/2019/jan/25/our-house-is-on-fire-greta-thunberg16-urges-leaders-to-act-on-climate.

2. See Prevention Web, International Recovery Platform, www.recoveryplatform.org/assets/tools_guidelines/GFDRR/Disaster%20Recovery%20Guidance%20Series-%20Building%20Back%20Better%20in%20Post-Disaster%20Recovery.pdf.

3. "Investment and Sustainable Development: Between Risk of Collapse and Opportunity to Build Back Better," Discussion Paper for Joint IC-CAC Session at the 2020 Roundtable on Investment and Sustainable Development, OECD, www.oecd.org/investment/Between-risk-of-collapse-and-opportunity-to-build-back-better.pdf.

4. See www.eval.org/About/Guiding-Principles.

5. "Guiding Principles for Evaluators," n.d., American Evaluation Association, www.eval.org/Portals/0/Docs/AEA_289398-18_GuidingPrinciples_Brochure_2.pdf.

6. See, for example, M. C. Alkin and J. A. King, "The Historical Development of Evaluation Use," *American Journal of Evaluation* 37, no. 4 (2016), pp. 568–79; C. A. Christie and M. C. Alkin, "An Evaluation Tree," in *Evaluation Roots: Tracing Theorists' Views and Influences*, edited by M. C. Alkin (Thousand Oaks, CA: SAGE Publications, 2013), pp. 11–57; D. M. Mertens, *Transformative Research and Evaluation* (New York: Guilford Press, 2009), www.equitableeval.org.

7. See American Evaluation Association, www.eval.org/Events/Evaluation-Conference/Conference-Theme.

8. M. Reeves and M. Deimler, *"Adaptability: The New Competitive Advantage,"* July–August 2011, *Harvard Business Review Magazine*.

9. See Impact Management Project website, https://impactmanagementproject.com/impact-management/impact-management.

10. See "What You Need to Know about Impact Investing," Global Impact Investing Network, n.d., https://thegiin.org/impact-investing/need-to-know/#how-big-is-the-impact-investing-market.

11. "ESG Assets May Hit $53 Trillion by 2025, A Third of the Global AUM," Bloomberg Professional Services, February 23, 2021, www.bloomberg.com/professional/blog/esg-assets-may-hit-53-trillion-by-2025-a-third-of-global-aum/.

12. "Impact Investing," Global Impact Investing Network, n.d., https://thegiin.org/impact-investing/#:~:text=Impact%20investments%20are%20investments%20made,impact%20alongside%20a%20financial%20return.

13. Particularly relevant to this paper, the former is more outcome-leaning and the latter is more output-leaning. K. P. Pucker, "Overselling Sustainability Reporting: We're Confusing Output with Impact," May–June 2021, *Harvard Business Review Magazine*.

14. See Impact Convergence website, http://impactconvergence.org/.

15. See *American Journal of Evaluation* 39, no. 3, September 2018, Sage Journals website, https://journals.sagepub.com/toc/aje/39/3.

16. See Impacting Responsibly, Salesforce.org website, www.salesforce.org/about-us/impacting-responsibly/.

17. And a 2020 comprehensive handbook about impact investing lifted up compelling questions about next generation evidence, including highly participatory and equity/justice-focused transformative evaluation and the nature of evidence, including the relevance of stakeholder voice and innovative perspectives about evidence for decision making.

18. Absent these two components, the evidence for successful initiatives aiming to do good (large or small) are valued solely through a funder, investor, or company-driven lens, favor quantitative metrics, and obscure potential negative impacts that can result from mission-driven activities. Additionally, the opportunity to manage for impact post interventions is severely hindered when stakeholders are not engaged in evidence development processes or when externalities are left out of the equation.

19. See "Stakeholder Engagement," Global Impact Investing Network, n.d, https://iris.thegiin.org/metric/5.0/oi7914/.

20. See Social Value International, May 2018, https://www.socialvalueint.org/principles.

21. Such evidence can be generated through processes involving co-creation of success measures and solutions, using technology assisted survey methods or storytelling to elicit stakeholders' perspectives, or embedding stakeholder engagement throughout management and governance decisions. See, for example, Strategy Development: Most Significant Change (MSC)," ODI, January 13, 2009, https://60decibels.com/; https://odi.org/en/publications/strategy-development-most-significant-change-msc/; https://keystoneaccountability.org/.

22. See "Promoting Social and Solidarity Economy Ecosystems," OECD, www.oecd.org/cfe/leed/social-economy/fpi-action.htm.

23. See Google definition, www.google.com/search?q=externalities+definition&oq=externalities&aqs=chrome.1.69i57j0i433l2j0l4j69i61.7324j0j4&sourceid=chrome&ie=UTF-8.

24. Janine Guillot, "Markets are Speaking, the IFRS Foundation is Listening," Value Reporting Foundation, December 16, 2020, www.sasb.org/blog/markets-are-speaking-the-ifrs-foundation-is-listening/.

25. Statement form the AEA Board of Directors Regarding Racism and Inequality in our Society, American Evaluation Association, March 31, 2021, www.eval.org/Full-Article/statement-from-the-aea-board-of-directors-regarding-racism-and-inequality-in-our-society.

26. See "Impact Washing: What It Is and How to Stop It" https://online.hbs.edu/blog/post/what-is-impact-washing.

THE PRIVATE IMPACT ECONOMY

THE GLOBAL MOVEMENT TO TRACEABLE SOCIETAL IMPACT

BRIAN KOMAR AND ANDREW MEANS

Companies love to celebrate the ways in which they are making the world a better place. They promote the good they are doing in the world in their ESG/sustainability reports, press releases, articles, and social marketing initiatives. This is not entirely different from nonprofit organizations, foundations, or even government agencies. While every sector has different motivations and incentives, all sectors increasingly want to be associated with making a difference. But can the stories we choose to tell about the good we are doing in the world be trusted?

The world is moving toward an impact economy. The future increasingly is one in which every industry in every sector will be held more accountable for the effect their work has on people, planet, and community. Producing traceable evidence of societal impact is among the most powerful levers to making the world a better place. Investing in the infrastructure, capacity, and technology necessary to trace the impact of our work is increasingly required by all organizations.

This is not to say that every kind of outcome or impact can be traced and proven. There are many interventions to which a purely quantitative, measurable determination of impact would be inappropriate. This is why we see evidence as such a powerful term. Evidence can include the quantitative and qualitative. It can adapt to the environment and context. What that evidence looks like, from using proven interventions to proving your particular intervention, can and should be different based on context.

A confluence of factors drive the impact economy across every sector. In the private sector, there is a growing recognition that previously considered externalities—nonfinancial impact on people, planet, community—have very real financial consequences and material value. Intangibles, the non-monetary aspects of a business such as brand, reputation, sentiment, etc., as a portion of total assets in the S&P 500 have reached unprecedented levels. As of 2020, 90 percent of all assets in the S&P 500 are intangible.[1] Evolving regulatory environments also are requiring companies to increasingly manage their environmental impacts. Many of these trends are accelerating in civil society as well.

For example, the Black Lives Movement and the renewed push in the private sector for advancing equality and racial justice have rightfully extended to nonprofits, foundations, and universities that are increasingly being held more accountable for sharing their diversity, equity, and inclusion efforts. Activists pushing for greater progress on climate change are driving changes in everything from elections to board rooms.

Recently, Salesforce.org had our annual social value (ASV) verified by a third party, something few, if any, organizations have done before. Annual social value is the monetary value of what we contribute to the community. It is not direct impact data; it does not speak to the changes those contributions have made for individuals and communities we seek to empower. It is, however, a demonstration that the stories organizations choose to tell about their impact should be rooted in trustworthy, traceable, ethical, and transparent data whenever possible.

Real value is created by how we treat our environment, our stakeholders, and our communities. The Business Roundtable, a collection of CEOs from the largest companies in the world, said last year that it was time for companies to be held accountable by all their stakeholders, not just their shareholders. Investment firms are acquiring impact analysis firms to help them evaluate the nonfinancial performance of their portfolios. Foundations and government agencies are increasingly engaged in performance-based contracting, and nonprofits are responding to these opportunities by working to invest more in evaluation and learning.

As we move toward this impact economy, we must work to increase the amount of our impact that is traceable. We must improve the quantity, quality, and interoperability of the information we use as evidence of impact. There are several things organizations of all types can do to improve the traceability of their impact.

INVEST IN STAKEHOLDER FEEDBACK

Any and every stakeholder impact conversation should begin with the people and communities whose lives are being impacted; that is, the stakeholders. This includes your customers or clients, your employees, your partners, and the communities impacted by your work.

The practice of soliciting feedback from those affected by your products and services is something that has long been practiced in the private sector, where there is an automatic closed feedback loop between buyers and sellers. Simply put, if you fail to listen to your customers, you go out of business.

Initiatives like Listen for Good[2] and Feedback Labs[3] are bringing these concepts into philanthropy and social service delivery and development, where such a feedback loop frequently does not exist. For example, the humanitarian sector uses direct feedback from beneficiaries to identify abuses by its field workers. The practice of Safeguarding allows for those receiving services to communicate directly with leadership to report misconduct. This kind of direct feedback has been shown to reduce abuses in large, distributed organizations.

Engaging stakeholders about impact is important but insufficient. You must be willing to not just collect that feedback but use it to drive decisions that can lead to profound changes. Listening is the start of an impact data journey that, when done right, produces insights and new understandings that drive learning, continuous improvement, and, ultimately, better impact performance.

INVEST IN IMPACT MANAGEMENT CAPACITY BUILDING

Technology and data collaboration investments can accelerate traceability of impact; however, the most important move is investing in our people. We can lead stakeholder engagement programs, purchase technology for our field staff, and invest in data sharing, but if we do not invest in the capacity of all organizations to measure and manage their nonfinancial performance, these efforts will fall flat.

In a world of traceable evidence of impact, we must ensure that all stakeholders are ready to consume and use that information. This can begin with our staff, but it must not end there. Yes, we must ensure that individuals across our organizations are able to track and use nonfinancial performance data. When we are making decisions, we must look at our financial measures, operational KPIs, and the impact of these efforts.

But we also must ensure that communities are able to access and use this information, that funders and donors have access to traceable impact, and that those served by our organizations also can consume our nonfinancial performance information.

INVEST IN ACTIVE IMPACT MANAGEMENT

Organizations must not just report on their impact but also engage in active impact management. Impact management is grounded in continuous improvement and learning theories. It is when organizations change how they are operating today to improve the impact they will see tomorrow by learning from what they did yesterday.

While the private sector is farther along when it comes to having systems to manage their core operations, even companies are nascent when it comes to the systems they use to manage their environmental, social, and governance (ESG) performance. We are seeing more and more investment into technology to monitor what has traditionally been seen as nonfinancial performance. Companies also are increasingly using new technologies, like blockchain, to track the societal impact of their operations through their supply chains.

Civil society must deepen its investments in active impact management. The sector has focused a disproportionate amount of its technology investment in fundraising operations as opposed to technology engaged in the delivery of services, most notably at the point of service delivery. When program and impact data are collected, far too often data is collected and aggregated at the central office, leaving those in the field with inadequate information and technology to help them do their jobs better.

As we seek to create more traceable impact, creating a digital trail of the interaction where impact is occurring is critical. Program management technology can be a great asset in helping people increase the amount of impact data about their work. We must ensure that everyone across the organization has access to technologies that demonstrate their performance.

Living Goods[4] is a great example of an organization that invests in technology at the point of service delivery. Working in three countries across Africa and Asia, Living Goods seeks to improve health outcomes in rural areas by giving community health workers better technology. This technology allows these health workers to provide better services and achieve greater impact in their work while simultaneously improving the evidence and traceability of that impact to the organization.

INVEST IN DATA COLLABORATION

Success in the impact economy will increasingly be a multistakeholder endeavor. The impact supply chains discussed above highlight how every organization's impact on the world is dependent on the impact of another organization. Organizations must think of themselves as parts of multiple data ecosystems and allocate greater investments into data sharing tools and infrastructure to support the sharing of their impact performance data across organizations and across sectors.

Supporting impact data harmonization efforts also is crucial. Efforts like the Impact Management Project[5] (IMP) are creating a cross-industry framework for understanding different types of impact. Thanks to initiatives like IMP, significant progress is being made to harmonize the alphabet soup of sustainability reporting standards for corporations. Cross-industry impact standards drive data interoperability even across similar programs in different sectors.

The World Economic Forum and the International Business Council are leading an effort to harmonize sustainability metrics and reporting for corporations.[6] Nonprofit and philanthropic organizations are increasingly seeing ground-up efforts for metric alignment, and groups like the Urban Institute are working to catalogue metrics by program area.[7]

Metric alignment is one necessary step. Once we have organizations collecting the same data, we need to support increasing reporting and sharing of impact information. Efforts like Guidestar's Platinum Seal,[8] requiring outcome data, and Charity Navigator's acquisition of ImpactMatters[9] demonstrate increased demand for outcome and impact information about nonprofit organizations' performance.

On the private sector side, investors are buying impact analysis firms and changing investment strategies to value nonfinancial performance. This is being driven by improvements in data sharing and reporting by companies against sustainability metrics.

To have traceability of impact, we must be able to share standard information with one another about our nonfinancial performance, and investing in our data sharing and infrastructure is a critical step on that journey.

CONCLUSION

Imagine what the world will look like when we trace a much larger amount of impact data across sectors. We will be able to hold one another more accountable for the effects of our actions. We can better direct resources to where they are able to do more good. We can ensure that all stakeholders' voices are heard and better influence the decisions of our business, government, and philanthropic organizations. In short, we will all deliver more societal impact together.

This future state is not a panacea. There are limits and risks. A lot of good work will not and cannot be traced back to some measurable impact. We must not allow easily traced outcomes to dominate more challenging, unmeasurable outcomes. We also must ensure that as we move toward an evidence- and data-driven impact economy we do so in a responsible and ethical way. We must not increase the vulnerability of already vulnerable populations through our impact data collection efforts. We just ground these efforts in frameworks that give people real agency over the collection and use of their data.

Evidence should be central to our understanding of the nonfinancial performance of all organizations. That evidence should be traceable and verifiable by outside sources, whether they be auditors or evaluators. Over the next ten years, we are sure to see more organizations investing in these efforts and billions of dollars will transact over the nonfinancial performance of all organizations.

NOTES

1. Aran Ali, "The Soaring Value of Intangible Assets in the S&P 500," Visual Capitalist, November 12, 2020, www.visualcapitalist.com/the-soaring-value-of-intangible-assets-in-the-sp-500/.

2. Fund for Shared Insight, Listen4Good, n.d., https://fundforsharedinsight.org/listen4good/.

3. See Feedback Labs website, https://feedbacklabs.org/.

4. See LivingGoods website, https://livinggoods.org/.

5. See Impact Management Project website, https://impactmanagementproject.com/.

6. "Measuring Stakeholder Capitalism: Towards Common Metrics and Consistent Reporting of Sustainable Value Creation," World Economic Forum, September 22, 2020, www.weforum.org/reports/measuring-stakeholder-capitalism-towards-common-metrics-and-consistent-reporting-of-sustainable-value-creation.

7. See Outcome Indicators Project, Urban Institute, website, www.urban.org/policy-centers/cross-center-initiatives/performance-management-measurement/projects/nonprofit-organizations/projects-focused-nonprofit-organizations/outcome-indicators-project.

8. See Candid website, https://learn.guidestar.org/seals.

9. See www.charitynavigator.org/index.cfm?bay=content.view&cpid=823.

BEYOND THE EVIDENCE ACT

DIANA EPSTEIN

THE EVIDENCE ACT OPPORTUNITY

The January 2021 Presidential Memorandum on Restoring Trust in Government Through Scientific Integrity and Evidence-Based Policymaking,[1] issued within the first week of the Biden-Harris administration, provided new momentum for evidence-based policymaking. The subsequent guidance issued from the Office of Management and Budget in response, OMB M-21-27,[2] stresses the importance of building a culture of evidence and embedding evidence into federal agency functions and processes. This new guidance not only unifies and reaffirms key principles such as the importance of rigor and transparency in evidence-building activities, it also elevates equity as a key consideration throughout the lifecycle of evidence building.

This renewed energy around evidence-based policymaking builds on the work of the bipartisan Commission on Evidence-Based Policymaking, which garnered new attention around improving how data and evidence are used through the federal government. The commission's report, issued in September 2017, offered a set of recommendations that illustrate both the challenges and enormous possibilities that a greater focus on evidence could bring in service of improving government effectiveness. The Foundations for Evidence-Based Policymaking Act of 2018, or the Evidence Act, addressed about half of the commission's recommendations, and served as a strong marker that the improved use of evidence must be a priority. This law includes the idea that evidence-based policymaking needs systematic planning, that we need strong data governance to use federal data assets effectively, and that we need coordinated support to share data effectively while protecting privacy and confidentiality. This chapter addresses progress toward executing on this first idea of more systematic evidence building and use in the federal government.

The Evidence Act builds on long-standing principles underlying federal policies and data infrastructure. While the government certainly is not starting from scratch, the law does create a new paradigm for federal agencies to think about how they build and use evidence. For one, it focuses on leadership by requiring agencies to designate three new senior officials (Evaluation Officers, Chief Data Officers, and Statistical Officials) who are responsible for leading implementation of the act's requirements. It also emphasizes collaboration and coordination across functions, recognizing that no single person or office can accomplish this work alone. It is designed to break down some of the long-standing siloes that have long stymied this work. It also puts in place a more strategic approach to building evidence as opposed to how it has traditionally happened in most agencies, which is very ad hoc, often in response to a particular mandate or driven by a specific group of motivated staff. And specific to the focus of this chapter, it elevates program evaluation as a key agency function. For years, the federal government had statutory systems for performance and statistics, but evaluation has been the missing link. While program evaluation was already happening in some places, it is not happening well in many other agencies, and the Evidence Act elevates evaluation as a key piece of the evidence-building enterprise. This is important because there are some kinds of questions—for example: Is a program or policy working as intended? Is it causing the intended changes?—that only evaluation can answer.

When a law like the Evidence Act passes, the Office of Management and Budget (OMB) is typically responsible for issuing guidance to agencies so they understand the details and how to implement the law. The Evidence Act is no exception, and it calls on OMB to issue guidance in a number of areas. Rather than issue guidance in many different pieces coming out of many different offices and then have agencies struggle with how to interpret and apply it, the OMB team decided to take a different approach. From the start, career staff have been coordinating internally, across OMB offices that don't always work closely together, to figure out how to issue guidance for agencies more cohesively. We decided to issue guidance in just four main phases (not necessarily in chronological order).[3] Phases 1, 2, and 3 roughly correspond to each of the three Titles of the Evidence Act, and the fourth piece of guidance is specifically related to the new program evaluation provisions in Title I. It is our hope that this coordinated approach makes it easier for agencies to understand what they are expected to do; at the same time, this intentionally reflects the kind of collaboration and coordination needed on the part of agencies and their partners both inside and outside government to implement this law meaningfully.

The Evidence Act is a rare opportunity for the government to take stock of past practices and try to create a more effective future. We know that federal agencies vary widely in their context and missions, as well as their history and current capabilities for evidence-based policymaking. To that point, the first guidance document issued by OMB provides key parameters around learning agendas, annual evaluation plans, and capacity assessments while allowing agencies flexibility to tailor these requirements to meet their specific needs.[4] While in some ways this creates new challenges by avoiding templates and standardized reporting, the intention is to allow agencies to drive this work themselves and do it in a way that makes sense for them. Time will tell if this approach bears fruit, but the effort is likely to fail unless agencies own this work and embrace it in a way they believe will lead to real change. The last thing we want is for the Evidence Act to be a compliance or reporting exercise where agencies put in minimal effort and check the boxes without anything really changing for the better.

LEARNING AGENDAS

Learning agendas are at the heart of this new approach to evidence building. The Evidence Act calls them "evidence-building plans," but the field had been using the term "learning agenda" for a number of years, so OMB decided to stick with that term, in part to emphasize the central learning and improvement function of this work. The two terms are synonymous, however, and using the phrase "strategic evidence-building plan" can sometimes be an equally useful if not better choice because this approach is designed to encourage deliberate and strategic planning of evidence-building activities. The learning agenda is a systematic plan for identifying and addressing priority questions relevant to the programs, policies, and regulations of an agency. If done well, multiyear learning agendas provide an evidence-building roadmap to support effective and efficient agency functioning. They provide a framework to use data in service of addressing the key questions an agency wants to answer to improve its operational and programmatic outcomes and develop appropriate policies and regulations to support successful mission accomplishment. A range of analytic methods and types of evidence can be used to answer the priority questions identified in a learning agenda. The important thing is to start with the question; the question should drive the method selected and not the other way around.

The Evidence Act requires learning agendas to be part of agency strategic plans. OMB expects that the learning agenda can function as a stand-alone document, but also that elements of the learning agenda should be woven through the strategic plan narrative. This alignment of evidence with strategic goals and objectives is an opening to bring the evidence builders and the strategic planners together from the outset. This has, typically, not been done in federal agencies, but the Evidence Act offers a new framework in which evidence-building priorities are aligned with strategy and envisioned together from the start. This elevates those important questions, both mission-strategic and operational, for which empirical answers will help agencies execute their missions more effectively and serve communities better. It also offers the opportunity for agencies to align their evidence-building questions to new priorities as they emerge; for example, for the Biden-Harris administration, this includes advancing racial equity, climate change, and economic recovery. Through this alignment, the learning agenda is an integral tool to building evidence that is more useful for decision makers and practitioners, in service of achieving better outcome for citizens.

STAKEHOLDER ENGAGEMENT AND TRANSPARENCY

Stakeholder engagement also is required as part of the learning agenda process (see figure 7.1-1), and, in fact, the Evidence Act specifies that a range of stakeholders, including the public, state and local governments, and nongovernmental researchers should be consulted. It is critically important for agencies to engage in wide-ranging and substantive stakeholder involvement from the beginning of the learning agenda process to identify the right priorities. To do this well is difficult, but the payoffs can be enormous. One element of stakeholder engagement is internal to the agency, with program administrations and implementors, those who understand how programs and policies truly function and are often best positioned to articulate the evidence they need to do their job better. Also important are stakeholders external to the agency, whether that is the state and local governments who receive federal funds and administer federal programs or the communities and members of the public who are the intended beneficiaries. It is these individuals who often have not been engaged but who may offer the most accurate assessment of the challenges they face and how they experience federal programs and policies. Nonprofits and other levels of government also are valuable generators of evidence and can push evidence up to the federal level through a meaningful stakeholder engagement process. Stakeholder engagement cannot begin and end with a posted Request for Information (RFI) or a checklist of token individuals consulted to meet the requirements of the law. It must be a sustained and iterative effort that occurs throughout the learning agenda cycle to ensure agencies are focused on the most salient priority questions and that the evidence they generate has the potential to be used by those who can benefit the most from this knowledge.

Strategic plans are public documents, which means that agency learning agendas are posted publicly (as are agency Capacity Assessments and Annual Evaluation Plans, two other planning documents required by the Evidence Act). This promotes transparency and accountability and also provides new opportunities for partnerships through more equitable and inclusive sharing of priorities. Learning agendas are an open broadcast to the world about an agency's evidence priorities. This offers a chance for academics, practitioners, think tanks, philanthropic foundations, and other researchers to align their own research portfolios with these priorities. Doing so allows researchers' work to be more policy-relevant and actionable, and it allows agencies to benefit from the added capacity, skills, and expertise these partners can bring to their efforts.

All this, of course, takes resources, but there are numerous ways in which agency budgets and those of their partners can incorporate evidence-building activities such as evaluation. Evaluation should be viewed as a core mission function and not something that detracts from service delivery; it is critical for program improvement and not a "nice to have" activity that happens on the side. From a federal award perspective, an evaluation cost is allowable and can be either direct or indirect, at the discretion of the federal awarding agency, unless prohibited by statute or regulation. As

FIGURE 5.5.1 **The Learning Agenda Cycle, Excerpted from OMB M-19-23**

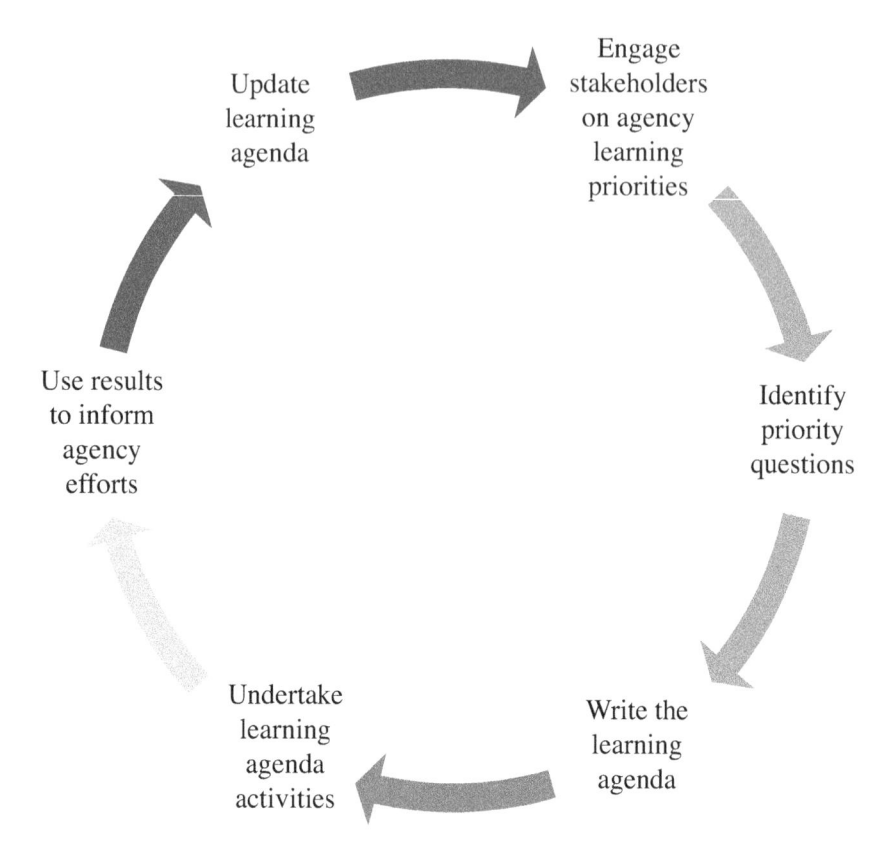

Update learning agenda → Engage stakeholders on agency learning priorities → Identify priority questions → Write the learning agenda → Undertake learning agenda activities → Use results to inform agency efforts → (back to Update learning agenda)

stated in 2CFR 200.413: "Typical costs charged directly to a Federal award are the compensation of employees who work on that award, their related fringe benefit costs, the costs of materials and other items of expense incurred for the Federal award. If directly related to a specific award, certain costs that otherwise would be treated as indirect costs may also be considered direct costs. Examples include extraordinary utility consumption, the cost of materials supplied from stock or services rendered by specialized facilities, program evaluation costs, or other institutional service operations." This language should serve as an invitation and opening for federal awardees to allocate a portion of their budgets toward evidence-building activities such as program evaluation, when appropriate.[5] Agencies also should consider available evidence when complying with the OMB uniform grants guidance on program design and also when using performance reporting to add to the body of evidence or determine new opportunities for learning.

DEDICATION, SUPPORT, AND PERSISTENCE

Building an evidence-based government is a long-term proposition; the cultural change needed to more systematically infuse evidence into decision making will not happen overnight. Nonetheless, many agencies have embraced this work and have made solid progress since the Evidence Act became law. For example, most agencies have named their designated officials—the Evaluation Officers, Statistical Officials, and Chief Data Officers—and those officials are collaborating and working together. The cross-agency councils for each of those officials meet regularly, and the councils are connecting on shared priorities and opportunities for collaboration. Agencies produced multiyear Learning Agendas, Annual Evaluation Plans, and Capacity Assessments and published them on their agency websites in the spring of 2022. Links to all these documents are also available at the new Evaluation.gov[6] website, which launched in September 2021 and provides a central online presence for federal evaluation and the Evaluation Officer Council.

Meanwhile, a vibrant community of career civil servants has been quietly pushing the work on learning agendas and evaluations forward, including the Evidence Team at OMB.[7] This team, which the author leads, is a small group of career staff that collaborates with other OMB offices and provides support and resources to agencies in a number of areas, including developing learning agendas, increasing agency capacity to build and use evidence, and providing expert advice and technical assistance on evaluation activities and initiatives for a broad range of federal agencies and functions. The team also partners with the Office of Evaluation Sciences at GSA to run a regular Evaluation and Evidence Training Series for federal staff, with thousands of attendees to date. Unknown to many on the outside, the team developed and maintains a comprehensive MAX (intranet) page with many different resources and tools as the anchor of a broader community of practice. In addition to the new Evaluation Officer Council, the Interagency Council on Evaluation Policy (ICEP) was rebooted and expanded, which provides a venue for evaluation experts across the government to work together to provide consultations, resources, events, and other peer support for the federal evaluation community.

LOOKING FORWARD

As this chapter noted at the outset, the 2021 Presidential Memorandum and associated OMB guidance gave new energy to federal evidence and evaluation efforts. The Memorandum specifically discusses agency learning agendas and annual evaluation plans, as well as other aspects of the Evidence Act. Its focus on transparency provides an additional opportunity to elevate the program evaluation standards and practices OMB issued in March 2020. This is the first time the government has had cross-agency standards for program evaluation, and these standards—relevance and utility, rigor, independence and objectivity, transparency, and ethics—are designed to improve the quality and use of evaluation across federal agencies.

The increased focus on program evaluation should help agencies prioritize answering more of these evaluative questions and with increased quality. By creating more demand for evaluation activities, it also may afford new opportunities for partnerships with state and local governments, nonprofits and community partners, and academics. The knowledge gained from these evaluations can help communities better understand which programs may work best in their own contexts and with their populations.

The Executive Order on Advancing Racial Equity and Support for Underserved Communities through the Federal Government[8] and its focus on data provides another important mechanism to not only advance evidence but ensure it is built and used in ways that advance equity for all. Priority questions around equity should be incorporated into agency learning agendas and the strategic evidence-building plans of their partners, and agencies must consider how they meaningfully involve all relevant stakeholders throughout the learning agenda process. This explicit focus on equity means the federal community and its many partners must consider how to put in practice what has often been an overlooked but critical consideration.

Getting these efforts right will take time, energy, and persistence, but this can be a turning point in building a stronger focus on evidence across the government. Federal agencies will need partners both inside and outside the government to create a culture where we collaborate, ask the tough questions, take risks, and share promising practices. If embraced and implemented thoughtfully, we can, together, make real progress in understanding how to best serve the American people.

NOTES

The views presented are those of the author and do not necessarily represent the views of the Office of Management and Budget, the Executive Office of the President, or the United States government, except where expressly stated as such.

1. Memorandum on Restoring Trust in Government through Scientific Integrity and Evidence-Based Policy, White House, January 27, 2021, www.whitehouse.gov/briefing-room/presidential-actions/2021/01/27/memorandum-on-restoring-trust-in-government-through-scientific-integrity-and-evidence-based-policymaking/.
2. See OMB M-21-27, White House, www.whitehouse.gov/wp-content/uploads/2021/06/M-21-27.pdf.
3. See OMB M-19-23, figure 1, White House, www.whitehouse.gov/wp-content/uploads/2019/07/M-19-23.pdf.
4. See OMB M-19-23, White House, www.whitehouse.gov/wp-content/uploads/2019/07/M-19-23.pdf.
5. See 2 CFR 200.202 Program Planning and Design and 2 CRF 200.301 Performance Measurement. For related information, see www.cfo.gov/wp-content/uploads/2021/Managing-for-Results-Performance-Management-Playbook-for-Federal-Awarding-Agencies.pdf.
6. See https://evaluation.gov.
7. I am indebted to my Evidence Team colleagues Erika Liliedahl, Erica Zielewski, and Danielle Berman, who are as much the intellectual owners of this chapter as I am. I would also like to acknowledge former teammate John Tambornino, who contributed valuable insights to OMB's Evidence Act work and former Evidence Team Lead Bethanne Barnes, who laid the foundation for so much of our current efforts.
8. See "Advancing Racial Equity and Support for Underserved Communities through the Federal Government, Executive Order," White House, January 20, 2020, www.whitehouse.gov/briefing-room/presidential-actions/2021/01/20/executive-order-advancing-racial-equity-and-support-for-underserved-communities-through-the-federal-government/.

REFERENCES

Evaluation.gov. Welcome Page. www.evaluation.gov/.
Foundations for Evidence-Based Policymaking Act of 2018. www.congress.gov/bill/115th-congress/house-bill/4174.
OMB. Evidence and Evaluation. White House. www.whitehouse.gov/omb/information-for-agencies/evidence-and-evaluation/.

OMB M-21-27. Evidence-Based Policymaking: Learning Agendas and Annual Evaluation Plans. White House, June 30, 2021. www.whitehouse.gov/wp-content/uploads/2021/06/M-21-27.pdf.

OMB M-20-12 Phase 4 Implementation of the Foundations for Evidence-Based Policymaking Act of 2018: Program Evaluation Standards and Practices. White House. March 10, 2020. www.whitehouse.gov/wp-content/uploads/2020/03/M-20-12.pdf.

OMB M-19-23 Phase 1 Implementation of the Foundations for Evidence-Based Policymaking Act of 2018: Learning Agendas, Personnel, and Planning Guidance. White House. July 10, 2019. www.whitehouse.gov/wp-content/uploads/2019/07/M-19-23.pdf.

Performance Management Playbook: Managing for Results: The Performance Management Playbook for Federal Awarding Agencies. April 27, 2020. www.cfo.gov/wp-content/uploads/2021/Managing-for-Results-Performance-Management-Playbook-for-Federal-Awarding-Agencies.pdf.

The views presented are those of the author and do not necessarily represent the views of the Office of Management and Budget, the Executive Office of the President, or the United States government, except where expressly stated as such.

BUILDING MORE HAYSTACKS, FINDING MORE NEEDLES

RYAN MARTIN

It is hearing day. For months, you have reviewed research, met with advocates, and called experts to get up to speed on what's happening.

A decade ago, Congress allocated tens of millions of dollars to start a new program focused on addressing a specific social problem. At the time, legislators—lacking evidence on what might work best—mandated the federal agency running this new program conduct an evaluation to see if it works.

Congress has waited a decade to know the answer, spent millions on studying the program, reviewed many interim reports, and now the final report is in. What did it find?

No significant impacts.

For the measures tracked by the study, there were some promising findings in the short term, but none of them lasted. It looks like there may have been some positive results for a subset of people in some places, but the report concludes with those four dreaded words: "more research is needed."

You, your colleagues, and members of Congress have met with dozens of groups who are lobbying for the program over the last six months—sharing anecdotes of success, highlighting the many thoughtful organizations working hard to address the issue, even giving your boss an award for championing the cause. You have received letters with hundreds of signatures calling for the extension and expansion of the program, seen op-eds placed in publications your boss reads, and heard how lobbyists at recent evening fundraisers reiterated the importance of this program to your boss. But does it work?

Today is the day Congress will decide. They are not voting on its fate, but the statements made at this hearing will set in motion a narrative that will harden as the program's profile rises and partisan viewpoints begin to take hold. Will key members of the committee support it and call for continued funding, hoping a few tweaks or another evaluation will show improvement? Or will they end it for good?

Millions of dollars are on the line, many reputations are at stake, and the fate of multiple nonprofits and dozens of local organizations hangs in the balance—not to mention the thousands of people participating in the program who hope it will make a real difference in their lives.

MOST THINGS WON'T WORK

I have seen this many times, and this scenario frequently plays out not only in Congress but in state legislatures and other decision-making bodies around the country.

Why? Addressing social problems is hard. There is no scientific formula or law of physics that says X action will cause Y reaction, guaranteeing less homelessness, fewer children in foster care, or higher earnings for those stuck in low-wage jobs. In fact, the only law related to the impact of a social program is that it is not likely to have any impact at all.

Known as *The Iron Law of Evaluation*, sociologist Peter H. Rossi argued in a 1972 paper[1]—updated and spread widely in the late 1980s—that "the expected value of any net impact assessment of any large scale social program is zero."[2] In making this claim, he notes "the Iron Law arises from the experience that few impact assessments of large scale social programs have found that the programs in question had any impact." He noted there were exceptions and some programs had demonstrated positive results but that these were unfortunately few and far between. Speaking in 2003 about the impact of his earlier paper, he noted the Iron Law was, thankfully, not as iron-clad as it had first seemed, saying, "I believe that we are learning how properly to design and implement interventions that are effective."[3]

The year 2003 was a long time ago. Is it really true that most social programs won't work? With all of our advances in understanding human behavior and the great leaps forward in technology and data analytics, haven't we cracked the code on how to change lives? Unfortunately, these efforts are still very much a work in progress.

In a July 2013 hearing held by the House Committee on Ways and Means, one expert noted how few studies of social programs had shown positive results. Summarizing the impacts of ninety randomized controlled trials in education, he noted about 90 percent found weak or no positive effects.[4] The same thing was found with employment and training programs—75 percent of thirteen programs that had been rigorously evaluated showed little or no positive impact.[5] In some cases, this has led to improvements in how evaluations are conducted to look more carefully for results. Unfortunately, however, this disappointment in finding that programs don't work sometimes means there is pressure to remove or weaken evaluation requirements, or to scale back evaluations because of the time and cost needed to conduct them (although easier access to administrative data is both speeding up and lowering the cost of such evaluations).

INEFFECTIVE PROGRAMS AREN'T HARMLESS

But remember, treating without testing can produce real harm. A program funded for a decade with the intention of helping low-wage workers move up the economic ladder is not benign if it doesn't work. Those who participated likely passed up other opportunities. They could have pursued a different education or training path, taken a new job, or even just stayed with the one they already had. They could have even ended up worse off than if they had never participated—in a lower-paying job or with less income, or having taken time away from work for a training that wasn't helpful. But even if the effects were not harmful to the individual themselves, there is still a large opportunity cost—what could have been done with those funds to truly help those needing a leg up, instead of spending on something that did not get them where they needed to go.

Is the answer, then, to avoid rigorously evaluating social programs? Or, since many things won't work, should funding be ended until we find the right answer? No. Instead, we need to fail with more certainty, more frequently, more cheaply, and much faster than ever before.

MORE HAYSTACKS = MORE NEEDLES

For some reason, the failure of a social program has been assigned an outsized burden of shame, ridicule, and finger pointing compared with failure in other disciplines. The witness at the 2013 Ways and Means hearing cited above pointed this out, trying to put the failure of so many social programs in context. As he noted at the time—and as the world has witnessed firsthand during the COVID-19 pandemic—the majority of medications under development turn out not to work, with a large share not yielding positive impacts in larger studies even when initial findings look promising.[6] In the business world, failures may be even more common, with thousands of studies conducted by Google and Microsoft on new products or strategies showing no significant effects.[7]

We need to get over it. Most things won't work, and that is ok. Given this, we have to think and work differently.

If we start from that point of view, we ask very different questions. If there are only a few needles in the haystack, how do we most quickly and effectively find them? Do we create one monolithic pile, take a sample, and say "on average, there were zero needles in this hay?" Do we create many different piles of hay but examine only one closely? No. We need lots of different haystacks, good detectives to look through them, and we need to do it over and over again.

As the former head of the Institute of Education Sciences put it,[8] "the probability of finding [an effective program] will be remote unless we search widely, frequently, and intelligently. In short, experiment, experiment, experiment."

We need to be all about finding the needles.

MAKING FAILURE A SUCCESS

If we know most social programs won't work, how can we make it so that finding something that doesn't work is acceptable and is seen as progress? Where program operators, funders, and evaluators aren't afraid to share results? In other words, how do we make failure a success? By treating the development of social programs the same way as we treat other disciplines.

The FDA approves the trial of a new drug to treat breast cancer. After promising results in an early study, the research is halted when a larger trial reveals no significant positive impacts and major side effects. Congress holds a hearing on the failure of the drug, noting research showed it didn't work but also highlighting anecdotes from some who seemed to benefit. The hearing ends with members at odds on whether the drug works or not, and a few months later funding to develop breast cancer treatments is eliminated.

No, of course not.

Yet this is how the world works for social programs. Congress and other entities provide funding for a specific program, review how it worked, then decide on whether it should continue, be changed, or end. This not only makes the learning process extremely slow; it also makes evaluation incredibly high-stakes. These factors create an environment where there are competing pressures to continue or discontinue programs for reasons other than their effectiveness.

There has got to be a better way, and there is.

LEGISLATE THE PRIORITY, NOT THE PROGRAM

In health care, Congress does not dictate which treatments are allowed or which medicines are approved. Instead, they created a process specifying the *priority* and not the *product*. This same approach can be taken with social programs.

At the federal level, progress has been made in recent years to do just that—specify what the goal is while leaving the selection of the specific program to the state or local entity. This also allows programs to change over time as new evidence is developed. Examples include the following:

- The Every Student Succeeds Act (P.L. 114-95) provides local education agencies with flexibility to select programs that best meet their needs, with programs being ranked in one of four levels as having strong, moderate, or promising evidence, or as demonstrating a good rationale for expecting positive impacts.[9]
- The Maternal, Infant, and Early Childhood Home Visiting program (MIECHV—section 511 of the Social Security Act) requires funding be spent to achieve outcomes listed in the law, using programs that meet evidence requirements delineated by the U.S. Department of Health and Human Services (HHS). States can also use up to 25 percent of their funds on programs that have not yet met the evidence requirements but that will undergo a rigorous evaluation.[10]

- The Family First Prevention Services Act (P.L. 115-123) directed HHS to review programs for mental health, substance abuse, and parenting focused on preventing children from entering foster care. Programs are rated as promising, supported, or well supported based on a rigorous review of evidence.[11]
- The Social Impact Partnerships to Pay for Results Act (P.L. 115-123) created a $100 million fund to tie payment directly to outcomes. In this case, instead of creating a list of evidence-based programs, an entity can sign an agreement to receive payment only if they produce the desired social outcome.[12] Similar language focused on tying funding to outcomes also has been included in the Workforce Innovation and Opportunity Act (P.L. 113-128), the Every Student Succeeds Act (P.L. 114-95), the Bipartisan Budget Act (social impact partnerships demonstration projects and the Maternal, Infant, and Early Childhood Home Visiting program [P.L. 115-123]), and the Carl D. Perkins Career and Technical Education Act (P.L. 115-224).

The next step in this effort is to not just evaluate programs to see if they achieve the priority but to help providers produce better outcomes over time. In many cases, program operators lack the resources—whether that be time, expertise, money, or data—to analyze what is working and what is not and improve their practices as a result. For example, many small programs or those serving a particularly disadvantaged group may not show positive results in an evaluation yet they may also lack opportunities to further learn from the evaluation and try something different. Funding continuous improvement can help overcome this challenge by increasing the number of interventions that can achieve the goal instead of relying solely on evaluations identifying programs that are already working.

By making this approach more commonplace—and by supporting continuous improvement—Congress and others can speed up the development of more effective programs, as well as redirect funding toward approaches that yield the best results. These efforts also can create a climate where failure is acceptable, evidence building is prioritized, and those running programs adapt based on what has been learned instead of fighting for the status quo.

In a world where Congress and others fund a priority and not a program, the failure of one intervention is no longer an existential threat with the potential to end investment in the issue. It just means it is time to learn what went wrong, improve that approach, or try another one instead. Funding to address the social problem does not go away. While it still may be difficult to identify what works and adapt to see what might be more effective, the priority still exists and support to address it continues.

With "tiered evidence" designs where financial support is provided to programs with varying levels of evidence, funding is not for a specific program but, instead, directed toward interventions that address the priority. If those who do not demonstrate results at first also receive support, this structure can help them identify improvements so they can become more effective over time, so that even a "failed" program can try again to achieve the goal.

Tying funding to outcomes can work in a similar way, shifting spending toward successful programs as well as potentially speeding the development of new ideas. "Pay for performance" often relies on real-time measurement of results as well as an evaluation of longer-term outcomes, so those providing services have strong incentives to monitor progress and adapt as needed to ensure they achieve the goal. This approach also has shown an ability to draw private investment and business expertise into social programs, which often brings with it a level of analysis and performance management not traditionally available to social service providers—allowing them to innovate and improve as they go.

A POSSIBLE FUTURE

It is hearing day. For months, you have reviewed the research, met with advocates, and called experts to get up to speed on what's happening.

A decade ago, Congress allocated tens of millions of dollars to address a social problem. At the time, legislators—lacking evidence on what might work best—mandated the federal agency running the new program help organizations build evidence of what works.

Congress has waited a decade to see what progress has been made, spent millions studying the impact, reviewed many interim reports, and now the final report is in. What did it find?

Progress. Specific programs (and certain features of other programs) have been shown to move the needle on improving people's lives. Those that work have been replicated and expanded, and new programs are in the pipeline that look promising. There are lots of failures, but some successes pointing the way to designing better programs.

Congress decides to continue the funding, clarify the goals they want to achieve, and let the thousands of people across the country working to solve this problem keep innovating. The message is clear: Build more haystacks, and keep looking for more needles.

NOTES

1. See the Welfare Reform Academy page at University of Maryland School of Public Policy website, Rossi's Remarks Iron Law Revisited, http://www.welfareacademy.org/rossi/Rossi_Remarks_Iron_Law_Reconsidered.pdf.

2. Peter H. Rossi, "The Iron Law of Evaluation and Other Metallic Rules," See this 1987 paper at www.gwern.net/docs/sociology/1987-rossi.pdf.

3. In Peter Rossi, "The 'Iron Law of Evaluation' Reconsidered," presented at 2003 AAPAM Research Conference, Washington DC., October 2003, www.welfareacademy.org/rossi/Rossi_Remarks_Iron_Law_Reconsidered.pdf.

4. Statement of Jon Baron, president of the Coalition for Evidence-Based Policy. House Committee on Ways and Means, Subcommittee on Human Resources Hearing on What Works, July 17, 2013, http://coalition4evidence.org/wp-content/uploads/2013/07/Testimony-before-Ways-and-Means-HR-subcommittee-7.17.13-Jon-Baron.pdf.

5. Ibid.

6. Ibid.

7. Ibid.

8. Russ Whitehurst, "Acceptance Remarks, November 9, 2007, www.welfareacademy.org/rossi/2007_whitehurst_speech.shtml.

9. See ESSA Tiers of Evidence: What You Need to Know page at REL Midwest website, https://ies.ed.gov/ncee/edlabs/regions/midwest/pdf/blogs/RELMW-ESSA-Tiers-Video-Handout-508.pdf.

10. Emily Sama-Miller, Julieta Lugo-Gil, Jessica Harding, and others, "Home Visiting Evidence of Effectiveness (HomVEE) Systematic Review. Handbook of Procedures and Evidence Standards: Version 2," p. 2, December 2020, HomVEE, https://homvee.acf.hhs.gov/sites/default/files/2021-02/HomVEE_Final_V2_Handbook-508.pdf.

11. See Welcome page at HHS clearinghouse website, https://preventionservices.abtsites.com.

12. See Treasury page, https://home.treasury.gov/services/social-impact-partnerships/sippra-pay-for-results.

TRANSFORMING GOVERNMENT

MICHELE JOLIN AND ZACHARY MARKOVITS

We are at a turning point for our public institutions. After forty years in which it became popular to demonize the government, and after four years in which that demonization metastasized into a physical attack on our government, the Biden-Harris administration is changing the paradigm. Rather than focusing on narrowing the size and scope of government, or making progress through incremental approaches, the new administration and its allies in Congress are unapologetically reinvesting in the physical, social, and civic infrastructure of the country.

The American Rescue Plan, with its $1.9 trillion in investments, has the potential to be the most effective social care package since the 1960s. Add onto that the transformative opportunities of the Infrastructure Investment and Jobs Act and the Inflation Reduction Act—which will pour trillions of dollars more into much-needed investments for cities, counties, states, and small businesses—and we have an opportunity to reshape American society, especially for residents and communities that have historically been left behind. But the ultimate success of this bold, progressive vision of government will not be determined in Washington, DC—it will depend on the actions of governors, state legislators, and agency leaders, and the thousands of mayors and city council members, county executives and commissioners, school superintendents and boards, as well as civil servants across the country.

With these new investments, we have a historic opportunity over the next three to ten years to remake public governance and restore trust in government. If we can harness the potential of these new federal dollars to help state and local leaders build new evidence—and accelerate their use of evidence and data—we can transform the way American government operates and begin to dismantle a legacy of racist policies and underinvestment in communities of color.

The foundation for this transformation is already in place. Over the last decade, a period defined by scarcity in the public sector, cities and states have made significant progress in how they invest taxpayer dollars. But the big bets and large-scale investments the federal government is now making provide an opportunity to accelerate progress on a wide range of challenges and build a government that works for all—and not just for the next three years, but for the next thirty.

In this chapter, we will explore how cities and states have spent the lean years of the 2010s building the capacity for this moment, the challenges governments face and how city and state leaders are meeting those challenges, and where there is an opportunity—with federal support and the right policies in place—to use this massive flow of funds to build and invest in the next generation of evidence.

CREATIVITY LOVES CONSTRAINT: PUBLIC SECTOR INNOVATION IN THE 2010S

Over the last decade, local and state governments have been clawing back from the deepest financial and staffing hole in the last fifty years. The strategies that many government leaders adopted during this period of austerity—which we call "Moneyball for Government"[1] to describe how innovative leaders are using evidence and data to make smarter investments to drive better results—have now set the stage for transformation in the public sector.

In the aftermath of the Great Recession, the federal government failed to step in to aid state and local governments. This underinvestment not only prolonged our nation's economic recovery by at least four years[2] but also forced many state and local governments to drastically reduce their workforce and to maintain only essential services. After hitting a trough in around 2013, many cities and states slowly began to hire again, but newly hired government staff had to take on a heavier workload. New practices around innovation, data governance, transparency, performance management, and program evaluation—areas that could help support a smaller staff and deliver better, more equitable services across government—began to take hold. By the time government employment reached its pre-recession levels—just a few months before it tanked again due to COVID-19[3]—local government effectiveness and efficiency had become redefined in a way that is driven by data.

Seizing on this demand among city leaders, in 2015, Bloomberg Philanthropies launched What Works Cities,[4] a national initiative led by Results for America in partnership with several other organizations that focused on helping cities use data and evidence to make better decisions and improve the lives of all their residents. It was founded on a bet that there was a fundamental gap between a mayor's acknowledgment that using the best available data can help manage a city well and their actual ability to create such a city. Prior to 2015, only a few U.S. cities had adopted a data-driven approach to improve decision making; many thought data-driven government was only for big, coastal cities.

Yet over this past half-decade, a quiet revolution has taken place as cities across the country have undergone a data-driven transformation. Local governments are changing how they do business, with a critical mass of cities helping staff improve their data skills and put in place critical data infrastructure that informs key decisions. This has enabled cities to operate more efficiently and effectively to better meet the needs of residents, which was underscored the past several years by their response to the COVID-19 pandemic. A recent report led by Monitor Institute at Deloitte[5] found that cities in the What Works Cities network made significant progress in foundational data practices. Monitor compared externally validated city responses in the 2020 What Works Cities Certification assessment to similarly reported results from 2015 to find that the percentage of cities:

- Monitoring and analyzing their progress toward key goals **has more than doubled** (from 30 percent to 75 percent).
- Engaging with residents on a goal and communicating progress **has more than tripled** (from 19 percent to 70 percent).
- With a platform and process to release data to residents **has more than tripled** (from 18 percent to 67 percent).
- Modifying existing programs based on data analytics **has more than doubled** (from 28 percent to 61 percent).

States also are making steady progress. Results for America's "Invest in What Works State Standard of Excellence,"[6] which sets a national benchmark for how state governments can consistently and effectively use evidence and data in budget, policy, and management decisions to achieve better outcomes for their residents, has found a dramatic growth of evidence-based and promising examples of impact and best practices over the past three years since the release of the standard.

USING DATA TO SOLVE AMERICA'S MOST INTRACTABLE PROBLEMS

While federal investments are increasing, so are the challenges state and local governments are facing. Governors, mayors, county executives, and other leaders now confront an overlapping set of economic, public health, climate, and racial justice crises that threaten their residents and communities.[7]

And yet we know that jurisdictions that invest in their data practices, that have systematically built up their ability to use data, are better prepared to deal with systemic challenges. They are able to use their data and evidence capacity to spotlight and understand root causes of community challenges, including racial disparities, and they can better target government investments to close racial gaps and accelerate economic opportunity.

This investment enables governments to respond quickly in the face of any crisis. In fact, 70 percent of cities in the Monitor study reported they are systematically using data-informed decision making to respond to the COVID-19 crisis. As Andy Berke, the former mayor of Chattanooga, Tennessee, said recently, "when a pandemic hits, [a data] culture is very important because the organization has got to respond. And it's only going to respond with data if that's the culture you built."[8]

Cincinnati, for example, spent years building up its system to better track and address the growing opioid epidemic in its backyard,[9] using data to tackle all parts of the problem, from mapping cases to tracking health response to improving emergency response to meet the national standard for call answer times. Within a few months, over 90 percent of emergency calls were being answered in less than ten seconds, up from a prior rate of 40 percent. This whole experience helped the city respond quickly when the pandemic struck; it was able to shift its entire performance and analytics team to support its planning and response to COVID-19, taking the same data-driven approach to COVID as it did to opioids. Similarly, Tulsa, Oklahoma, is now better using data to repurpose dollars or defund ineffective programs. By focusing on addressing critical equity concerns, city leaders shifted $500,000 of federal funding away from wealthier communities and to the city's poorest neighborhoods after an analysis showed that existing processes in which every neighborhood received funding regardless of need were not helping the city's most vulnerable communities.

These long-term investments in strengthening the evidence and data capacity of governments may not be as flashy or headline-grabbing as announcing a new program or policy, but they are just as critical. We cannot address the most intractable problems of tomorrow if we don't build a strong data and governmental infrastructure today.

STATE AND LOCAL GOVERNMENTS CONTRIBUTING TO THE NEXT GENERATION OF EVIDENCE

Building on the gains made by local and state governments over the last decade, the new infusion of federal dollars is an opportunity for state and local governments to spur innovation, prioritize learning, and improve evidence building and use. To do this, we think jurisdictions should focus their attention on four main areas:

Invest in Infrastructure to Build and Use Data and Evidence to Create a Culture of Learning

To ensure this one-time influx of funds is used in a way that leads to long-term, sustainable impacts,[10] state and local governments need to use federal dollars to build critical data and evidence infrastructure. To do this, state and local government need to:

- **Understand where they are, so they can map where to go next.** What Works Cities provides a tool[11] to measure how well cities are using data to manage, and it can be used as a diagnostic for local governments to determine the data and evidence infrastructure most important to invest in next.
- **Focus on people, policy, and process before technology.** Leaders will have to make real choices and have the opportunity to invest real dollars[12] in new positions and systems to lead this work. Technology is just a tool to enable progress—city leaders must determine the personnel, training, and technical assistance needed

to gather, analyze, and understand their data to help solve their most intractable challenges.
- **Put equity first.** For governments to address the legacy of underinvestment in communities of color and close racial gaps, they need to understand the scope of the problem before they act. This requires disaggregating data by race, setting specific equity goals, and building integrated data systems to help cities, counties, states, and school districts act with full knowledge of who is succeeding and who is being left behind by the current system.

Orient toward Learning, Testing, and Improving

State and local governments need to use this influx of funds to continue to foster a culture of learning, testing, and improving. This means both using the best evidence-backed programs to help residents and building on the existing programmatic evidence-base so jurisdictions can continue to implement what works best. For example, Dayton, Ohio, is building on previous research to test new strategies to reduce the racial kindergarten readiness gap[13] and is working with scholars from the University of Dayton to evaluate the program's effectiveness. In more than two dozen cities across the country, mayors are building on the evidence derived from the pilot in Stockton, California,[14] to test new ways of providing a guaranteed income[15] to residents.

State and local government leaders continuously need to make decisions based on the best information available. It is critical that government decision makers are given resources, skills, and data and analytic tools to test the impact of decisions and investments, learn what is working, and improve over time. Further, the basis of this research and data capacity ought to be to improve decision making—and, thus, residents' lives—in all sub-federal governments rather than as a compliance mechanism to report to the federal government or grant makers. Federal American Rescue Plan dollars can be used to build this and test, learn, and improve infrastructure that will improve state and local decision making for decades to come. In fact, Results for America is working with Mathematica to track how cities, counties and states are expanding their evidence and data capacity in a new American Rescue Plan Data and Evidence Dashboard[16] that enables better sharing and shows how these dollars are being used to deliver real results.

Incorporate a Critical Source of Evidence into Decision Making—Community Voices

It is not enough for governments to invite residents to community meetings, or meet with groups of stakeholders, or survey their population. These processes are helpful tools, but by themselves they reinforce inequities in ways that are antithetical to the opportunities governments have to reshape the future.

Rather, we must challenge cities and states to use this flow of federal dollars to pursue a new model of understanding the public need—finding a multimodal way of quantifying "community intelligence"—and then building those insights into the decision making process in government.

For example, in Racine, Wisconsin, city, county, school, and civic groups are thinking about how they can deliver impact collectively. Working with a local collective impact partnership, Higher Expectations for Racine County, this collaborative is developing performance measures to evaluate the impact of its Rental Empowerment and Neighborhood Tenant Services Initiative, as well as think about ensuring more residents earn their high school diploma.[17] Here, they have not created something brand new but, rather, have brought together different community groups with diverging viewpoints and representing different voices and, based on a common interest, have focused their attention around a set of common, intractable problems and brought new common data and resources to the table to try and solve those problems together.

LOOKING TO THE FUTURE

Looking back to today thirty years from now, we hope that few—beyond those of us in good government circles—will even remember a time when government was not results-driven. We hope residents will simply feel the results of living in cities and states that offer easy access to high-quality social services, invest in programs that work, and actively listen to their input. At a time when too many people experience government as a headache—while navigating complex websites or waiting in long lines for jobless benefits or nutrition assistance—let's imagine a future when the mundane technical and technocratic capability cities and states build today sets off a new age of efficient and effective governance. No one may remember the painstaking work to deliver this dream, but the more equitable, sustainable, and prosperous society that grows out of it will be felt for generations.

NOTES

1. "Moneyball for Government: The Book," https://moneyballforgov.com/moneyball-for-government-the-book/.

2. Josh Bivens, "A Prolonged Depression is Guaranteed without Significant Federal Aid to State and Local Governments," Working Economics Blog, Economic Policy Institute, May 19, 2020, www.epi.org/blog/a-prolonged-depression-is-guaranteed-without-significant-federal-aid-to-state-and-local-governments/.

3. "October Marks Continued Loss in Local Government Jobs," National Association of Counties, www.naco.org/resources/featured/october-jobs-report.

4. See the What Works page at the Bloomberg Philanthropies website, https://whatworkscities.bloomberg.org/about/.

5. Rhonda Evans, Tony Siesfeld, and Mario Zapata Encinas, "Closing the Data Gap: How Cities are Delivering Better Results for Residents," Deloitte, June 30, 2021, www2.deloitte.com/us/en/blog/monitor-institute-blog/2021/closing-the-data-gap.html.

6. "Results for America Releases the 2020 Invest in What Works State Standard of Excellence," August 12, 2020, Results for America, https://results4america.org/press-releases/results-america-releases-2020-invest-works-state-standard-excellence/.

7. Michele Jolin and David Medina, "How Philanthropy Can Help Governments Accelerate a Real Recovery," Sanford Social Innovation Review, August 11, 2020, https://ssir.org/articles/entry/how_philanthropy_can_help_governments_accelerate_a_real_recovery.

8. Rhonda Evans, Tony Siesfeld, and Mario Zapata Encinas, "Closing the Data Gap: How Cities are Delivering Better Results for Residents," Deloitte, June 30, 2021, www2.deloitte.com/us/en/blog/monitor-institute-blog/2021/closing-the-data-gap.html, p. 7.

9. Brandon E. Crowley, "Data's Critical Role in Reversing the Opioid Epidemic," Governing, June 14, 2017, www.governing.com/gov-institute/voices/col-cincinnati-app-track-reverse-opioid-heroine-overdoses.html.

10. Zachary Markovits, "How the New Stimulus Can Strengthen Local Governments," Governing, April 9, 2021, www.governing.com/finance/how-the-new-stimulus-can-strengthen-local-governments.html.

11. What Works Cities Certification, Bloomberg Philanthropies, What Works Cities, https://whatworkscities.bloomberg.org/certification/.

12. Molly Daniell and Zachary Markovits, "Moving from Vision to Action: The Case (and Tips) for Investing in Data and Evidence to 'Build Back Better,'" What Works Cities, May 29, 2020, https://whatworkscities.medium.com/moving-from-vision-to-action-the-case-and-tips-for-investing-in-data-and-evidence-to-build-44c22c0f0d67.

13. Alison Gardy, "Dayton Doubles Down on Preschool Attendance," What Works Cities, September 18, 2020, https://medium.com/what-works-cities-economic-mobility-initiative/dayton-doubles-down-on-preschool-attendance-f10cb4b6a095.

14. Annie Lowrey, "Stockton's Basic-Income Experiment Pays Off," *The Atlantic*, March 2021, www.theatlantic.com/ideas/archive/2021/03/stocktons-basic-income-experiment-pays-off/618174/.

15. Sarah Wray, "Wave of US Cities to Pilot Guaranteed Income Programmes," Cities Today, March 16, 2021, https://cities-today.com/wave-of-us-cities-to-pilot-guaranteed-income-programmes/.

16. American Rescue Plan Data and Evidence Dashboard, Results for America, https://results4america.org/tools/arp-dashboard/.

17. Alison Gardy, "Recharging Racine," What Works Cities, February 4, 2020, https://medium.com/what-works-cities-economic-mobility-initiative/recharging-racine-226eabb1ee22.

DEMOCRATIZING EVIDENCE

VIVIAN TSENG

THE OLIGARCHY OF EVIDENCE

It is not an exaggeration to say that the evidence-based policy enterprise in the United States resembles an oligarchy more than a representative democracy. It is an enterprise shaped by elites: evidence for the public, shaped by the few.

For over two decades, through Republican and Democratic administrations, and in systems as varied as education, human services, criminal justice, and international development, the federal approach to developing and using evidence has been top down. Evidence-based policy was dominated by support for randomized controlled trials to test the impact of social programs and then the leveraging of federal dollars to incentivize states and localities to adopt those programs (Haskins and Margolis 2014). At first glance, it is hard to argue with this strategy. The logic is tidy: *Fund more of what works and less of what doesn't* (Orzag 2010). But closer scrutiny reveals the shortcomings of a system that privileges the perspectives of federal policymakers over that of system leaders, front-line practitioners, and communities. In this light, federal evidence-based policy initiatives have too often suffered the folly of paternalism, presuming to know what practitioners and communities need better than they do.

Practitioners' and communities' distrust of evidence does not stem only from federal policymakers' actions. Researchers also have been complicit.

Academics have long been critiqued for "drive-by" research, in which they enter a poor or racially marginalized community to collect data for their studies and then exit without engaging the community in ways that could enhance its welfare. Researchers too rarely even circle back to share their findings with communities. Universities reward academics when their research impacts their fellow researchers but fail to appreciate—or sometimes outright disdain—when research impacts communities and practitioners (Hart and Silka 2020). In education circles, teachers and families have characterized this phenomenon as research done *to them* rather than *with them*. Others simply label it as extractive.

DEMOCRATIZING EVIDENCE

A more equitable approach to producing and using evidence to support policy would embrace democratic principles. Stakeholders across civic and professional roles and positions in society would have meaningful roles in identifying what evidence is needed and deciding how it should be used (Tseng, Fleischman, and Quintero 2018; Democratizing Evidence in Education 2022). *Democratizing evidence* calls for an inclusive process to determine the purpose evidence should serve. Whereas research questions often arise from researchers' conversations with each other, a more democratic approach would pursue research agendas that arise from vibrant back-and-forth exchanges between researchers, practitioners, and communities as they tackle the real-world problems most important to them. Program evaluations would be driven not by policymakers seeking thumbs up/thumbs down judgments but by practitioners seeking to improve their work and by the beneficiaries of public services who want programs to better meet their needs. Under a democratized evidence agenda, setting research goals and priorities would be less an academic exercise, and evaluation would not be a check-the-box compliance exercise to satisfy policymakers. Instead, diverse stakeholders would deliberate, negotiate, and compromise over what evidence is needed and for what purposes. The agenda-setting process would likely be messier, take longer, and be more resource intensive, but evidence initiatives would yield meaningful work that serves the public interest.

Democratizing evidence also means communities, practitioners, and the broader public have access to evidence and are equipped to use it to advocate for the policies and services that would benefit their communities.

People will, of course, continue to disagree about their values and the proper role of government, but greater access to evidence and well-designed opportunities for public deliberation over evidence can foster a more evidence-informed citizenry. Moreover, research and data can help forge a shared public understanding of the major problems facing society and the range of potential solutions for them. Perhaps most importantly, evidence can be a stronger tool for democracy: communities can hold the government accountable for its use, nonuse, or misuse of evidence.

DEMOCRATIZING EVIDENCE IN ACTION

Research initiatives that embody democratic principles already can be found in communities across the country. For example, the local Children and Youth Cabinet in Providence, Rhode Island, has brought together fifty-five cabinet and community members from two neighborhoods to set data-driven priorities for kids, select programs to address those priorities, and develop a plan to finance and implement the programs (Annie E. Casey Foundation 2022). In Broward County, Florida, girls and young women conducted youth participatory action research to identify, and then advocate for, ways to improve the juvenile justice system—a project supported by the county's Children Services Council (Gallagher 2019). And for years, community members and education researchers have jointly designed science education curricula that integrate Indigenous ways of knowing with Western science by teaching about plants and animals alongside students' Indigenous cultural practices, histories, and stories about the environment (Meléndez and others 2018).

Examples of democratizing data include the Rockefeller Foundation and Mastercard Center for Inclusive Growth's investments in data science for social impact, which dovetail with grassroots efforts such as Discriminology,[1] an initiative that enables Black and brown communities to use school data to advocate for educational equity. Data for Black Lives[2] is another organization that unites activists, organizers, and mathematicians in the mission of "using data science to create concrete and measurable change in the lives of Black people." These efforts share the underlying principle that those who are most harmed by society's racial and economic inequalities must be able to "have a greater say over their future" (Pacetti 2016). As data is leveraged for social impact, we must be sure community members are active participants. Wielding data allows communities to exercise self-determination to ensure that policies and programs serve them *in the ways they want to be served* (Gallagher 2019).

LOOKING AHEAD

A Democratizing Evidence initiative would fit well within President Biden's goals of "bringing science back," while fostering racial equity. On January 20, 2021, Biden issued the Executive Order on Advancing Racial Equity and Support for Underserved Communities through the Federal Government, which required agencies to conduct equity assessments and develop plans for redressing long-standing inequities across the federal government. A week later, his Memorandum on Restoring Trust in Government through Scientific Integrity and Evidence-Based Policymaking called for equitable delivery of programs across all areas of the federal government. To fulfill these ambitious goals, the communities meant to benefit from government policies and programs should have access to the evidence. They should have a say in identifying which problems require more evidence. And they should have a seat at the table in interpreting the evidence and determining what it means for government action and spending.

In short, the Biden administration must democratize evidence. Incorporating the basic principles of democracy into federal evidence initiatives would overturn the oligarchy of evidence and leave an enduring legacy for generations to come. To get there, the administration could: 1) require science agencies and research and evaluation offices to meaningfully engage communities and practitioners in establishing research priorities; 2) set aside funding to equip community-based organizations to participate in evidence initiatives from the agenda-setting to the implementation and monitoring stages; and 3) ensure community-based organizations have equitable access to federally funded research and evaluation findings and well-designed opportunities to deliberate over those findings and their relevance for future policy action. Democratizing evidence in these ways would usher in a new era of equity-centered and evidence-informed policymaking.

NOTES

1. See the Discriminology website, www.discriminology.org.
2. See the Black Lives Matter website, http://d4bl.org.

REFERENCES

Annie E. Casey Foundation. 2022. Evidence2Success. www.aecf.org/work/evidence-based-practice/evidence2success/.

Democratizing Evidence in Education. *Statement of Principles*. 2022. http://democratizingevidence4.us.

Gallagher, S. *Integrated Data Systems and Youth Voice: Co-Creating Racial Equity*. Presentation at the Ready by 21 Annual Meeting. Seattle, WA. 2019.

Hart, D., and Silka, L. "Rebuilding the Ivory Tower: A Bottom-Up Experiment in Aligning Research with Societal Needs." *Issues in Science and Technology*, 2020.

Haskins, R., and Margolis, G. *Show Me the Evidence: Obama's Fight for Rigor and Results in Social Policy*. Washington, DC: Brookings. 2014.

Meléndez, José, Marin, Ananda, Bang, Megan, and others. Community-Based Design Partnerships: Examples from a New Generation of CHAT/DBR. 2018.

Orzag, P. Memorandum for the Heads of Executive Departments and Agencies: Evaluating Programs for Efficacy and Cost-Efficiency. Office of Management and Budget. July 29, 2010.

Pacetti, E. G. The Five Characteristics of an Inclusive Economy: Getting Beyond the Equity-Growth Dichotomy. December 2016. www.rockefellerfoundation.org/blog/five-characteristics-inclusive-economy-getting-beyond-equity-growth-dichotomy/.

Tseng, V., Fleischman, S., and Quintero, E. "Democratizing Evidence in Education." In *Connecting Research and Practice for Educational Improvement: Ethical and Equitable Approaches*, edited by B. Bevan and W. R. Penuel. New York: Routledge. 2018.

STANFORD REGLAB-SANTA CLARA COUNTY

ACADEMIC-PUBLIC HEALTH COLLABORATION FOR RAPID EVIDENCE BUILDING

SARA H. CODY AND DANIEL E. HO

In March 2020, in conjunction with five other Bay Area counties, one of us issued the first shelter-in-place order in the country in response to the emerging COVID-19 pandemic. As the county health officer of Santa Clara County, California, home to roughly 1.9 million residents, San Jose, and Silicon Valley, I (Cody) had the benefit of long-standing trust and collaboration with other Bay Area health officers. Collaboration, iteration, and rapid information sharing were critical at a time when public health infrastructure was strained to the max. What is less known is that, through the crisis, the Public Health Department (PHD) and Emergency Operations Center (EOC) also developed partnerships with several groups at Stanford, including Stanford's RegLab[1] (directed by Ho) that shaped key aspects of COVID-19 response.

In this chapter, we describe some of the elements of the RegLab partnership and articulate what we have learned about academic-public health partnerships.[2] We emphasize that the problems we faced were profound. Many lessons will be drawn from a once-in-a-generation crisis, spanning far beyond the scope of this chapter. Yet our collaboration has persuaded us that one important set of lessons is about getting academic-government collaborations right. How can health departments and researchers partner most effectively to tackle the most vexing problems, when the current ecosystem often impedes such collaborations?

ORIGINS

At the beginning of the pandemic, there was already a long-standing history of collaboration and consultation between the county and Stanford. PHD, for instance, had consulted extensively with faculty engaged in infectious disease modeling to understand the spread of COVID-19 (James and others 2021). Our specific collaboration began when a PHD epidemiologist attended a virtual talk about the use of mobility information to understand disease spread, based on joint work with the city of San Jose (Ouyang and others 2020). The RegLab began a series of conversations with PHD staff and EOC leadership on the potential use and limitations of mobility information for situational awareness. The Stanford RegLab team built out a mobility dashboard that enabled the county to ascertain: a) which areas exhibited lower (apparent) social distancing compliance; b) business activities; and c) intercounty travel patterns. Such information helped to inform, for instance, public health order revisions in advance of fall holidays. Similar situational awareness came from wastewater sampling, also developed by a Stanford group (Graham and others 2020).

EVOLUTION

Beyond that initial connection, however, the Stanford RegLab (and its sister lab, the Future Bay Initiative) engaged in a series of exploratory conversations, mindful of the extreme demands on time, with a range of EOC/PHD stakeholders. We identified an immediate need around data science for health equity (see, for example, Krass, Henderson, and Ho 2020). While Latinx individuals are roughly 25 percent of county residents, they represented over 50 percent of COVID-19 cases, due to long-standing structural sources of inequality. As a result, we examined how a partnership could augment pandemic response to address health inequities. This resulted in three areas of investment:

1. *Contact Tracing.* The Stanford RegLab team built out a language matching algorithm to enable over 900 contact tracers to be matched to predicted language of incoming cases, using census data. Previously, because laboratory reports have only spotty information about language and ethnicity, cases were effectively assigned blindly, requiring many contact tracers to dial in for third-party translation. In a randomized trial, this intervention reduced time to interview cases by nearly fourteen hours per case and increased the likelihood of interview completion (Lu and others 2021).
2. *Testing.* After a series of in-depth focus groups with community members, the county and Stanford RegLab partnered with community health workers (*promotores de salud*) to launch a novel door-to-door COVID-19 testing program that utilized both local knowledge and machine learning. The trial increased the proportion of tests administered to Latinx individuals from 49 percent at the closest neighborhood site to *88 percent*; and it yielded an 11 percent positivity rate, dramatically expanding testing resources in the most vulnerable communities (Chugg and others 2021).
3. *Supporting Services.* Quarantine and self-isolation can be profoundly challenging for more marginalized communities. To address this, the county built a specialty team of contact tracers offering "high-touch" support services. This team matched diagnosed cases with social support services, such as rental assistance, grocery delivery, cleaning supplies, and hotel accommodations. Stanford RegLab helped design the rollout with an impact demonstration in mind, showing that high-touch services improved the take-up rates of such services by up to 16 percent.

In later periods, the collaboration has pivoted toward vaccine distribution (for example, mobile vaccine siting and outreach efforts) and variant tracking based on a similar data-driven approach.

In normal times, each of these interventions might have taken months, if not years, to deploy. The pandemic, however, required rapid iteration within days. Such agility demonstrates what government could be and yet so often is not: innovative, evidence-driven, and fast-moving.

LESSONS

What lessons can we learn from this case study of rapid innovation? For public health and the public sector, we think there are three:

1. *Build trust, relationships, and capacity.* Critical to the pandemic response were relationships of trust, within the county, with community stakeholders, and across the county-academic divide. We were aided here by many informal ties between the groups, but without such preexisting relationships, it will be key to foster open exchanges around ideas and opportunities. Increases in public health funding can improve this kind of capacity for historically understaffed departments.

2. *Find champions and empower them.* Departments should identify the individuals within the organization who have the vision and desire to do things differently. Who are the "operational nerds" who spot process improvements and can identify places where external partners can help? Who are the evidence champions? Critical to the RegLab partnership were these champions inside the EOC (for example, Greta Hansen, Pamela Stoddard, Sarah Rudman, Anandi Sujeer, Analilia Garcia, and Alexis D'Agostino) who could help quickly identify "win-wins" (that is, projects that would not get done but for an academic partner) and key stakeholders to be involved.

3. *Assign barrier-busters.* Academic-public sector collaborations can fail in many different steps. For contact tracing, there was initial resistance to changing a process that had been painstakingly built. (In Assistant Health Officer Dr. Sarah Rudman's words: "We were building the plane as it was taking off.") This might have made routing cases to specialty language teams impossible. But Dr. Rudman busted these barriers. For testing, one barrier was how to deliver private health information to *promotores* in a way that protected the privacy of individuals. Within days, we figured out, with the exceptional help of compliance and legal counsel, how to provide county-issued devices that were subject to public health security restrictions. Assigning specific individuals the role to bust these barriers is critical.

This then leads us to the lessons for the academy. Academics can play pivotal roles for the future of public health. But barriers need to be busted in universities, as well. Contract review, for instance, can blindly fixate on risks, and it took escalating the matter up to the Stanford provost to get sign-off on our initial data use agreement. The future of public health will depend on a significant transformation of how academic researchers organize themselves:

1. *Escape silos and building teams.* University units are organized by specialization. Academics are, hence, sometimes perceived as "hammers in search of nails" or as engaged in "extractive research" (take the dataset, publish, and run). Instead, curiosity about the world should include curiosity about things we know nothing about. COVID-19 response does not stay neatly confined in an infectious disease department, as evidenced by profound social disparities. Epidemiologists, data scientists, engineers, social scientists, and lawyers all have critical roles to play, but need to do so together, in defiance of conventional academic units. What this invariably will mean is building collaborative teams without regard to academic methodology, conventions, and hierarchy.

2. *Center the real problem.* Curiosity should entail learning first about the most important problems. There was much hype at the beginning of the pandemic about what artificial intelligence (AI) can do to fight COVID-19 (Krass and others 2021). But when major health departments were still receiving droves of lab reports by *fax machine*, off-the-shelf AI may be entirely inapposite. Of course, AI did prove critical in specific respects, but it first took an understanding of the human, community, and institutional challenges to know what algorithms, if any, might help. For instance, extensive engagement around design and weekly check-ins with community health workers helped develop a shared sense of the motivation, constraints, and goals of the approach. This kind of "human-centered" approach will be critical to adapt state-of-the-art tools for actual problem solving. Researchers and academic journals will need to recognize the unique value of community-embedded, institutionally-grounded, and problem-oriented research collaborations.

3. *Solve first, publish later.* Conventional academic models posit influence through publication. (Step One: Publish. Step Two: Question Mark. Step Three: Influence!) Our model was distinctly different. In pandemic times, publication cycles largely cannot respond to the moment, and so we addressed problems first and developed publications later, when there was time to catch our breath. For instance, one of the early things we noticed was that widely used mobility data exhibited demographic bias. We were mindful of this bias when presenting data for operational insights but wrote up the general implications for algorithmic bias audits later (Coston and others 2021). Universities need to recognize these collaborations in promotions and tenure decisions. Publishing later does, ultimately, involve publication, the currency for academics, but merely on a different timeline.

4. *Follow through in practice.* Our theory of impact was to directly help embed data-driven interventions into COVID-19 response. Often, that meant solving a range of practical problems on the way, as operational systems often are not built to facilitate research. For instance, the county had developed an elaborate system for case intake on top of the state system for contact tracing. We realized after extensive deliberation that it would be much better to automate the process entirely, enabling iterative assignment and any refinements of the process. Our team, hence, built out the automated process that saved time and enabled interventions that were, otherwise, operationally infeasible. For many academics, this would be seen as a distraction. For us, it was part of mutual problem solving and building trust in the partnership.

Last, we turn to some broader policy implications. For the first time in decades, public health has seen the increase in public investment it deserves. Controlling COVID-19, preventing the next pandemic, and reducing the social disparities of health will be critical for ensuring health equity going forward. Several reforms could ensure that academic–public health collaborations can thrive.

1. *Invest in information infrastructure.* During this collaboration, our teams built a data infrastructure on tests, cases, mobility, housing units, and demographics largely from scratch. One of Stanford's on-premises servers for health research, luckily not used for this

work, went down for over six months during the pandemic. The basic public health data and information system used for surveillance and situational awareness in California, CalREDIE, went down several times during the course of the pandemic, leaving the PHD essentially blind. This is not the future. Policymakers need to invest in public health data infrastructure (DeSalvo and others 2021; Maani and Galea 2020) and initiatives like the National Secure Data Service[3] and the National Research Cloud[4] to ensure that secure data and computing infrastructure is in place to engage in this kind of work.

2. *Intergovernmental Mobility for States and Localities.* Federal agencies can easily assign academics to function as agency employees under a somewhat obscure statute, the Intergovernmental Personnel Act[5] (IPA). The IPA has been used to great success to streamline access under government security standards to sensitive data and information. Yet states and localities lack such a vehicle for bringing academics in. We addressed this in part by having Stanford students and researchers work as part-time employees or volunteers so they could quickly understand county systems, subject to full security protocols. But such authority needs to be established more generally; we need model state IPA and wide adoption to enable academic-local government partnerships.

3. *Open Systems.* Proprietary systems can be major blockers for innovation. If the contact tracing system had not been controllable by code (that is, by application programming interface), many of the improvements to contact tracing would have required intensive manual workarounds at a time with no FTEs to spare. Such technical systems need to be opened up to facilitate the ability to work and extend such systems effectively.

4. *Funding Models.* Much of this work would not have been possible without core funding. All the Stanford work was done on a pro bono basis without a prespecified grant deliverable, which enabled rapid iteration and adaptation. Conventional grant cycles simply do not work in this timeframe, and both government and philanthropic communities need to recognize that project-specific funding may crowd out some of the most innovative work. Instead, funders should sponsor partnerships with built-in space to explore,

iterate, and pivot where necessary. One of Stanford's newest initiatives, the Stanford Impact Labs,[6] where one of us (Ho) is on the advisory board, for instance, is an important step in this direction, as are initiatives like FDA's Centers of Excellence in Regulatory Science and Innovation[7] that partner with universities.

CONCLUSION

We each bring different perspectives to the table. From the perspective of the County Health Officer, I (Cody) have seen the challenges of getting academic partnerships to work, and want to promote this kind of collaboration that moves from lab to field. From the perspective of an academic who has partnered with many government agencies, I (Ho) have seen many initiatives fail because one barrier or another was not busted.

We make the recommendations above in the spirit of genuine excitement about what is possible when academics focus on problems and when government is agile. Ensuring that such innovation happens is critical to government programs and mitigating what Michael Lewis vividly coined the "Fifth Risk" (Lewis 2018). With such collaborations, we have an opportunity to shape, transform, and revitalize public health and government.

NOTES

1. See Stanford website, https://reglab.stanford.edu/.

2. See PHF website, www.phf.org/programs/AHDLC/Pages/Academic_Health_Departments.aspx.

3. Nick Hart and Nancy Potok, "Modernizing U. S. Data Infrastructure: Design Considerations for Implementing a National Secure Data Service to Improve Statistics and Evidence Building," Data Foundation, July 2020, www.datafoundation.org/modernizing-us-data-infrastructure-2020.

4. See National Research Cloud page at Stanford University website, https://hai.stanford.edu/policy/national-research-cloud.

5. See Policy, Data, Oversite page at OPM.gov website, www.opm.gov/policy-data-oversight/hiring-information/intergovernment-personnel-act/.

6. See Partnership Helps Oakland Students Thrive after Juvenile Detention page at Stanford website, www.opm.gov/policy-data-oversight/hiring-information/intergovernment-personnel-act/.

7. See Centers of Excellence in Regulatory Science and Innovation (CERSIs) at USFDA website, www.fda.gov/science-research/advancing-regulatory-science/centers-excellence-regulatory-science-and-innovation-cersis.

REFERENCES

Chugg, Ben, Lisa Lu, Derek Ouyang, Benjamin Anderson, and others. "Evaluation of Allocation Schemes of COVID-19 Testing Resources in a Community-Based Door-to-Door Testing Program." *JAMA Health Forum* 2, no. 8 (2021).

Coston, Amanda, Neel Guha, Derek Ouyang, Lisa Lu, and others. "Leveraging Administrative Data for Bias Audits: Assessing Disparate Coverage with Mobility Data for COVID-19 Policy." In *Proceedings of the 2021 ACM Conference on Fairness, Accountability, and Transparency* (2021): 173–84.

DeSalvo, Karen, Bob Hughes, Mary Bassett, Georges Benjamin, and others. "Public Health COVID-19 Impact Assessment: Lessons Learned and Compelling Needs." *NAM Perspectives* (2021).

Graham, Katherine E., Stephanie K. Loeb, Marlene K. Wolfe, David Catoe, and others. "SARS-CoV-2 RNA in Wastewater Settled Solids Is Associated with COVID-19 Cases in a Large Urban Sewershed." *Environmental Science & Technology* 55 (December 2020): 488–98.

James, Lyndon P., Joshua A. Salomon, Caroline O. Buckee, and Nicolas A. Menzies. 2021. "The Use and Misuse of Mathematical Modeling for Infectious Disease Policymaking: Lessons for the COVID-19 Pandemic." *Medical Decision Making* 41, no. 4 (2021): 379–85.

Krass, Mark, Peter Henderson, and Daniel E. Ho. "Prioritizing Public Health Resources for COVID-19 Investigations: How Administrative Data Can Protect Vulnerable Populations." *Health Affairs* (blog). April 22, 2020. www.healthaffairs.org/do/10.1377/hblog20200420.729086/full/.

Krass, Mark, Peter Henderson, Michelle M. Mello, David M. Studdert, and others. "How US Law Will Evaluate Artificial Intelligence for Covid-19." *The BMJ* (2021): 372.

Lewis, Michael. *The Fifth Risk: Undoing Democracy*. UK: Penguin, 2018.

Lu, Lisa, Benjamin Anderson, Raymond Ha, Alexis D'Agostino, and others. "A Language Matching Model to Improve Equity and Efficiency of COVID-19 Contact Tracing." *Proceedings of the National Academy of Sciences* 118, 2021.

Maani, Nason, and Sandro Galea. "COVID-19 and Underinvestment in the Public Health Infrastructure of the United States." *The Milbank Quarterly* 98, no. 2 (2020): 250.

Ouyang, Derek, Cansu Culha, Neel Kasmalkar, Maeve Givens, and others. "Stanford Future Bay Initiative Covid-19 Projects: Social Distancing Compliance." 2020. http://bay.stanford.edu/covid19.

CAMDEN COALITION

HEALTHCARE AND PUBLIC HEALTH DATA INTEGRATION DURING COVID

AARON TRUCHIL, CHRISTINE MCBRIDE, AUDREY HENDRICKS, NATASHA DRAVID, AND KATHLEEN NOONAN

The Camden Coalition, a multidisciplinary nonprofit working to improve care for people with complex health and social needs, has been addressing multiple health disparities in southern New Jersey and beyond for the past two decades.[1] Through this work, we have learned that technology, no matter how sophisticated, can go only so far. What *is* required: a judicious, collaborative, and hands-on patient- and partner-centered approach to using and integrating technology.

The Camden Coalition launched its Health Information Exchange (HIE) in 2010 to connect the siloed data of our regional health systems and improve care delivery in the city of Camden.[2] The HIE is a web-based application that links individual-level data from providers across Camden and the region to enable real-time access to a holistic picture of an individual's clinical data. Since its launch, hospital electronic medical record (EMR) systems have evolved to allow for expanded capacity to share data across hospitals, yet there are still major gaps in access for other key providers. The Camden Coalition's HIE helps remedy this gap, by enabling the same level of visibility to a broad array of health and social service providers that are equally critical for improving wellbeing. These additional providers include federally qualified health centers, the jail's healthcare provider, skilled nursing facilities, and social service organizations such as shelters, senior living facilities, and medically indicated meal services, among others.

COVID-19 created a host of challenges, particularly for low income, urban communities like Camden City, where high rates of community transmission exacerbated existing structural issues and inequities. The pandemic shined a spotlight on the need to integrate data across institutions and sectors. The Camden Coalition and our partners recognized that the Camden Coalition HIE was uniquely positioned to help bridge this data divide and could serve as a critical support tool for front-line agencies in their efforts to support a community that saw its healthcare and social needs grow exponentially in the wake of COVID-19. Four core use cases for the HIE emerged:

1. Providing comprehensive insight into the impact of COVID-19 on our community
2. Ensuring providers had greater access to lab results
3. Enhancing contact tracer's ability to identify and engage individuals
4. Standing up cross-agency and provider workflows to support vulnerable populations

PARTNERS

As was the case in most of the country, much of the front-line response to the pandemic in New Jersey fell on county health departments, which were tasked with standing up testing and contact tracing operations, establishing safe quarantine options for individuals with unstable housing, and, eventually, deploying vaccinations. Prior to the COVID-19 pandemic, the Camden Coalition had begun preliminary discussions with the Camden County Department of Health about becoming an HIE participant, but plans had not been finalized to onboard them before the pandemic was declared a national emergency on March 13, 2020. One week later, the Camden Coalition received a call from the county requesting an immediate connection to the HIE. We granted their request and quickly finalized the necessary agreements to ensure privacy, security, and proper consenting procedures so we could start training county staff right away and also determine their reporting needs.

Other key partners during this period were our local health systems, including emergency departments, primary care and population health teams, and a local housing provider whose shelter staff utilized the HIE to coordinate care and medication management services for their most complex patients. Health systems leadership also reached out with requests to train additional users on the HIE to facilitate greater use of the evolving COVID-19 data points we were integrating into patient- and population-level reports.

LEVERAGING DATA TO ADDRESS COVID-19 NEEDS

To address the variety of COVID-19 needs articulated by the county health department and other partners, the Camden Coalition worked quickly to develop new functionality for the HIE. These expanded capabilities were done with limited additional funding (under $20,000) and capitalized on much of the existing infrastructure of the HIE.

1. **Real-time population health dashboards:** Recognizing that different stakeholders had different components of the data picture and no single entity held a comprehensive picture, the Camden HIE could provide a holistic view of our community and the impact of COVID-19. The coalition's data team constructed a data warehouse to include daily extracts of all COVID-19-related HIE data. Incorporating feedback from different stakeholders, the Camden Coalition constructed numerous Tableau dashboards to monitor COVID-19 trends, such as daily new cases, lab positivity rates, hospitalization and ICU utilization, geographic hotspots, and other relevant metrics.

2. **Expanding access to COVID-19 lab data:** While all COVID-19 test results were mandated to be shared with the state, clinicians were limited to seeing only labs their organizations had ordered. As testing sites were rapidly being established and patients were scrambling to get tested wherever possible, the Camden Coalition worked closely with providers and lab companies in our region to ensure as much lab data as possible made its way into the HIE and that the data would, then, be available to providers in user-friendly formats. To ensure the information was front and center, alerts were created in the HIE that high-

lighted the most recent lab date and result. Existing population management reports already being utilized by providers were expanded to include additional fields that indicated recent test results to avoid duplicative testing.

3. **Enhancing contact tracing efforts with additional contact information:** Recognizing that the HIE had longitudinal demographic data on a large subset of the region's population and that one of the primary barriers to contact tracing was the ability for tracers to successfully call and get through to individuals, the Camden Coalition worked closely with the county to create a contact tracer user role with tailored access to relevant information in HIE. Contact tracer users could look up an individual in the HIE and access phone number and address data across all our data contributors without the liability of seeing HIPAA-protected clinical data. Every additional phone number or address was an opportunity to re-engage someone who might have had incorrect or out-of-date contact information.

4. **Flagging vulnerable patients:** The Camden Coalition recognized that the HIE also could flag individuals at highest risk for developing a severe response to COVID-19. Using the CDC's risk criteria, the coalition constructed variables to identify these high-risk populations and incorporated them into existing provider reports. With these report additions, providers could look at their populations and prioritize patients at higher medical risk for telephonic check-ins, educational outreach, and appointments.

5. **Cross-agency workflow development**: As partners realized the need to stand up new multiagency interventions, a common data platform became necessary to facilitate these workflows. One such intervention deployed in Camden and other communities across the country were quarantine hotels, which provided access to temporary housing for individuals without stable housing to prevent further exposures in the community. These interventions needed to launch quickly with limited time and resources to build out the data systems to support them. The HIE, with its ability to quickly stand up data collection and workflow tools across partners, helped fill the gap.

The housing nonprofit and county managed the administration, intake, and onsite support to hotel residents, while Camden Coalition staff provided care coordination. The Camden Coalition team worked closely with partner agencies to quickly understand and map out the workflow—from referral to intake and all the way through care coordination and exiting of the program—and translate it to provide all the information necessary at each step of the process. We then converted the workflow into a set of forms and documentation steps, and took a minimum viable product (MVP) approach to turn around a prototype we could share with our partners for their input. With the MVP developed, we were able to demo the key functionality to partners and solicit rapid feedback critical to refining the tool. We continued to co-design all the forms with partners to ensure critical information moved with the patient throughout each step of the process to alleviate the patient's need to repeat information about their situation to multiple providers and to allow providers to more efficiently communicate with one another.

CHALLENGES AND RESPONSES

From a data perspective, the foremost challenge to the COVID-19 response was that neither the state and county health departments nor our social service providers had sufficient preexisting data and analytics support. Given the unprecedented nature of the pandemic, everyone—from the state and county health departments to our local health systems and social service providers—was scrambling to move quickly, and with limited, and sometimes conflicting, guidance. These conditions produced a somewhat chaotic environment that made it difficult at times to step back and think about opportunities to use data innovatively and more holistically rather than just focus on the immediate crisis at hand. As a result, there was a general lack of bandwidth by partners, and getting momentum on determining how the HIE could support the work was sometimes a challenge.

There also was an onslaught of opportunistic technology companies pitching their solutions as the silver bullet to combat the pandemic as they eyed emergency funding allocations as a new gold rush. Even though the HIE was already funded and the coalition was not seeking additional funds to support the work, there was a general perception that we were just one of a plethora of vendors trying to sell new products which, at times, seemed to stymie the work.

Compounding this was a lack of clarity over who had the decision making authority to make data and technology decisions. While we scrambled to expand the HIE's lab data to the extent we could, we recognized a direct connection with the State's Communicable Disease Reporting and Surveillance System (CDRSS) would provide us the most comprehensive lab picture possible to our provider community. After multiple conversations with the state, we were unable to gain traction with establishing an interface between the HIE and the CDRSS system. We also were unable to secure permission to obtain data extracts from the CDRSS that, while not directly feeding into the HIE, would complement our dashboarding and population health efforts.

To counter these challenges, the Camden Coalition tried to anticipate the needs of its partners and move forward with building out new functionality regardless of whether there was clarity around funding and/or statewide consensus on whether to move forward with using the HIE for a given use case. This meant that time was invested developing tools that were not, ultimately, needed in certain cases, but it also meant that as partners solidified their needs for the HIE, we were ready to support them as rapidly as possible. In the long term, we anticipate working with the county to build their data and analytics capacity, and offer similar support to surrounding counties in South Jersey.

RESULTS

In the urgency of responding quickly to the pandemic and to understand the potential for the HIE to support our partners, the Camden Coalition was able to leverage the HIE for a variety of new purposes and onboard the Camden County Health Department in a short period of time. Some of the tools and functionality developed had immediate benefit and impact—the population dashboards allowed us to establish additional, better located testing sites; the contact tracing role within the HIE allowed county staff to have more accurate outreach information; the multiagency care coordination and workflow tool for the quarantine hotel was critical for standing up a brand new, time-limited program that served one of the city's most vulnerable populations.

Other tools we developed did not provide as much utility. We were not successful in activating the provider community to use the clinical and social vulnerability flags, as providers were primarily focused on treading water in a radically new landscape. However, our internal teams used this variable as a prioritization strategy when conducting outreach to schedule vaccination appointments for Camden residents at a pop-up FEMA vaccination site in February and March 2021.

The longer-term implications of these collaborations were that the Camden Coalition was able to further reinforce the HIE as a critical and nimble resource within our community. We were able to onboard the county health department and showcase the HIE's ability to serve as a uniquely situated, inexpensive, cross-sector, cross-agency tool, which has helped initiate conversation about broader uses for the HIE to support non-COVID-19 programming and to create a more robust ecosystem of care in our region. The efforts to quickly build a variety of new functionality and collaborate across agencies strengthened our muscles for how the HIE can react and adapt to future needs as they emerge

REFLECTIONS

The collaborative, data-driven efforts of the Camden Coalition and its partners to leverage the HIE to respond to the COVID-19 crises showcases the value of ongoing investments in community data infrastructure. With very limited additional funds, the Camden Coalition was able to quickly stand up a variety of new functionalities that supported the diverse needs of our partners. This would not have been possible without many years of prior investment in the underlying data and analytics infrastructure and strong preexisting partnerships on the ground.

Given the need to react quickly to the crisis, our early efforts focused on rapid-cycle development of functions that were urgently needed. As we continue to tackle the ongoing challenges of COVID-19, the early work we did with the county and a local housing nonprofit created an opportunity to more systematically build a cross-sector data capacity in our region that we are already expanding today, and to develop ways to more effectively respond to subsequent challenges.

NOTES

1. See Camden Health website, https://camdenhealth.org/about/.
2. See Camden Health, https://camdenhealth.org/connecting-data/hie/.

UNITEDHEALTHCARE COMMUNITY & STATE

USING HOUSING TO IMPROVE HEALTH–A SOCIAL IMPACT INVESTING STRATEGY

ANDY MCMAHON AND NICOLE TRUHE

The mission of UnitedHealth Group, a diversified Fortune 500 health and well-being company and UnitedHealthcare Community & State's parent company, is to help people live healthier lives and help make the health system work better for everyone. At UnitedHealthcare Community & State, we live that mission by providing high-quality public sector health benefits to low-income individuals and families, people with disabilities, and seniors.

Through our role as a managed care organization (MCO), we manage health care for nearly 6 million beneficiaries or members in thirty-one states plus Washington, DC, across a variety of programs, including Medicaid, the Children's Health Insurance Program (CHIP), Aged, Blind, and Disabled (ABD) plans, and Special Needs Plans (SNP). The primary populations we cover include children from low-income families, low-income adults, people with disabilities, and seniors with limited income. More than 2.7 million UnitedHealthcare Community & State members are children.

In the United States, Medicaid and CHIP cover nearly one in five Americans and 40 percent of all births. The most prevalent health conditions among children who are Medicaid and CHIP beneficiaries are asthma and ADHD or ADD. Among nonelderly adult Medicaid beneficiaries, hypertension, arthritis, asthma, and diabetes are the most common health conditions, with mental disorders the most reported condition for which adults receive care. Social and economic factors, health behaviors, and the physical environment account for 80 percent of an individual's health status, a greater proportion of an individual's health than medical care. These factors, or social determinants of health (SDOH), impact Medicaid beneficiaries—who are primarily low income—more than the general population.

As a result, we take a whole person approach to care and ensure that medical, social, and behavioral and addiction services and supports work hand-in-hand. Members are screened for social barriers, and partnerships with community-based organizations are developed to help members address their identified social needs. We cover services such as transportation, pilot new initiatives, engage communities, provide grants, volunteer in the community, and find other ways to support the members and communities we serve. In addition, we have developed a social impact investment strategy, in conjunction with the UnitedHealth Group Treasury team, that is making investments to build the capacity of organizations and programs that improve community health and reduce unnecessary healthcare utilization.

OUR APPROACH TO BUILDING THE EVIDENCE

UnitedHealthcare Community & State aims to deliver on the Institute of Healthcare Improvement's "Triple Aim"—improving the patient care experience, improving the health of populations, and reducing the per capita costs of care. Some interventions, such as supportive housing, which offers affordable housing linked to intensive case management, have a strong evidence base. Reports and studies have shown supportive housing improves housing stability and mental and physical health and reduces substance use. There are many other programs addressing social determinants of health where the evidence is emerging, including medically tailored meals and home remediation for asthma. However, there are very few rigorous studies that focus on alignment between healthcare and social services, particularly among a broad set of Medicaid beneficiaries.

Building the evidence base of the impact of housing, nutrition, and other interventions that have an impact on health care is essential to ensuring we most effectively serve our members. This also plays an integral role in our social impact investment strategy and decision making as we look for opportunities to address the social determinants of health.

As we develop partnerships and execute our social impact investment strategy, we aim to catalyze change and build pathways toward sustainability for successful programs. Building a strong evidence base is critical for strengthening interest, engagement, and investment by government, health plans, and others in programs that demonstrate an ability to impact healthcare outcomes and utilization. We are an active partner and investor throughout projects to offer support during challenges and to ensure success. Our approach involves:

- **Identifying and prioritizing organizations that have a commitment and focus on outcomes**—Organizations do not need top-notch expertise in evaluation and outcomes; rather, we look for organizations that have an interest in partnering to build the evidence.
- **Building evaluation plans collaboratively**—Many organizations are looking to improve their approach to evaluation but have not partnered with a Medicaid managed care plan. We work hand-in-hand to determine whether our partnership can help strengthen their approach to evaluation by offering our perspective, ideas, and, potentially, data.
- **Providing concessionary capital or grant funds for evaluation**—We understand that organizations often have a difficult time funding evaluation. We explore opportunities to concede return on investments to finance not only services but also evaluation of implementation of those services. Through the grant funding we do, we also will partially allocate funds toward the development and implementation of evaluation.

Example: Community Catalyst

In 2018, we launched an innovative partnership with the Council of Large Public Housing Authorities (CLPHA) aimed to improve the health outcomes of Medicaid managed care beneficiaries living in publicly assisted housing. By leveraging the capacity, resources, and expertise of local public housing authorities, our local Medicaid health plans, and other community partners, we believed we could improve both member and community health. Our aim was to develop these partnerships, implement healthcare interventions and strategies, and measure the impact the interventions have on healthcare outcomes and utilization.

We launched the partnership by convening some of UnitedHealthcare's Medicaid health plans and the Public Housing Agency (PHA) in six communities—Akron and Columbus, OH; Seattle and King County, WA; and Austin and Houston, TX. Initially, the UnitedHealthcare health plans and public housing agencies shared their respective challenges and priorities. We built on these early conversations by establishing data sharing agreements to match public housing authority data with our claims data. The data dashboards helped the teams understand the number of shared members or residents and their most prevalent healthcare conditions. With this information, each local team identified an initial health challenge to address in their community (for example, lead poisoning or asthma among children, diabetes, nonessential emergency utilization, and mental health).

At the start of 2020, teams were collaboratively developing engagement strategies to address these health priorities. Unfortunately, the COVID-19 pandemic disrupted further progress on the initiative. Currently, we are reengaging in the work and including federally qualified health centers (FQHC) and community-based organization (CBO) partners in the collaboratives. FQHCs are clinics that provide primary and specialty care to underserved populations. They serve one in three people in poverty and one in five Medicaid members, on average nationally. CBOs provide vital services in communities. Both FQHCs and CBOs lend new insights into

FIGURE 5.11.1 Proposed Interventions by Site

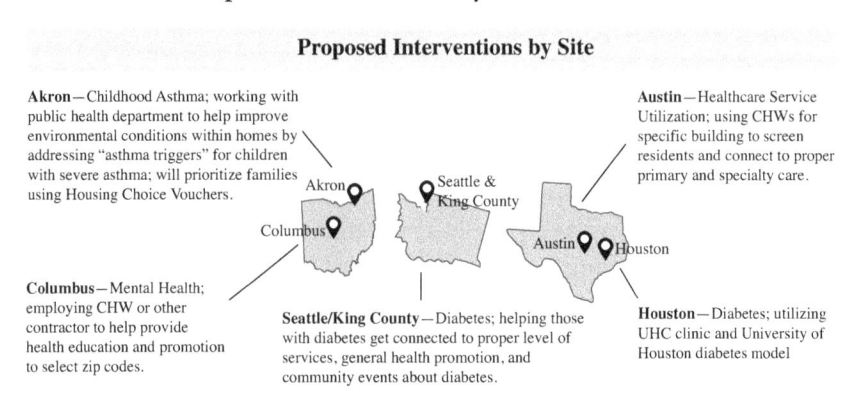

Proposed Interventions by Site

Akron—Childhood Asthma; working with public health department to help improve environmental conditions within homes by addressing "asthma triggers" for children with severe asthma; will prioritize families using Housing Choice Vouchers.

Austin—Healthcare Service Utilization; using CHWs for specific building to screen residents and connect to proper primary and specialty care.

Columbus—Mental Health; employing CHW or other contractor to help provide health education and promotion to select zip codes.

Seattle/King County—Diabetes; helping those with diabetes get connected to proper level of services, general health promotion, and community events about diabetes.

Houston—Diabetes; utilizing UHC clinic and University of Houston diabetes model

FIGURE 5.11.2 **Community Catalyst Approach**

Community catalyst approach

Community Needs

- Community-Based Organization
- Public Housing Agency
- FQHC
- UHC Medicaid Plan

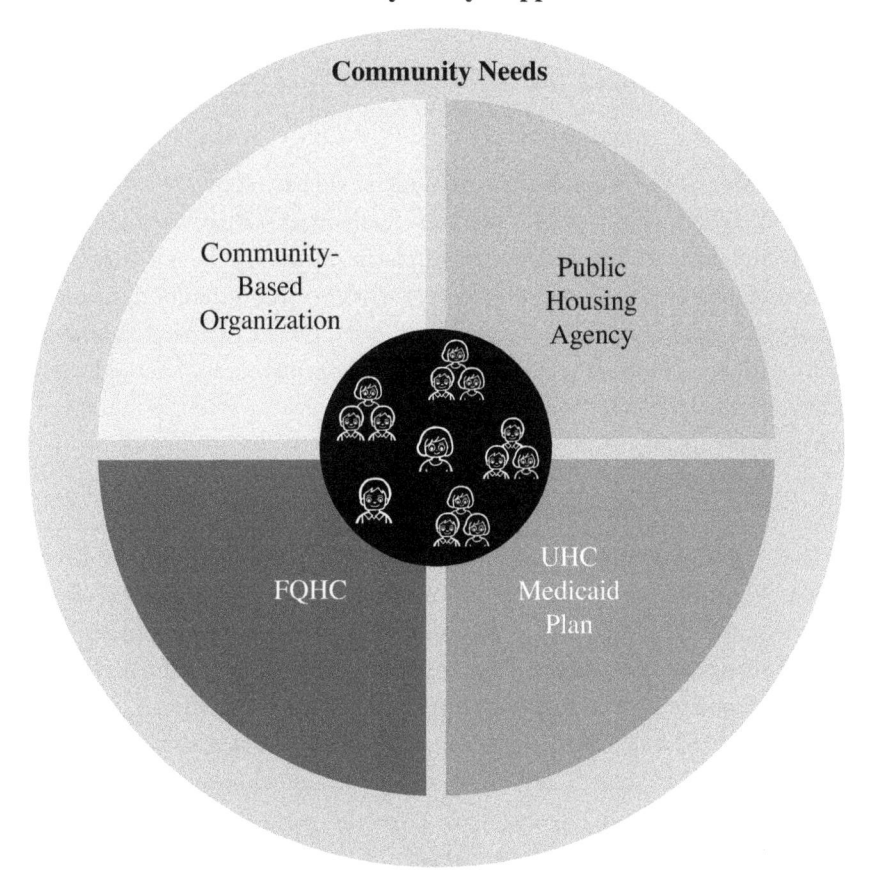

community health challenges and will be important collaborators as we develop and implement local healthcare interventions.

We are in the process of expanding this approach to five more communities through what we are calling our *Community Catalyst* initiative. This initiative aims to foster community-based collaborations that improve community health. We believe that working together with organizations across sectors and individuals with deep community knowledge and experience is the most effective way to make progress toward solutions that will positively impact the health of our communities.

Decisions in the *Community Catalyst* work are made collaboratively, including decisions about measurement and evaluation. Partner PHAs, FQHCs, CBOs, and health plans will have different perspectives on evaluation and access to diverse types of data and information. We think this diversity will contribute to holistic and robust findings and learnings. Through this process, we also will build the evaluation capacity among the participating organizations.

From our early *Community Catalyst* work, we have found that partners are excited about working across sectors. Each organization has traditional partners and ways of navigating those relationships (CBO and funder, insurer and provider, etc.) and cross-sector work creates different dynamics. This work is stretching both UnitedHealthcare and our partnership leaders to work in new ways. It takes time for organizations to understand each other and how to work together, but this is a critical step to the success of the partnership.

Example: Health and Housing Fund

In June 2020, UnitedHealthcare announced the creation of a $100 million investment fund that will create 1,000 new apartments with integration of housing and health care present in every project, including on-site health-related services for residents. The Health and Housing Fund (the Fund) was developed in partnership with Stewards of Affordable Housing for the Future (SAHF) and National Affordable Housing Trust (NAHT), two leading affordable housing organizations. The Fund aims to expand access to affordable housing in communities across the county, a need that has intensified with the COVID-19 pandemic.

To ensure we could track outcomes, we are investing in SAHF's "Housing as a Platform" evaluation tool to assess changes in resident health outcomes. Through a collaborative effort to develop an evaluation plan, SAHF will measure improvements in core measures—physical and mental health, access to primary care, and food insecurity. Sites also must select one to two additional measures from a menu with indicators for financial stability, education, health and wellness, housing stability, and community and safety.

The Fund also includes investments in housing developments to provide health-related services for residents. UnitedHealth Group conceded some return on investment to invest $1 million in an array of services and strategies to integrate housing and health care. Housing developers have the ability to request grant funding of between $25,000 to $75,000 toward providing services and supports to enhance the health and wellness of their residents. Potential use of those grant funds includes financial or health coaching, broadband to enable educational programming and telehealth, and development of community spaces with the option to use them as clinical spaces for residents to receive care onsite.

Our hope is that the use of the Housing as a Platform evaluation tool to evaluate the enhanced onsite services will add to the evidence base on the connection between housing and health and provide insights into the impact of targeted interventions. Seven housing developments are underway with Fund financing, and two developments are finished and open to residents. Services have become and will continue to be an important part of our evaluation criteria for affordable housing investments. In the first two completed residences, housing managers are using UnitedHealthcare grant funding for an onsite food pantry and a community health worker and peer health coach program to enhance resident connectivity to primary care.

We are committed to identifying and supporting innovative approaches to finance programs that improve health and wellness. When programs can demonstrate a positive impact on health outcomes, we can more easily build the case to concede returns on our social impact investment capital. We believe that conceding capital is a method of enhancing impact and contributing further evidence for programs.

REFLECTIONS AND THE PATH FORWARD

Clear and simple evidence is essential for replicating and scaling successful programs. But there remains a large majority of interventions that lack the data and outcomes tracking needed to assess impact. While randomized controlled trials (RCTs) are considered the gold standard of clinical research, much social sector work is not conducive to RCT design, and most service providers do not have the time or expertise to engage in such rigorous research. A disciplined process for learning, testing, and improving provides the basics for building the evidence. There is value to well-designed RCTs, but we can move much more quickly if they are supplemented with other, more accessible and easily implemented evaluation techniques.

That is why we seek to leverage the available evidence and work to build on it, using our own healthcare data where possible. We do this by bringing together health care, housing, and social services with a focus on concretely measuring the impact of these efforts so we can improve community health and help scale what works. Evidence of impact and success helps us make the case to expand our efforts, bring other partners to the table, and advocate for policy change that will support these types of cross-sector partnerships. And we are looking for partners who are eager to collaborate on and further this effort to build the evidence.

While we are a very large and complex organization, we also come to this work as a humble partner. UnitedHealthcare can contribute funding, data, and expertise, but we prioritize listening to and learning from our community partners. When it comes to evaluating impact and advocating for increased investment in efforts that work, we will succeed if we work together.

Finally, in addition to contributing our own resources and expertise for building the evidence, we actively encourage other MCOs, health systems, and all levels of government to do the same. Building the evidence is an investment in longer-term, more sustainable, systemic change. And it is the only path forward to ensure that, as a society, we are investing in the right things at the right time, to improve the health and well-being of people and communities. Together, we can achieve it.

REFERENCES

Centers for Medicare and Medicaid Services. "Medicaid and CHIP Beneficiary Profile: Characteristics, Health Status, Access, Utilization, Expenditures, and Experience." February 2020. www.medicaid.gov/medicaid/quality-of-care/downloads/beneficiary-profile.pdf.

Corporation for Supportive Housing. Supportive Housing 101. www.csh.org/supportive-housing-101/.

County Health Rankings and Roadmaps. County Health Rankings Model. 2014. www.countyhealthrankings.org/explore-health-rankings/measures-data-sources/county-health-rankings-model.

Institute for Healthcare Improvement. The IHI Triple Aim. 2021. www.ihi.org/Engage/Initiatives/TripleAim/Pages/default.aspx.

Spencer, A., Freda, B., and McGinnis, T. "Measuring Social Determinants of Health among Medicaid Beneficiaries: Early State Lessons." *Center for Health Care Strategies, Inc.* December 2016. www.chcs.org/media/CHCS-SDOH-Measures-Brief_120716_FINAL.pdf.

UnitedHealthcare Community & State. Partnering with Health Centers. www.uhccommunityandstate.com/articles/partnering-with-health-centers.html.

ABOUT THE CONTRIBUTORS

Lola Adedokun is the executive director of the Aspen Global Innovators Group at the Aspen Institute and cochair of the Aspen Forum on Women and Girls, where she leads a portfolio of programs that expand opportunities for and access to health and prosperity. Prior to joining the institute, she administered almost $200 million as director of both the African Health Initiative and the Child Well-being Program at the Doris Duke Charitable Foundation. She earned dual BA degrees with honors in health policy & society and sociology from Dartmouth College and an MPH from Columbia University's Mailman School of Public Health.

Mandy A. Allison is an associate professor in the Department of Pediatrics at the University of Colorado. Before medical school, she taught public school, where she saw the effect of structural inequities on health and education outcomes. She completed her residency at the University of Utah. Dr. Allison sees patients and teaches residents and students at Children's Hospital Colorado, serving a racially, culturally, and linguistically diverse, mainly low-income population. She joined the team at the Prevention Research Center for Family and Child Health (PRC) in 2016 and has been the codirector of the PRC with Dr. David Olds since June 2019.

Plinio Ayala The Per Scholas staff is led by CEO and president Plinio Ayala. Plinio was born and raised in the South Bronx, just as it was growing into a national emblem for urban poverty and disinvestment. His experience instilled a lifelong passion for creating economic opportunity, and shortly after graduating from Wesleyan University with a degree in

American studies, he devoted his career to building win-win solutions to social and economic problems, first at Jobs for Youth and then at SoBro.

In 2003, Plinio became president and CEO of Per Scholas and since that time has been instrumental in all the organization's achievements—from evolving its original mission, which was to bridge the digital divide by refurbishing end-of-life computer equipment, to leading its accelerating national growth. In the process, Plinio has incubated strong organizational capacities to respond to changing market conditions, pursue entrepreneurial opportunities, and embrace rigorous measurements of impact. He often says that he can imagine no greater satisfaction than seeing overlooked people—many of whom have struggled with educational and public systems that seem designed to stymie rather than uplift them—finally channel all their passion and curiosity into transformative careers.

Plinio sits on the boards of the Workforce Professional Training Institute, Economic Mobility Partners, and SoBro. He has also served on the New York State Workforce Recovery Strategy committee since 2020. He has received numerous honors, including the Hispanic Chamber of Commerce Community Leader Award in 2019, Newscorp's Murdoch Community Hero Award in 2018, and Hispanic Community Leader of the Year by Crain's New York in 2016, among others.

Tamar Bauer is an attorney with expertise in harnessing policy strategies to drive better and more equitable outcomes for communities. She focuses on actionable innovation in the public and private sectors, with demonstrated success in driving federal and state policy change. As chief policy officer at Nurse-Family Partnership from 2006 to 2017, she helped secure $1.5 billion in federal funds to create the Maternal, Infant and Early Childhood Home Visiting Program and mobilized $30 million in public and private funding to expand services for families in South Carolina's Pay for Success initiative. Prior to NFP, Tamar helped launch the New York Academy of Medicine's Child Health Forum and advanced policy work at the New York March of Dimes and American Academy of Pediatrics.

Jake Bowers is associate professor of political science and statistics at the University of Illinois Urbana-Champaign. He is a fellow in the Office of Evaluation Sciences in the General Services Administration of the U.S. Federal Government and has served as a fellow in the Policy Lab at Brown

University and as methods director for the Evidence in Governance and Politics network. He cofounded and codirects the Causal Inference for Social Impact Lab at the Center for Advanced Study in the Behavioral Sciences at Stanford University. He also cofounded Research 4 Impact, an organization devoted to connecting academia with practice.

Jessica Britt serves as senior director of research and evaluation at Year Up, one of the nation's leading workforce development organizations. Jess has over ten years' experience in program management and evaluation both in the United States and internationally. In addition to her work at Year Up, Jess serves as a member of the board of directors for Safe Passage / Camino Seguro in Guatemala City. To connect on LinkedIn: https://www.linkedin.com/in/jessica-britt/.

Jennifer L. Brooks helps philanthropic organizations, nonprofits, and governments strengthen their impact through evaluation, metrics, and evidence-based practice. Brooks held senior positions at the Bill and Melinda Gates Foundation, the National Governors Association (NGA), and the Administration for Children and Families. At NGA, Brooks oversaw technical assistance on human services, workforce, and economic development programs and led Governor Hickenlooper's NGA Chair's Initiative, Delivering Results. At the federal government, she led a research and evaluation portfolio for the federal Head Start program. Brooks holds a PhD and MS from Penn State University and an MA from the University of Chicago.

John Brothers currently serves as the president of the T. Rowe Price Foundation and president of T. Rowe Price Charitable. Brothers comes to T. Rowe Price from Quidoo, an international consulting firm he started and led for over a decade, merging the firm in 2016.

Brothers served as a management and social policy professor for over a decade at NYU and Rutgers and served as a visiting fellow at the Hauser Center at Harvard. He is currently serving as an honorary professor of practice at Queen University in Belfast, Northern Ireland, and works with the China Global Philanthropy Institute in Beijing.

Brothers has been a writer with the *Stanford Social Innovation Review*, *Nonprofit Quarterly*, and the *Huffington Post* and is an author of several books. He has been interviewed, referenced, or quoted in dozens of local, national, and international media outlets including the *New York Times*, *Washington Post*, *Newsweek*, ABC News, and the *Wall Street Journal*. Brothers has spoken to thousands on nonprofit and philanthropic effectiveness.

Brothers, who grew up in deep poverty, began his work in the local community, serving as a community organizer and family case manager in urban neighborhoods in the Midwest, then moved to leadership positions, including CEO, with local and national organizations on the East Coast. Brothers is proud that this work leaves a legacy of innovative efforts that every day continue to serve a wide network of children and families. These efforts include emergency services for homeless women and children in Northern Virginia, after-school programs for children in the housing projects in South Brooklyn, and transitional housing options for immigrant families in Boston who are suffering from domestic violence.

Laurie Miller Brotman, PhD, is the Bezos Family Foundation Professor of Early Childhood Development and professor of population health and child and adolescent psychiatry at the NYU Grossman School of Medicine and director of the Center for Early Childhood Health and Development. Dr. Brotman is a clinical developmental psychologist whose scholarship focuses on culturally responsive family engagement, social emotional learning, and scaling programs to reduce racial and income disparities. Dr. Brotman is the founding director of ParentCorps, a family-centered enhancement to pre-K programs serving racially and culturally diverse families in historically disinvested neighborhoods.

Daniel J. Cardinali is president and CEO of Independent Sector, the only national membership organization that brings together a diverse set of nonprofits, foundations, and corporations to advance the common good.

Before joining IS in 2016, Dan served on the IS Board of Directors and several IS member committees. He also led IS member, Communities In Schools, the nation's largest and most effective dropout prevention organization, for twelve years after working in other positions at the organization.

As a thought leader in the field of public education, Dan was credited with fostering the growing national trend toward community involvement in schools through partnerships with parents, businesses, policymakers, and local nonprofit groups. As the president and CEO of IS, he believes strongly in the power of nonprofits, foundations, and other organizations to work collaboratively to improve life and the environment for individuals and communities around the world. Dan is known for his commitment to performance management to drive evidence-based programs and high-impact organizations.

Early in his career, Dan worked as a community organizer in Guadalajara, Mexico, organizing a squatter community to secure land rights, running water, and public education. He then returned to Washington, DC, for a research fellowship at the Woodstock Theological Center at Georgetown University.

Helen I. Chen As founder and principal of HC Consulting LLC, Helen I. Chen has a passion for improving opportunities for youth through education. Her consulting practice focuses on program evaluation, curriculum and teacher professional development, and technical assistance. She provides thought partnership to leaders intent on delivering high-quality, evidence-based programs, at scale, with the goal of reducing gaps in opportunities. She guides organizations in program evaluation, project management, and coaching to improve their direct services and internal capacity for scale and sustainability.

Carrie S. Cihak leads evidence-informed practice and partnerships for the regional government of the twelfth largest county in the United States. Cihak has served as sponsor of the county's work on equity and social justice and is the architect of several county initiatives, such as Best Starts for Kids. Cihak previously served as chief of policy for the King County Executive Office, senior policy staff for the King County Council, and as staff economist on President Clinton's Council of Economic Advisers. Cihak is a research affiliate at the Center for Advanced Study in the Behavioral Sciences at Stanford University, local government fellow at Results for America, and board member for the Society for Causal Inference.

Sara H. Cody has worked in governmental public health for over twenty-five years. She is currently the health officer and public health director in Santa Clara County, the community where she grew up. Dr. Cody is a graduate of Stanford University, Yale University School of Medicine, and Stanford Internal Medicine Residency program. She is best known for her response to the COVID-19 pandemic, including leading a regional group of health officers to implement the first shelter-in-place order in the country. Her early and decisive action is estimated to have saved thousands of lives and was informed in part by trusted academic partners.

Kevin Corinth is the staff director of the U.S. Congress Joint Economic Committee. Previously, he was the executive director of the Comprehensive Income Dataset Project at the University of Chicago, chief economist at the Council of Economic Advisers in the Executive Office of the President, and a research fellow at the American Enterprise Institute. His research focuses on poverty, income measurement, tax policy, housing, and homelessness. He obtained a PhD in economics from the University of Chicago, and a BA in economics and political science from Boston College.

Tracy E. Costigan serves as senior director in the Executive Vice President's office at the Robert Wood Johnson Foundation. In this role, Costigan partners with foundation leadership to implement strategies that promote fair and just opportunities for health and well-being in the United States by addressing the intersection of structural racism, other forms of discrimination, and social conditions that impact health. Previous roles include leading large-scale complex research and evaluation at the American Institutes for Research and the Children's Hospital of Philadelphia. Costigan holds a PhD in clinical psychology from Drexel University and a BA in biology-psychology from Tufts University.

Spring Dawson-McClure, PhD, is a ParentCorps manager, overseeing the research strategy as the program scales nationally. As a white scientist-practitioner, with opportunities for deep learning from families, educators, and colleagues in the context of school-based randomized controlled trials for nearly two decades, Dr. Dawson-McClure brings a strong commitment to advancing health and education equity and deepening her practice of antiracist and community-engaged research.

Natasha Dravid serves as senior director for care management and redesign initiatives at the Camden Coalition and has been with the organization since 2013. She oversees a portfolio of interventions designed to improve care for patients who face the systemic barriers of racism, poverty, and limited access to care. Her current projects cover maternal health, behavioral health, vaccine promotion, and cancer screening and are anchored in an acknowledgment of the social determinants of health. She also oversees the organization's care management programs, including the high-touch Camden Core Model, Housing First, and Horizon Neighbors in Health programs. Natasha was instrumental in setting up the coalition's Medicaid Accountable Care Organization demonstration project, including activating a real-time data infrastructure for patient triage, launching a citywide quality improvement plan rooted in the primary care system, and developing and overseeing contracts with managed care organizations to improve healthcare delivery for Medicaid patients. Natasha continues to build on the lessons learned from clinical redesign projects to integrate successful strategies into the wider healthcare delivery system through the recently established Regional Health Hub structure in New Jersey. She also oversees the user-facing operations of the Camden Coalition Health Information Exchange, which drives regional workflow enhancements in clinical delivery. Natasha is passionate about working alongside care teams to activate real-time data in service of healthcare innovation. Natasha holds an MBA from the Yale School of Management and a BA in English from Haverford College.

Brad Dudding has served as chief impact officer at the Bail Project (TBP) since 2019. His work is focused on scaling TBP's impact nationwide, leveling up program quality and performance, and championing the organization's learning and research agenda. Prior to joining TBP, he worked for two decades at the Center for Employment Opportunities (CEO), a nonprofit committed to enriching the lives of returning citizens through employment. At CEO, Brad held several senior positions focused on building capacity to scale and delivering desired results.

Dylan Edwards has worked in the development sector for over a decade, including on projects in public health, community safety, youth empowerment, and affordable housing. Dylan joined AMP Health as a management partner and was embedded in the community health team at the Zambia Ministry of Health for two years before taking up his current position as deputy director, business development and communications.

Diana Epstein is the evidence team lead at the Office of Management and Budget. She was previously a research and evaluation manager at the

Corporation for National and Community Service and a program evaluator and policy analyst at Abt Associates, the American Institutes for Research, and the RAND Corporation. She has a PhD from the Pardee RAND Graduate School, an MPP from the Goldman School at UC Berkeley, and a bachelor's degree in applied math-biology from Brown University.

Katy Brodsky Falco served as executive director of Crime Lab New York, a criminal justice research organization that partners with civic leaders to identify, test, and scale programs and policies with the greatest potential to improve lives. She also served as executive director of assessments and reentry services at the New York City Department of Correction, where she designed the city's first performance-based reentry services contract targeting inmates at highest risk of recidivism, and piloted the use of evidence-based assessment tools. She also worked as a staff attorney at Legal Aid Society. Falco received a JD from NYU Law School and a BA from Harvard University.

David Fein As a principal associate at Abt Associates, David Fein has over three decades of experience leading rigorous evaluations of innovative programs aiming to improve the well-being of low-income adults and their families. These studies span multiple antipoverty initiatives sponsored by the federal Administration for Children and Families (ACF) at the U.S. Department of Health & Human Services, ranging from welfare reforms to healthy marriage initiatives to career pathway approaches. Trained as a demographer, he brings a strong multidisciplinary orientation to his work. Fein's recent work has focused on workforce training. He codeveloped the first randomized controlled trial evaluation of career pathway strategies—the nine-site, sixteen-year ACF-sponsored Pathways for Advancing Careers and Education (PACE) project. As an outgrowth of PACE, Fein has partnered with Year Up to build evidence on multiple program generations on this exemplary program for low-income youth. This partnership has generated a rich array of findings—ranging from the "improve" to the "prove" ends of the evaluation spectrum—as well as examples of best practices in researcher-practitioner collaboration. Current topics of interest include disparities in program effectiveness for participants with varying characteristics; the role of skills and employer connections in producing program impacts; and how the COVID-19 pandemic is reshaping the design, delivery, and effects of workforce training. Fein's longstanding interest in challenging measurement problems extends from his dissertation research on census undercount to a recent paper applying sequence analysis (a data-mining technique originating in DNA analysis) to discern the impacts of workforce training on whole career pathways.

Kelly Fitzsimmons is a committed social innovator. Before founding Project Evident in 2017, Kelly served as vice president / chief program and strategy officer at the Edna McConnell Clark Foundation (EMCF), where she led policy innovation, evaluation, grantmaking, and the early capital aggregation pilot. Prior to EMCF, she cofounded Leadwell Partners and New Profit Inc., held senior leadership positions in nonprofit organizations, and served on several foundation and social sector boards and advisory committees. Kelly currently serves as a Leap of Reason ambassador and is a member of Results for America's Invest in What Works Federal Standard of Excellence Advisory Committee. A graduate of McGill University in Montreal, Fitzsimmons holds an MBA from Boston University.

Gary Glickman is an entrepreneurial, global senior executive with over thirty years of providing leadership in both the public and the private sectors.

Gary has worked at the highest levels of the federal government, including the Executive Office of the President and later as a senior policy advisor for the Department of the Treasury. He was the coordinator of the Partnership Fund for Program Integrity Innovation, which was charged with bringing together a diverse group of state, local, not-for-profit, and philanthropic stakeholders to seek and test innovative approaches to improve efficiency and integrity in social service programs.

Gary founded and led as president and CEO a pair of successful consulting firms with expertise in helping state and local governments in the integration of delivery services across banking, electronic commerce, and manufacturing. He has served as president and CEO of a U.S. subsidiary of a German manufacturing company and as president and chief marketing officer of an NHSE listed enterprise that focused on assisting government agencies in driving customer service and enhancing citizen relationships.

Gary is a sought-after speaker and recognized thought leader. He has written and spoken extensively on social impact bonds, electronic benefits (which totally revolutionized how food stamps were delivered and monitored), cybersecurity, access to health care, and identity management. He has a BA in American studies and sociology from Brandeis and an MBA in finance and economics from the Stern School of Business, NYU. Gary donates his time and expertise to various professional and not-for-profit boards. He resides in the greater Washington, DC, area.

Makeda Mays Green is senior vice president, digital and cultural consumer insights at Nickelodeon. In her role, she evaluates the most effective ways to reach diverse target audiences through innovative research methodologies across Nickelodeon's platforms. Green is also a proud advisory board member of Raising Good Gamers, an initiative developed to create positive change in the culture and climate of online video gaming for youth, and of Determined to Educate, a nonprofit designed to support underserved youth through mentoring programs. She holds a BA from Wesleyan University and an MA and EdM in psychology from Teachers College, Columbia University, and resides in Stamford, Connecticut, with her husband and three children.

Shanika Gunaratna, MPP, is a ParentCorps manager, overseeing external engagement and strategic partnerships as ParentCorps works to scale to early childhood education settings nationwide. She brings more than a decade of experience in media, policy, and early childhood research and innovation to this role. Gunaratna lives in Brooklyn.

Ron Haskins is a senior fellow emeritus in the Economic Studies program at the Brookings Institution, where he was formerly codirector of the Center on Children and Families. He is formerly a senior consultant at the Annie E. Casey Foundation and was the president of the Association for Public Policy Analysis and Management in 2016. Haskins previously cochaired the Evidence-Based Policymaking Commission appointed by Speaker Paul Ryan. He is the coauthor of *Show Me the Evidence: Obama's Fight for Rigor and Results in Social Policy* (2015) and the author of *Work over Welfare: The Inside Story of the 1996 Welfare Reform Law* (2006). Beginning in 1986, he spent fourteen years on the staff of the House Ways and Means Committee and was subsequently appointed as senior advisor to President Bush for welfare policy. Haskins currently sits on the boards of MDRC, UNC Chapel Hill School of Education Foundation, and Power to Decide (formerly the National Campaign), as well as the Smith Richardson Foundation grants advisory board.

Audrey Hendricks As the senior program manager for data-driven workflows, Audrey Hendricks manages the development of workflow, documentation, and reporting tools in the Health Information Exchange for various population health programs within the Clinical Redesign Initiatives team. She also manages a portfolio of maternal healthcare delivery initiatives and oversees the data-driven patient identification process (triage) for Camden Coalition programs. Since joining the Camden Coalition in 2012, Audrey has served in various roles performing patient outreach for the Camden Core model, developing and implementing patient-centered care delivery initiatives in partnership with local providers, and coordinating technical assistance. She is passionate about empowerment-based engagement strategies and holds a BA in anthropology from Haverford College.

Daniel E. Ho is the William Benjamin Scott and Luna M. Scott Professor of Law, professor of political science, and senior fellow at the Stanford Institute for Economic Policy Research at Stanford University. He serves as associate director of the Stanford Institute for Human-Centered Artificial Intelligence, faculty fellow at the Center for Advanced Study in the Behavioral Sciences, and director of the Regulation, Evaluation, and Governance Lab. He received his JD from Yale Law School and PhD from Harvard University and clerked for Judge Stephen F. Williams on the U.S. Court of Appeals, District of Columbia Circuit.

Betina Jean-Louis, PhD, is principal consultant at Arc of Evidence, an evaluation company with expertise along the full evidence spectrum. Arc of Evidence works with social change agents to use data in strategic ways and to create, research, and continuously improve interventions that promote equity and social justice. Jean-Louis currently serves as senior advisor for equity and evidence at Project Evident; previously, she created and led Harlem Children's Zone's evaluation department for eighteen years. Jean-Louis has partnered with practitioners and funders to support the pursuit of equitable outcomes. A first-generation immigrant and college student, she earned undergraduate and graduate degrees from Columbia and Yale.

Michele Jolin is the CEO and cofounder of Results for America. Michele has held several leadership roles in the White House, including as a senior advisor for social innovation under President Obama (where she designed and launched the first social innovation fund), and as chief of staff for President Clinton's Council of Economic Advisers for CEA chairs Janet Yellen and Joseph Stiglitz. Michele was also part of the presidential transition teams for Obama/Biden and Biden/Harris. In 2007, Michele led the Presidential Transition Project at the Center for American Progress and coedited the book *Change for America: A Progressive Blueprint for the 44th President*. Earlier in her career, Michele was a senior vice president at Ashoka, a global foundation that invests in social entrepreneurs in more than fifty countries around the world, and worked for Senator Barbara Boxer (D-CA) on the Senate Banking, Housing, and Urban Affairs Committee.

Archie Jones has spent more than twenty years leading and maximizing the impact of high-growth, innovative enterprises in the private and social sectors. Jones currently serves as the chief financial officer of NOW Corporation. In the social sector, Jones has served as a partner at New Profit and a board member of Year Up National. He currently serves as a board member of the Taly Foundation, the Mickey Leland Kibbutzim Foundation, and First Choice Credit Union and is also a founding board member and vice chairman of Year Up Greater Atlanta. Over the past two decades, Jones has led private equity, privately held, and publicly traded companies and has served on the board of directors of several corporate and nonprofit organizations. Jones is a certified public accountant. He holds an MBA from Harvard University and is a graduate of Morehouse College.

Heather King is an expert in structuring evidence to unlock its potential for data-driven decision making in the social impact sector. She has done this work in a variety of sectors, including education, financial health, social capital, youth development, food security, housing, obesity prevention, and more. She holds a PhD in evolutionary biology from the University of Chicago.

Chris Kingsley works to create stronger and better-integrated public data systems as a resource for those working to improve the lives of children, families, and communities. Prior to joining the Annie E. Casey Foundation's evidence team, he led the Data Quality Campaign's local policy advocacy and consulted on matters related to data privacy, ethical use of predictive analytics, and collective impact. Chris served as the principal associate for data initiatives with the National League of Cities Institute for Youth, Education, and Families, and he has authored reports on performance management, municipal social media strategies, citywide information systems design, and economic development. As a Watson fellow, Chris studied telecommunications policy and development in Africa, India, and China. He is a graduate of Haverford College and the University of Pennsylvania.

Brian Komar is an impact institution builder whose career includes executive leadership roles at Salesforce, the Center for American Progress, and the Leadership Conference on Civil Rights. Brian's experience spans the public, private, and philanthropic sectors, and his areas of expertise include impact/ESG/sustainability, marketing, external affairs, and coalition building. He currently serves as vice president, global impact management at Salesforce, where he helps bring the full force of Salesforce to help its customers realize their potential to be platforms for positive social and environmental change.

Heather Krause, PStat, remains unconvinced. As a mathematical statistician with decades of global experience working on complex data problems and producing real-world knowledge, she has developed the Data Equity Framework to address equity issues in data products and research projects. We All Count, a project for equity in data, is working with teams across the globe to align their work with their equity goals in their data products, from funding to data collection to statistical analysis and data visualization. Her emphasis is on combining strong statistical analysis with clear and meaningful communication.

Erin Lashua-Shriftman, MA, is a ParentCorps manager, overseeing a team dedicated to ParentCorps programmatic data collection and management. With more than fifteen years' experience coordinating and managing longitudinal research and programmatic data, Lashua-Shriftman is a champion of continuous improvement and innovation of data collection and utilization, and interrogating data practices to bring them into alignment with the organization's value for racial equity. Lashua-Shriftman lives outside NYC with her family.

Michael H. Levine has a track record of driving early childhood and education reform transformation at public and private companies, foundations, and government agencies. He has more than twenty years of experience researching educational, developmental, and socioeconomic implications of the emerging media and learning landscape to inform policy, fuel innovation, promote educational equity, and influence professional practice. He is an author of more than forty publications and policy briefs, including *Tap, Click, Read: Growing Readers in a World of Screens*, and a board member for nonprofit and double bottom-line social venture organizations. He was named one of the United States' most influential leaders in family policy by *Working Mother* magazine.

Matt Levy As VP of evaluation and learning, Matt Levy is responsible for the measurement and evaluation of programs and new interventions implemented by First Place for Youth in California and with national partners in support of transition-age foster youth (ages eighteen to twenty-four). He also oversees and implements the organization's data systems from AWS to Power BI, creating state-of-the-art dashboards, ensuring automated ETL processes function, and driving the implementation and sale of a new proprietary and predictive analytics tool: the Youth Roadmap Tool. He also plays a key role in steering the organization's innovation agenda, leveraging test+learn techniques to pilot interventions, surface learnings, evaluate success, and when successful, support scaling. He is expert in data visualization, scripting in R and SQL, and leveraging human-centered design to cocreate dashboards with users to ensure data drives action. Increasingly, he is leveraging an equity and participatory lens in his work, supporting the organization's transition to become an antiracist organization.

Christopher Lowenkamp received his PhD in criminal justice from the University of Cincinnati. He has developed numerous assessments for use in the criminal justice system. Lowenkamp's research focuses on bridging the gap between research and practice.

Rhett Mabry is president of the Duke Endowment, a Charlotte-based private philanthropic foundation. A native of Greensboro, North Carolina, he joined the endowment in 1992 as associate director of health care.

He became director of the Child Care program area in 1998 and was named vice president of the endowment in 2009. He became president on July 1, 2016. Mabry holds a master of health administration from Duke University and a bachelor's degree from UNC Chapel Hill. Before joining the endowment, he held managerial positions at Ernst & Young and HCA West Paces Ferry Hospital. Mabry has served on the North Carolina governor's Early Childhood Advisory Council and is a past board chair of the Southeastern Council of Foundations. He also serves on the board of Candid, a national organization that compiles and evaluates philanthropic data.

James Manzi is a cofounder and managing partner of Foundry.ai, an artificial intelligence technology studio. He was founder, CEO, and chairman of Applied Predictive Technologies, which became the world's largest cloud-based AI software company. Jim is the author of several software patents, as well as the 2012 book *Uncontrolled*. He received a BS in mathematics from MIT.

Zachary Markovits is the vice president and local practice lead at Results for America, where he is focused on helping all local governments use data and evidence to make real and more equitable change in the lives of residents. Before joining Results for America, he worked at the Pew Charitable Trusts, where he led Pew's elections performance portfolio, as well as the Voting Information Project. Previously, he worked at the University of California's Survey Research Center and served as a community organizer on the south side of Providence, Rhode Island.

Ryan Martin is the deputy director of the National Governors Association's Center for Best Practices, where he assists in the center's work to help states develop effective solutions to public policy challenges. Prior to joining NGA, Ryan spent ten years working with members of the U.S. House Committee on Ways and Means and U.S. Senate Finance Committee to develop and advance legislation to reduce poverty, protect children, improve maternal and child health, and ensure social programs achieve results. Prior to working for Congress, Ryan was the executive officer for the Office of Family Assistance, U.S. Department of Health and Human Services.

Mary Marx is the president and CEO of the Pace Center for Girls and over the past decade has led the organization through an extensive period of growth. Since its founding in 1985, Pace has positively impacted the lives of more than 40,000 girls, and its advocacy work over the past decade has contributed to a more than 60 percent decrease in the number of girls referred to Florida's juvenile justice system. In 2019, Pace embarked on a national expansion strategy using a community participatory action model grounded in the needs, issues, and strategies of communities to achieve social change.

Rebecca A. Maynard is University Chair Professor of Education and Social Policy Emeritus, University of Pennsylvania. She is an expert in randomized controlled trials and rapid-cycle evaluation. Her research focuses on population groups from infants and toddlers to un- or underemployed adults. It addresses a range of policies and practices including childcare access and quality, teen pregnancy prevention, K-12 school reform, career and technical education, and social welfare policies. She is an advocate of open science and of strategic application of multiple methods of research in service of better and more equitable outcomes for all.

Michael McAfee is the president and CEO of PolicyLink, a national research and action institute focused on advancing racial and economic equity: just and fair inclusion for everyone living in America. He brings over twenty years of experience as a leader who has partnered with organizations across the public, philanthropic, and private sectors to realize this vision. Before joining PolicyLink, Michael served as senior community planning and development representative in the Chicago Regional Office of the U.S. Department of Housing and Urban Development. He earned his doctorate of education in human and organizational learning from George Washington University and completed Harvard University's Executive Program in Public Management.

Christine McBride has eight years of experience consulting and building partnerships with health information exchanges (HIEs) and state agencies to share and activate data. Her work has focused on strategic planning to enhance HIEs' functionality, develop service offerings for participants, and collaborate with stakeholders and community partners. Her experience working closely with medical and social service providers has shown the value of using data to improve patient outcomes and patients' experiences navigating the healthcare system.

Raymond McGhee Jr. joined the Robert Wood Johnson Foundation (RWJF) in 2020, bringing his expertise—in research and evaluation studies, policy analysis, and program design—to the foundation's Research-Evaluation-Learning team. At RWJF, he manages grantmaking to nonprofit organizations and academic research evaluating program investments. A key role McGhee plays is using evidence from the results of funded research and evaluations to support organizational learning that informs foundation investment strategy. McGhee also serves as an equity lead as a part of RWJF's Equity Leadership Group, collaborating with staff and foundation leaders to promote equity and inclusion within the foundation.

Andy McMahon is an entrepreneurial professional with more than two decades of experience in the fields of affordable housing, health care, human services, and the integration among them. Andy has a proven record of cultivating and executing collaborations across government agencies, philanthropy, healthcare entities, and community partners. Andy has strong and diverse government relations experience in affordable housing, health care, and criminal justice sectors and has a time-tested commitment and successful track record integrating public systems and private partners to better serve individuals and families with complex health needs.

Andrew Means is a serial social entrepreneur who has dedicated his career to helping all organizations measure, manage, and report their impact.

Tatewin Means is Sisseton Wahpeton Dakota, Oglala Lakota, and Inhanktonwan. An advocate for human rights, children, and families, she served as attorney general for the Oglala Sioux Tribe (2012–17) and in 2018 sought the democratic nomination for South Dakota attorney general—the first Indigenous woman to seek the office of state attorney general in the United States. She holds a BS in environmental engineering, an MA in Lakota leadership, and a JD with a concentration in human rights law. Tatewin is the executive director of Thunder Valley Community Development Corporation, an Indigenous organization seeking liberation for Lakota people through language, lifeways, and spirituality.

Bruce D. Meyer has been the McCormick Foundation Professor at the University of Chicago's Harris School of Public Policy since 2004. He is also a research associate at the NBER and a nonresident senior fellow at AEI. He has published on poverty, inequality, tax policy, survey accuracy, and government safety net programs in the major economics journals. Meyer received his BA and MA in economics from Northwestern University and his PhD in economics from MIT in 1987. Meyer served on the Commission on Evidence-Based Policymaking and cochaired the Federal Interagency Technical Working Group Exploring Alternative Measures of Poverty.

Kevin Miklasz is the senior director, learning analytics and insights at Noggin, where he leads efforts in learning analytics and data-driven assessments. Kevin has worked in the fields of game design and education for over ten years, gaining a variety of diverse experiences, from designing science curriculum and games, to teaching after-school science programs and game design jams, to conducting data analyses to improve EdTech products. Kevin is also the author of the book *Intrinsic Rewards in Games and Learning*. Kevin has a BA in physics from the University of Chicago and a PhD in biology from Stanford University.

Katie Smith Milway is founder and principal of MilwayPLUS social impact advisors, which works with clients as partners, focusing on philanthropic research, content development, influence strategies, and nonprofit innovation and growth. With a professional background in journalism, nonprofit management, strategy consulting, and governance, she is a frequent speaker at convenings on research-related themes.

Katie is also adjunct faculty at the Lilly Family School of Philanthropy, Indiana University, and a senior advisor at the Bridgespan Group, where for a decade she served as head of the knowledge practice. Prior to Bridgespan, she spent fourteen years at Bain & Company, consulting to global clients and becoming the firm's founding editorial director and publisher. She began her journalism career at the *Wall Street Journal*, and nonprofit service at Food for the Hungry.

Jordan Morrisey has worked with AMP Health for five years and for the past three years has served as AMP's deputy director for global operations. Prior to AMP, Jordan served as a community health and HIV/AIDS prevention volunteer, embedded with the Namibia Ministry of Health with the U.S. Peace Corps. He has experience in human capital development, grassroots mobilization, and supporting communities to reduce poverty and increase opportunity and access to health care. Jordan holds an MS in development management from American University's School of International Service and a BA in international affairs from George Washington University's Elliott School of International Affairs.

Neal Myrick is the vice president of transformative philanthropy at Salesforce. He leads philanthropy innovation to help the company's philanthropic efforts meet today's most complex challenges. Before Salesforce, Neal was the founding head of Tableau Foundation, leading efforts to donate more than $100 million globally over eight years. Neal is a former global IT leader at pioneering software companies and was a climate-focused nonprofit executive director. Neal is a clean-tech angel investor and philanthropist. He is also on several international advisory committees focused on global health and development, innovation, and the United Nations' sustainable development goals.

Dallas M. Nelson Aš'ápi (Dallas M. Nelson) was born and raised on the Pine Ridge Reservation in South Dakota and is a citizen of the Oglála Lakȟóta Nation. Dallas is the director of the Lakota Language and Education Initiative at Thunder Valley Community Development Corporation. He received his bachelor's degree in sociology and American Indian studies at Black Hills State University and his master's degree in Lakota leadership and management at Oglala Lakota College. Dallas is a longtime advocate for Indigenous education, Lakota language reclamation and revitalization, social change, and social justice for all Indigenous children and families.

Dusty Lee Nelson Wi Pxehin Ji Win / Dusty L. Nelson was born and raised in the Red Cloud community of Pine Ridge, South Dakota. She is a graduate of Oglala Lakota College and Montessori Center of Minnesota and is also an Oglala mother of 3 children. Dusty has spent her professional career devoting her efforts towards educating all ages of youth in various types of language and cultural programs, schools, and youth camps. In 2021 Dusty founded a home-based Montessori immersion Program called Lakota Children's House. In her free time she mentors young women and participates in community organizing focused on social justice and liberation.

Robert Newman is a pediatrician with thirty years of experience in global health and development. He is currently executive director of AMP Health, working with African governments to develop visionary and effective public sector teams. Previously, Dr. Newman held roles as Cambodia country director for U.S. CDC; managing director for policy and performance at Gavi; director of the Global Malaria Programme at the World Health Organization, CDC team lead for the President's Malaria Initiative; and Mozambique country coordinator for Health Alliance International.

Kathleen Noonan is president and CEO of the Camden Coalition of Healthcare Providers, a multidisciplinary nonprofit in Camden, New Jersey, established in 2002 as a citywide alliance of health and social services organizations, as well as community representatives with the goal of delivering better care to individuals with complex health and social needs. In 2008, Kathleen cofounded PolicyLab at the Children's Hospital of Philadelphia to connect clinical research with real-world health policy priorities and solutions. She received her JD from Northeastern University School of Law and her BA from Barnard College, Columbia University.

Amy O'Hara is a research professor in the Massive Data Institute and executive director of the Federal Statistical Research Data Center at the McCourt School for Public Policy. She also leads the Administrative Data Research Initiative, improving secure, responsible data access for research and evaluation. O'Hara addresses risks involved with data sharing by connecting practices across the social, health, computer, and data sciences. Her research focuses on population measurement, data quality, and record linkage. O'Hara has published on topics including the measurement of income, longitudinal linkages to measure economic mobility, and the data infrastructure necessary to support government research.

Veronica Olazabal is chief impact and evaluation officer at the BHP Foundation, president of the American Evaluation Association, and a teacher at Columbia University's School of International and Public Affairs. Her professional background ranges about twenty years and six continents and includes designing, implementing, and leading global programs, research, and evaluation for the Rockefeller and MasterCard Foundations. Veronica has served on various funding and advisory boards including, most recently, the World Benchmarking Alliance and the World Bank's Center for Learning on Evaluation and Results. She is the recipient of several industry awards and has published in the *American Journal of Evaluation*, *Evaluation*, and the *Stanford Social Innovations Review*. Olazabal holds a BA in communications and a master's degree in urban policy and planning from Rutgers, the State University of New Jersey.

David Olds is professor of pediatrics at the University of Colorado, where he codirects the Prevention Research Center for Family and Child Health. He has conducted randomized trials of Nurse Family Partnership (NFP), the only prenatal/early childhood program to meet evidence-based programs' "Top Tier" of evidence. NFP is identified as having the strongest evidence in the world that it prevents child maltreatment. Today, NFP serves over 50,000 families in the United States and 18,000 per year in seven other countries. David has received numerous awards, including the Charles A. Dana Award for Pioneering Achievements in Health and the Stockholm Prize in Criminology.

Nisha G. Patel has more than two decades of cross-sector experience leading and implementing initiatives to create community-centered economic opportunity. Previously, she served as executive director of the U.S. Partnership on Mobility from Poverty, in the Obama administration as director of the Office of Family Assistance, and deputy director and part of the founding team of Ascend at the Aspen Institute. Nisha has designed and launched multiple place-based philanthropic initiatives, including as a program officer at the Bill & Melinda Gates Foundation and director of programs at Washington Area Women's Foundation. She resides in DC, the future fifty-first state.

Marika Pfefferkorn As an interdisciplinary and cross-sector thought leader and community advocate, Marika Pfefferkorn is a change agent working to transform systems and scale successes across educational ecosystems, focusing on emerging technologies. Pfefferkorn works along the continuum from community to theory to practice, integrating collective cultural wisdom and applying a restorative lens to upend carceral conditions in education and to reimagine education through a liberatory lens. She has successfully co-led campaigns to end discriminatory suspension practices in Minnesota schools, to remove the presence of police in Twin Cities schools, and to increase investment in Indigenous restorative practices in education and community settings.

Jane Reisman bridges the worlds of impact management and the evaluation profession. As founder of the evaluation firm ORS Impact, which she led for over twenty-five years, Jane engaged in new frontiers to scale impact. Her current work as social impact advisor focuses on field-building and design efforts that strengthen impact measurement and management practices for impact investing. She is active in networks and boards that seek convergence of evaluation, impact measurement and management, and ESG practices and writes and presents regularly about developments and best practices.

Jason Saul is a leader in the field of social impact measurement. He is cofounder of the Impact Genome Project, a publicly funded initiative to standardize social impact data.

Brian Scholl is an economist, practitioner, and thought leader in evaluation, evidence systems, institutional design, organizational capacity, and public policy. Scholl previously served as chief economist of the United States Senate Budget Committee, where he managed the committee's Economics Unit, advised members, promoted evidence practice and use in the federal government, and worked tirelessly to integrate deep research and insights into public policy design. He helped to develop a broad range of economic policies to aid recovery from the Great Recession with particular attention to issues in labor, macroeconomic policy, household finance, international finance, and financial markets. He has previously worked in U.S. policymaking institutions in foreign affairs, financial regulation, and economic policy. He has been awarded the Federal Evaluation Innovator Award by the Evaluation Officer Council for his work designing compact and cost-effective rapid-cycle evidence initiatives.

Since founding the boutique consulting firm Global Innometrics in 2001, Scholl has worked with hundreds of clients as a direct provider of evaluation, evidence, program implementation, and organizational development services, as well as in policy design and evaluation. He has worked with an extremely diverse range of global organizations from direct service provider civic organizations and firms, to business and civic associations, to financial institutions, to local and national governments, developing a unique perspective of all aspects of the evidence value chain in varied cultural and institutional contexts.

Scholl earned his PhD in economics and MA in statistics at the University of California at Berkeley. He has conducted extensive research in household finance and behavioral decision making, political economy, development economics, public sector capacity, and macroeconomics and finance. Much of his recent research has focused on government capacity to serve the public interest and to use evidence for effective policymaking.

Jane Schroeder is the chief policy officer at First Place for Youth, where she leads the organization's policy advocacy and systems change initiatives at the federal, state, and local levels, advancing policies that remove barriers for foster youth transitioning into independence, and helping to create a policy environment where impact-driven nonprofits can thrive.

Prior to joining First Place in November 2016, Jane worked in government relations for the California Nurses Association / National Nurses United, where she advocated for legislation and regulation to protect patient safety and advance the nursing profession. Jane earned her JD degree from the University of Washington School of Law, and her bachelor's degree from McGill University in Montreal, Quebec.

Michael D. Smith is the eighth CEO of AmeriCorps, the federal agency for service and volunteerism formerly known as the Corporation for National and Community Service. Smith was nominated by President Biden in June 2021, confirmed by the U.S. Senate in December 2021, and officially started in January 2022.

Smith has dedicated his career to social justice and public service in underserved communities like those where he grew up. Most recently, he served as executive director of the My Brother's Keeper Alliance and director of Youth Opportunity Programs at the Obama Foundation. In these roles, Smith led the foundation's efforts to reduce barriers and expand opportunity for boys and young men of color, their families, and other underserved youth. Smith was part of the team that designed and launched the My Brother's Keeper initiative in the Obama administration and was appointed special assistant to President Obama and senior director of cabinet affairs, managing the initiative and interagency task force at the White House. My Brother's Keeper led to new federal policy initiatives and grant programs; tens of thousands of new mentors; more than 250 MBK communities in most states, DC, Puerto Rico, and nineteen tribal nations; and more than $1 billion in private sector and philanthropic investments.

Before this, Smith was a political appointee in the Obama administration serving as director of the Social Innovation Fund, a key White House initiative and program of the Corporation for National and Community Service. He reinvigorated the initiative, managed its largest funding competition, introduced its first Pay for Success grant program, and oversaw a portfolio of more than $700 million in public-private investments in support of more than 200 nonprofits. Before this, Smith served as senior vice president of social innovation at the Case Foundation, where he oversaw the foundation's domestic giving and program strategy and guided numerous global public-private partnerships. Earlier in his career, he helped build national initiatives aimed at bridging the digital divide at the Beaumont Foundation of America and PowerUP, served as a senior staff member at the Family Center Boys & Girls Club, and was an aide to U.S. congressman Richard E. Neal.

Smith is a Senior Atlantic Fellow for Racial Equity and a member of Boys and Girls Clubs of America's Alumni Hall of Fame, the highest honor bestowed by the organization. Prior to his government service, he served on the board of directors of Results for America, Venture Philanthropy Partners, Public Allies, Idealist.org, and Philanthropy for Active Civic Engagement. Smith earned his bachelor's degree in communications from Marymount University and resides in Springfield, Virginia.

Christopher Spera is president and CEO of Arbor Research Collaborative for Health. Arbor Research conducts studies that lead to improvements in patient care, clinical practices, and health-related public policy in the United States, Europe, and Asia. Prior to his current role, Spera was the division vice president for health and environment at Abt Associates. He also served as the director of research and evaluation at the Corporation for National and Community Service (AmeriCorps), a $1 billion federal agency, and served as a vice president at ICF International. In these roles, he directed groundbreaking studies, program evaluations, and survey research practices. Outside of his work at Arbor Research, he continues to serve as a professor of public policy and management at Carnegie Mellon University's Heinz School of Public Policy, where he teaches program evaluation. He has more than twenty peer-reviewed publications and technical published reports and earned a PhD in human development and quantitative methodology from the University of Maryland.

Kathy Stack is a senior fellow at the Tobin Center for Economic Policy and an independent consultant who advises nonprofit organizations, foundations, research organizations, and government officials on strategies to advance cross-program innovation and evidence-based decision making. She spent twenty-seven years at the White House Office of Management and Budget, where she oversaw federal education, labor, and major human services programs.

Stephanie Straus (she/her) helps governmental and administrative agencies increase their data use for research and evaluation purposes, across education and civil justice. Using the most appropriate and current privacy-enhancing technologies, Straus also advocates for secure data governance models that address the legal and regulatory risks involved with data sharing.

Kiribakka Tendo was born and raised in Uganda. A statistician by training, he began his professional career in investment banking. Passionate about management, he got his MBA in the United States and worked in retail management at Amazon. Keen to build managerial capacity, he joined AMP Health, where he was a management partner at the Ministry of Health and Sanitation in Sierra Leone for three years. Kiribakka is now the deputy director for country support at AMP Health, overseeing its operations in Africa, and is based in Johannesburg.

Aaron Truchil serves as the senior director of analytics at the Camden Coalition, where he oversees the organization's applied data and research activities aimed at improving care for individuals with complex health and social needs. Aaron earned an MS in social policy from the University of Pennsylvania and a BA from Wesleyan University.

Nicole Truhe is the senior director of policy, Medicaid at UnitedHealthcare Community & State. Nicole leads the development of policy positioning, advocacy, and thought leadership strategies related to traditional and complex Medicaid populations. Previously, Nicole led policy and advocacy efforts on Pay for Success / evidence-based policy, workforce development, and social innovation at a national social innovation advocacy organization. Nicole also worked for over a decade at a national child welfare and children's mental health nonprofit, where she advocated for policy changes in the child welfare, juvenile justice, children's mental health, and innovative financing policy areas.

Vivian Tseng is president and CEO of the Foundation for Child Development. She is recognized for her leadership in building an interdisciplinary field of research on the use of research in policy and practice and expanding research-practice partnerships nationwide. She publishes and speaks internationally on evidence-informed policy and practice. Her abiding commitment to racial equity is reflected in her mentoring of young professionals, board service, academic publications, advocacy work, and development of programs to support researchers of color and nonprofit leaders from racially minoritized and LGBTQ communities. She received her PhD from NYU and her BA from UCLA.

Gregory Tung is an associate professor in the Colorado School of Public Health's Department of Health Systems, Management & Policy. His research interests relate to how scientific evidence is incorporated into policy and program decision making, with a special emphasis on injury prevention. Dr. Tung works on a diverse range of injury topics, including the prevention of youth violence, suicides, poisonings, and child abuse. His research interests also include the integration of health services and public health systems, with a focus on nonprofit hospital community benefit activities.

Erika Van Buren is the founder and CEO of Line of Sight, a consulting firm that supports leaders, organizations, and systems in strengthening their capacity for improvement and equitable impact. She has served as a seasoned independent consultant in service to the nonprofit, philanthropic, and government sectors for over twenty years. During her professional tenure, Dr. Van Buren has cultivated expertise in the areas of evaluation capacity building, applied research, program and systems change design strategy, evaluation, and performance management for human service delivery systems. She received her doctorate in clinical child psychology from the University of California, Los Angeles, and has dedicated her career to studying and improving community-based services for populations of color within and across mental health, child welfare, justice, and other public systems. Prior to Line of Sight, Dr. Van Buren served as the chief innovation officer for First Place for Youth, an organization based in Oakland, California, and nationally recognized for its focus on learning, data use, evidence generation, and the delivery of results-based care in education and employment services for transition-age foster youth. At First Place, she was responsible for leading the design, utilization, and maintenance of the organization's performance management systems and structures, and providing leadership and oversight of the organization's national scaling efforts. She crafted and implemented the internal and external evaluation agenda for the agency, disseminated knowledge that was leveraged for policy advocacy and reform, and worked closely with program and system leadership to identify and roll out best and evidence-supported strategies to improve practice and the child welfare system's impact on transition-age foster youth. As a thought leader in this space, Dr. Van Buren has been recognized as a Ford Foundation fellow, a LEAP of Reason ambassador, and an Annie E. Casey Foundation Leadership fellow.

Bi Vuong is an experienced education policy professional who is committed to a practical approach to building evidence and improving outcomes and opportunities for students. She is the author of *Strategic Budgeting: Using Evidence to Mitigate the "COVID Slide" and Move Toward Improvement"* (2020), a contributor to *Opportunity and Performance: Equity for Children from Poverty* (2021), and one of the leaders featured in *Taking Charge of Change* (2021), by Paul Shoemaker. Before joining Project Evident, Bi was the director of Proving Ground at the Center for Education Policy Research at Harvard University, where she worked with states and districts across the country to implement a continuous improvement framework built on meaningful, measurable outcomes. She also launched the National Center for Rural Education Research Networks, bringing evidence-building capacity to districts in rural New York and Ohio. Prior to Proving Ground, Bi served as the deputy chief financial officer at the School District of Philadelphia; she has also held positions at the Data

Quality Campaign and EducationCounsel, LLC. Bi currently serves on the board of the Academic Development Institute. A graduate of Kenyon College, Bi also holds an MPA from the Woodrow Wilson School at Princeton University.

Garrett Warfield and his team oversee all studies designed to test and improve the impact of Year Up programs and strengthen business operations, often in partnership with leading research experts across the country. Before joining Year Up in 2014, Garrett spent over ten years as a researcher, evaluator, teacher, performance manager, and all-around data nerd for government agencies, nonprofits, and universities. He holds a BA in psychology and statistical methods from Boston University, an MSc in criminology with forensic psychology from Middlesex University in England, and a PhD in criminology and justice policy from Northeastern University.

Tara Watford is the chief data officer at the Bail Project. Prior to joining TBP, she was the senior director of research and evaluation at the Youth Policy Institute, where she measured the collective impact of programs designed to empower students and families in high-poverty communities throughout Los Angeles. A passionate advocate for social justice, Tara believes that data—especially when derived from the voices and experiences of those most marginalized—are fundamental building blocks of progress and an essential tool in creating equitable policy and a just society. She received her PhD from UCLA.

Ahmed Whitt leads the Learning+Impact Unit at the Center for Employment Opportunities. For more than ten years, Whitt has led federal, state, and privately funded evaluation projects ranging from public health to criminal justice. His academic research has focused on the influence of neighborhood contextual factors on individual economic, mental health, and behavioral outcomes. He is an alumnus of the University of Pennsylvania and the University of North Carolina at Chapel Hill.

Carina Wong has spent her career redesigning education and training systems to enable young people from vulnerable communities to reach their full potential. She has worked at the intersection of policy, practice, and philanthropy for over three decades. She holds advanced degrees in education and policy from Stanford and George Washington Universities. She is a trustee at the CA College of the Arts, where she earned an MBA in design strategy. A former Peace Corps volunteer and proud mother of three young children, she is passionate about the arts, food, and cooking.

David Yokum, JD, PhD, is director of the Policy Lab at Brown University and host of the *30,000 Leagues* podcast. He was previously the founding director of the Lab @ DC in the DC Mayor's Office and, before that, a founding member of the White House's Social & Behavioral Sciences Team and inaugural director of the U.S. Office of Evaluation Sciences (OES). Over one hundred field experiments have now been completed under the Policy Lab, the Lab @ DC, and OES. David's expertise draws on the cognitive foundations of judgment and decision making and, in particular, how that knowledge and associated methodologies can be extended into applied settings.

Peter York, principal at BCT Partners, is a national social impact measurement and big data analytics leader. He has worked with nonprofits, foundations, and government agencies for over twenty-five years to help them plan, evaluate, and improve their performance. This includes spending the past ten years developing and refining precision analytics, a machine learning approach to rigorously evaluate social programs and produce actionable evidence to front-line practitioners. He has authored numerous research papers and articles for academic and professional journals and, most recently, a case study on the application of precision analytics in the child welfare and mental health sectors.

INDEX

Figures are indicated by "f" following the page number.

Abstinence Only Education Evaluation program, 38–39
Abt Associates, 39, 46, 47, 268
Abuse of evidence, 257–59, 296–98
Academic-public health collaboration for evidence building, 14, 413–21; evolution of, 414–15; lessons from, 416–20; origins of, 414
Accountability: community learning and, 203; data sovereignty and, 138; evaluation norms and, 158; learning frameworks and, 11, 265, 267–68; no-evidence trap and, 260; stakeholder engagement and, 385; standards-based assessments and, 107–09
Accountability for Equitable Results Framework (Annie E. Casey Foundation), 328
Actionable evidence: evaluation design and, 105–07; evidence-based policy and, 35–45, 357–59; examples of, 35–45; generation of, 335–44; importance of, 2; local governments and, 146; practitioner-research partnerships for, 53; user-friendly language for, 267; value of evidence and, 252
Actionable Evidence Initiative, 41

Actionable Intelligence for Social Policy (AISP) of University of Pennsylvania, 116, 130, 132
Adedokun, Lola, 7, 161
Administration for Children and Families, 47
Administrative Data Research Facility, 281
Administrative Data Research Network in the United Kingdom (ADRN-UK), 127
Adoption placement agencies, 209–10
Advancing Racial Equity and Support for Underserved Communities through the Federal Government, Executive Order on (2020), 7, 388, 411
Affordable Care Act (2015), 280
AI (artificial intelligence), 171, 194, 195, 417
AIR (American Institutes for Research), 221
AISP (Actionable Intelligence for Social Policy) of University of Pennsylvania, 116, 130, 132
Alaska Permanent Fund, 172
Alfred P. Sloan Foundation, 134
Allison, Mandy A., 4, 56
All of Us (NIH), 128

American Evaluation Association (AEA), 365, 367, 369–70
American Humane (organization), 129
American Institutes for Research (AIR), 221
American Journal of Evaluation, 367
American Rescue Plan (2021), 14, 176, 280, 399, 404
American Rescue Plan Data and Evidence Dashboard, 404
AmeriCorps, 266, 268
Anatole, Manzi, 166
Annie E. Casey Foundation, 75, 116, 328
Annual Evaluation Plans, 385, 387
Annual social value (ASV) verification, 375
Anti-evidence culture, 260
Arches Transformative Mentoring program (Arches), 312
Artificial intelligence (AI), 171, 194, 195, 417
Ayala, Plinio, 4–5, 80

Baelen, Rebecca, 47
The Bail Project (TBP), 4, 67–73
Ballmer, Connie, 22
Baltimore City Public Schools, 4, 74–79
Baltimore Fund for Educational Excellence, 75
Baron, Jon, 265
Barrow Street Consulting (BSC), 82–85
Bauer, Tamar, 1
BBB (building back better) movement, 364–65, 368, 370
BCT Partners, 336–37, 340
Behavioral health issues assessment. *See* Criminal Justice Lab of New York University
Belmont principles, 127
Benitez, Lymari, 234, 235–37
Berke, Andy, 402
Best practices: Career and Technical Education programs, 76; data governance, 6, 133–34; data transparency, 102; evaluation design, 105; governance transparency, 130, 131; learning impact evidence, 319, 321, 323, 325; pandemic response, 288; project management, 298–99; public sector innovation, 402; reading instruction, 219
Best Starts for Kids (BSK) property tax levy, 145, 149
Biased data: causal inference methods and, 257; evidence-based policy and, 5–6, 94–102, 95–98*f*; publication bias and, 298–301; reducing likelihood of, 60; in surveys, 272
Biden, Joe: data integration and, 280–81; democratizing evidence and, 411; Evidence Act and, 381; Executive Order on racial disparities and inequalities, 7, 388, 411
Big data: data justice and data sharing risks, 194, 197–99; decision making and, 120; ethics and access to data, 126–36; health goals and programming, 123; narrative analysis and, 104; research and evaluation use of, 340; standards for, 116
Bill and Melinda Gates Foundation, 41, 81
Biomarkers, 128
Bipartisan Budget Act (2015), 396
Black-led organizations, 354
Blockchain, 377
Bloomberg Philanthropies, 401
Bowers, Jake, 12, 295
Boyd, Clinton, 165
Bradley, D., 41
Bridgespan Group, 69, 354
Brighthive, 341
Britt, Jessica, 46, 47
Bronx Defenders, 68
Brooks, Jennifer L., 13, 356
Brothers, John, 9, 201, 206
Brotman, Laurie Miller, 180
Bryan, Tara Kolar, 203
BSC (Barrow Street Consulting), 82–85
BSK (Best Starts for Kids) property tax levy, 145, 149
Building back better (BBB) movement, 364–65, 368, 370

Burkhauser, S., 41
Bush, George W., 22
Business Roundtable, 375

California Policy Lab, 281
Camden Coalition, 14, 252, 282, 422–28
Cancel, Yessica, 235
Capacity Assessments, 385, 387
Cardinali, Daniel J., 8, 189
Career and Technical Education (CTE) programming, 4, 74–79; reflections, 78–79; resources, 79; results, 78; transparency and stakeholder input, 77–78
Career training: Career and Technical Education programming, 4, 74–79; for low-income parents, 170–71; Per Scholas, participant feedback during COVID-19 pandemic and, 4–5, 80–86; tying funding to outcomes in, 396; Year Up Professional Training Corps, 4, 46–55
Carl D. Perkins Career and Technical Education Act (2006), 396
"Carrot and Stick Philanthropy" (Brothers), 206
Carter, Charles, 23
Cash distributions, 7, 170–79; context for, 170–71; COVID-19 pandemic and, 172; culture of learning and, 404; economic changes and child poverty, 171; evidence for, 172–73; next generation evidence and, 173–75; recommendations for, 175–76; structural racism and gender inequity, 171–72
Causal evidence, 322
Causal inference, 357–58
Causal Inference for Social Impact Lab, 150
CBDDM (Community-Based Data for Decision Making) strategy of Ethiopia, 122
CBO (community-based organization) partners, 432–33
CDC (Centers for Disease Control and Prevention), 425

Census Bureau (U.S.), 272, 273, 275
Center for Advanced Study in the Behavioral Sciences at Stanford University, 150
Center for Employment Opportunities (CEO), 10, 311–16; challenges and responses, 312–13; Credible Messenger Initiative, 312; employment success, reducing barriers to, 313–15; reflections, 315–16
Center for Innovation through Data Intelligence (CIDI) of New York City, 114
Centers for Disease Control and Prevention (CDC), 425
Centers for Medicare and Medicaid Services (CMS), 266
CEO. See Center for Employment Opportunities
CEPEI, 291–92
Chan Zuckerberg Initiative, 360
Charity Navigator, 378
Chen, Helen I., 218
Chicago Beyond Initiative, 112, 157–58
Child First, 211–12
Children. See Public schools and students
Children and Youth Cabinet, Providence, 410
Children's Defense Fund, 90
Children's Health Insurance Program (CHIP), 429–30
Child tax credit (CTC) payments, 176
Churchill, Winston, 255
CID. See Comprehensive Income Dataset Project
CIDI (Center for Innovation through Data Intelligence) of New York City, 114
Cihak, Carrie S., 6–7, 144
Civic Eagle, 200
Civil court data, 133–34
Civil Justice Data Commons, 133–34
Civil Legal System Modernization project of Pew Charitable Trusts, 132–34
CLPHA (Council of Large Public Housing Authorities), 431

CMI (Credible Messenger Initiative), 312–16
CMS (Centers for Medicare and Medicaid Services), 266
Code for America, 113, 115
Coding of data, 358–62
Cody, Sara H., 413
Coleridge Initiative, 281
Colonialism, 137–42, 353
Colorado Evaluation and Action Lab, 281
Columbia University, 176
Commission on Evidence-Based Policymaking, 381–82
Communities for Just Schools Fund, 198
Communities in Schools, 190
Community-Based Data for Decision Making (CBDDM) strategy of Ethiopia, 122
Community-based organization (CBO) partners, 432–33
Community-based research, 133, 174
Community Catalyst, 431–34, 432–33*f*
Community engagement, 130–31, 147, 300–302
Community learning, 202–04
Community organizers and leaders, 164–65, 201–02
Community Release with Support, 68
Community trust. *See* Public trust
Community voice, 8–9, 187–240; behavioral health and criminal justice, 225–30, 228*f*; community trust and philanthropy, 9, 209–17; data justice and data sharing risks, 194–200; overview, 187–88; participant-centered research and evaluation, 231–39; philanthropy's role in evaluation, 9, 201–08; power of community voice and, 189–93; Summer Literacy Initiative and, 218–24
Compliance vs. learning frameworks, 265, 267–68
Comprehensive Child Welfare Information Systems, 280
Comprehensive Income Dataset (CID) Project, 11, 270–77; evidence generated by, 275; existing data, problems with, 272; future of, 275–76; overview, 270–72; purpose of, 273
Confidentiality and privacy issues: actionable evidence and, 340–41; anonymized identification codes for, 273; COVID-19 pandemic information sharing and, 416, 423; data economy and, 115–16; ethics and access to data, 126–36; Evidence Act and, 271, 282; health data and, 131, 416, 425; improving access to data and, 281–82; public school and student information, 32; public trust and, 198–99
Connected Classrooms, 81
Consultants, 252–54
Continuous evidence building. *See* Evidence building
Continuous improvement: culture of evidence and, 255; evidence-based policies vs., 213; evidence-building strategies and, 180–81; evidence engagement and, 31; federal funding for, 396; impact management and, 377; next generation of evidence and, 262–63; Pace Center for Girls and, 233–34, 237; Project Evident and, 25; real-time research methods and assessment of, 48; Rural Church Summer Literacy Initiative and, 218, 224; teams for, 329; Year Up's research methods for, 48
Cook, Thomas D., 203
Corinth, Kevin, 11, 270
Corporate Social Responsibility (CSR), 366
Correlational evidence, 321–22
Costigan, Tracy E., 7, 153
Costs of program evaluations, 48, 49, 250–52, 266, 393. *See also* Funders
Council of Large Public Housing Authorities (CLPHA), 431
Council on Foundations, 293
Court data, 133–34
Courtney, Mark, 331

COVID-19 Data and Innovation Centre, 292
COVID-19 pandemic: academic-public health collaboration and, 14, 413–21; cash distributions during, 7, 170–79; cross-agency workgroups and response to, 114; data economy and response to, 113; evidence and impact in post-COVID-19 world, 13, 364–73; healthcare and public health data integration during, 14, 422–28; homelessness and, 113–14; local and state governments, effect on, 401, 402–03; nonprofit sector and creativity during, 12, 287–94; Pace Center for Girls, remote services of, 236; Per Scholas, participant feedback during, 4, 80–86; pre-existing social and health crises, exacerbation of, 162–63; racial disparities and, 352; school closures and evidence-building interruptions, 184–85; The Bail Project and, 72; Year Up online coaching delivery and, 52
Cradle to Prison Algorithm Community Summit (2018), 196
Credible Messenger Initiative (CMI), 312–16
Criminal Justice Lab of New York University, 9, 225–30; diversion from criminal justice system, 229–30; HealthLink Diversion Tool, creation of, 226–28, 228f
Critical Race Theory (CRT), 155
Cross-program data infrastructure and analytics, 278–86
CSR (Corporate Social Responsibility), 366
CTC (child tax credit) payments, 176
CTE. *See* Career and Technical Education programming
Cultural awareness, 57
Cultural humility, 222
Culture, organizational, 233–35, 255–56, 260
Culture of evidence, 255–56, 381
Culture of learning, 3, 403–04

Current Population Survey Annual Social and Economic Supplement, 272
Customer loyalty, 82

Dashboards: of American Rescue Plan, 404; for COVID-19 information, 414, 424, 427, 432; daily actionable evidence and, 339–42; evidence building and integration of, 249; to support outcome attainment, 330-32
Data: anonymization of, 131–32, 273; practitioner use and understanding of, 29–32; for public good, 127–30, 198–200; risks from information gaps in, 24; standardized coding of, 358–62; standards for, 215. *See also* Biased data; Big data; Equity in data and evidence; Health data; Research and development-like use of data and evidence; *other headings starting with "Data"*
Data analytics capacity, 11–12, 278–86; data access and, 281–83; facilitation of, 284; funding for, 280–81; meta-analyses and, 358–60; staff capacity and, 283–84
Data collaboration, 378–79
Data collection investments, 250–51
Data economy, 6, 111–18; infrastructure for, 115; overhauling, 112–14; sharing and building power for, 115–17
Data entrapment, 194
Data ethics and access, 6, 126–36; Civil Justice Data Commons and, 133–34; equitable practices for, 132–34; federal government and, 129–30; harm, assessing, 134; philanthropy and, 131–32; public interest, trust, and transparency, 126–29; state and local governments and, 130–31
Data extraction, 6, 119–25; COVID-19 pandemic and, 417; data sovereignty and governance, 6, 137–43; democracy and, 124–25; democratizing evidence and, 409; grassroots capacity and, 121–22; transparency and, 122–24

Data for Black Lives, 116, 410
Data for Public Good (D4PG), 198–200
Data infrastructure improvements, 14, 399–407; future of, 405–06; next generation evidence and, 403–05; for problem solving, 402–03; for public health, 419; public sector innovation and, 400–402
Data intermediaries, 115
Data justice and data sharing risks, 8–9, 194–200; community summit on, 196–98; example of, 195; future of, 200; public good, data for, 198–200
Data literacy, 101, 194, 199
Data myths, 94–102, 95–98f
Data Saves Lives campaign, 131
Data security. *See* Confidentiality and privacy issues
Data sovereignty and governance, 6, 137–43
Data trusts, 341
Dawson-McClure, Spring, 180
DDCF (Doris Duke Charitable Foundation), 162–63, 165–68
Dead-end studies, 246, 248–49, 267, 298
Dean-Coffey, Jara, 92
Deceptive evidence, 257–59, 296–97, 370
Decision making: data extraction and, 119–25; power imbalances and, 109, 120; transparency in, 77–78
Decolonizing Wealth (Villanueva), 353
Deloitte, 401
Democracy, 124–25, 261–62
Democratizing evidence, 14, 408–12; in action, 410–11; future of, 411; oligarchy of evidence and, 408–09; policies for, 409–10
Desousa, Kevin C., 290
Digital Promise, 319
Dignity in Schools Campaign, 198
Directional evidence, 321
Disconfirmability principle, 297
Discrimination: discriminology, 410; in employment, 171–72; health equity and, 154; researchers of color and, 167. *See also* Racial disparities and inequalities

Diversity, equity, and inclusion efforts, 375
Doris Duke Charitable Foundation (DDCF), 162–63, 165–68
Dravid, Natasha, 422
Dubois, W.E.B., 9, 233
Dudding, Brad, 67
The Duke Endowment (TDE): community trust and, 9, 209–17; Summer Literacy Initiative, 9, 214, 218–24
Dunleavy, Angela, 288–89

Eastern Band of Cherokee, 172
EBP. *See* Evidence-based policy and practice
Echoing Green, 354
Economic mobility. *See* Career training
Edin, Kathryn, 270
Edna McConnell Clark Foundation (EMCF), 25, 232, 234
Education. *See* Public schools and students
Education Department, U.S., 212, 280, 282, 307
Education Partnership Coalition, 199
Edwards, Dylan, 6, 119
EdWeek, 236
EEF (equitable evaluation framework), 156–58, 156f
EEI (Equitable Evaluation Initiative), 92, 150, 156
Effect sizes, 298–301
Elevate Data for Equity project of Urban Institute, 132
EMCF (Edna McConnell Clark Foundation), 25, 232, 234
Employment: discrimination, 171–72; First Place for Youth program and, 327–34; Great Recession and, 400; guaranteed income programs and, 173; of homeless individuals, 275; poverty and, 170–71; unemployment rates, 171–72, 173. *See also* Career training; Center for Employment Opportunities
Engineered consent, 199

Enright, Kathleen, 293
Environmental, social impact, and governance (ESG), 13, 374–80
EPF (European Patients' Forum), 131
Epstein, Diana, 13, 381
Equitable evaluation framework (EEF), 156–58, 156*f*
Equitable Evaluation Initiative (EEI), 92, 150, 156
Equity in data and evidence, 5–8, 87–186; biased data and, 94–102; cash distributions and, 170–79; data economy and, 111–18; data ethics and access, 126–36; data extraction and, 119–25; diversity and, 375; funder evaluation norms and, 153–60; Indigenous data sovereignty and, 137–43; justice, evidence as, 89–93; just philanthropy and, 103–10; leadership to advance, 7, 161–69; local government and equity, 144–52; overview, 87–88; Parent-Corps and, 180–86; participant-centered measures and, 233–36
ESG (environmental, social impact, and governance), 13, 374–80
ESSA (Every Student Succeeds Act, 2015), 318–22, 395–96
Ethics: deceptive evidence and, 257–59, 296–97; impact washing and, 370; randomized controlled trials and withholding treatments, 221–22, 231, 256. *See also* Data ethics and access
Ethiopia, Community-Based Data for Decision Making (CBDDM) strategy of, 122
European Institute for Innovation through Health Data, 131
European Patients' Forum (EPF), 131
Evaluation.gov, 387
Evaluation Officer Council, 387
Evaluation practices, 4, 21–27; actionable evidence and, 35–45, 53, 105–07, 335–44; challenges in, 266–70; collaboration and, 150, 431; dead-end studies and, 246, 248–49, 251, 267, 298; evaluation questions and, 220; Evidence Act and, 382; evidence-informed decision making and, 365; funders and norms centered on equity, 4, 153–60, 156*f*; housing and health outcomes, 434–35; jargon and, 205–06; just philanthropy design principles, 6, 103–10; lack of, 350; Next Generation of Evidence and, 26–27; participant-centered measures and, 232, 233–36; philanthropy's role in, 154–55, 201–08; for place-based investments, 214–15; privilege and racism in, 92–93; Project Evident and, 25–26; racial disparities and research design, 167; rating systems for summits and, 191; research questions and design, 98–102, 167, 221–22, 247, 257–59, 296–98, 330; standards for, 215. *See also* Evidence building; Randomized controlled trials
Every Student Succeeds Act (ESSA, 2015), 318–22, 395–96
Evidence Act (2018). *See* Foundations for Evidence-Based Policy-Making Act
Evidence-based policy and practice (EBP): actionable evidence and, 35–45, 357–58; community voice and, 191–92; data sharing risks and, 196–97; democratizing evidence and, 408–12; end users and, 361–62; Evidence Act and, 381–82; evidence building for, 244–46; future of, 362; Impact Genome Project and, 359–61; implementation challenges and, 211–12; increase in, 22, 29–30, 35; justice and, 89–93; legacy of racism, systemic change to address, 352–53; local governments advancing racial equity and, 144–52; next generation evidence and, 13, 356–63; ParentCorps early childhood intervention and, 182–83; racial disparities and, 163–64; RCTs and, 22, 211, 357–58; unintended consequences of replication, 212–13

Evidence-Based Policymaking Commission, 271
Evidence building, 11, 243–64; abuse of evidence and, 257–59, 296–98; academic-public health collaboration for, 14, 413–21; costs of, 250–52; culture of evidence and, 255–56; housing and health improvements, 430–31, 435–36; infrastructure for, 252–55; local governments and, 147–51; management integration and, 249–50; methodological fundamentalism and, 256–57; next generation of evidence and, 261–63; no-evidence trap and, 259–60; ParentCorps early childhood intervention and, 181–83; for place-based investments, 214–15; power of working backward, 246–49; promise of evidence and, 244–46; strategies for, 180–81; valuing evidence investments and, 251–52. See also Evaluation practices; Practitioner-centered evidence building
Evidence-building plans. See Learning agendas
Evidence fears, 11, 265–69
Evidence fluency, 254
Evidence reimagined. See Reimagined evidence
Executive Order on Advancing Racial Equity and Support for Underserved Communities Through the Federal Government (2021), 7, 388, 411
Externalities, 251, 368–69, 375
Extractive data. See Data extraction

Falco, Katy Brodsky, 225
Family Connects, 213
Family First Prevention Services Act (2018), 396
Family search techniques, 209–10
FareStart, 288–89
Federal government: data analytics capacity, strategies to strengthen, 278–86; data ethics and access, 129–30; equity agenda, 7, 388, 411; impact evaluations and program funding, 391–98; standards for data security, 128. See also Evidence-based policy and practice; *specific acts*
Federally qualified health centers (FQHC), 432–33
Federal strategies for data analytics, 11–12
Feedback Labs, 376
Fein, David, 46, 47
The Fifth Discipline (Senge), 203
First Place for Youth (FPFY), 11, 327–34; actionable evidence and, 41; continuous quality improvement for, 329; goals of, 327–29; metrics, refining, 330–31; on-demand actional evidence generation, 335–37; policy change agenda and, 329–30; *Raising the Bar* policy brief of, 332, 342; results and reflections, 332–33; scalability of programming, 343; sustainable change strategies of, 331–32
Fitzsimmons, Kelly, 1, 3, 4, 21
FiveThirtyEight website tool, 297
Ford Foundation, 293
Foster care: aging out of, 327–34, 342–43; locating relatives and, 209–10
Foundations for Evidence-Based Policy-Making Act (Evidence Act, 2018), 381–90; data analysis, 280; data security and availability, 271, 282; dedication, support, and persistence, 387; future of, 388; learning agendas and, 146, 383–88, 386f; National Secure Data Service and, 116; overview, 381–83; racial equity and, 7; stakeholder engagement and transparency, 384–87, 386f
FQHCs (federally qualified health centers), 432–33
Frett, Latanya Mapp, 294
Friedman, Mark, 112
Funders: community data, learning from, 203–04; COVID-19 pandemic, creativity during, 287–94;

evaluation norms and equity, 7, 153–60, 156*f*; evaluation outcomes and, 22–23, 85; evidence-based policy and, 30, 213; leaders of color, support for, 166–68; social impact investment strategies and, 431. *See also* Philanthropy
Fund for Shared Insight, 232–34

Gates-MacGinitie Reading Test (GMRT), 220
Gemma Services, 11, 335–44; actionable evidence and, 41; challenges and responses, 340–41; data readiness and, 337; practitioner buy-in and, 336–37; precision modeling process and, 337–40; real-time insights from, 341–42; reflections, 343
Gender inequity, 171–72
Generalization problem, 307–10
General Services Administration, 284
Glickman, Gary, 11, 278
Global Reporting Initiative (GRI), 366
GMRT (Gates-MacGinitie Reading Test), 220
Government. *See* Federal government; Local and state governments
Government Performance Lab of Harvard, 283
Government surveys, 271–72
Grant applicants: COVID-19 pandemic and, 293; data analytics, federal strategies to strengthen, 278–84; grant cycles, limits of, 419–20; Pay for Success program, 278–79, 283; prior evidence, including, 30
Grassroots nonprofit organizations, 213
Great Recession, 400
Green, Makeda Mays, 317
Guaranteed income programs. *See* Cash distributions
Guidestar, 378
Gunaratna, Shanika, 180

HARKing (Hypothesize after the Results are Known), 297

Harlem Children's Zone (HCZ), 30, 31–32, 91, 181
Harvard's Government Performance Lab, 283
Haskins, Ron, 1, 2–3, 29
Health and Housing Fund, 434–35
Health and Human Services Department, U.S., 280–81, 395–96
Health and sex education programs, 38–39
Health data: electronic health records, 128; extraction and use of, 119–25; funding surveys for, 145–46; housing and health improvements, 429–36; safeguards for, 131, 416. *See also* COVID-19 pandemic
Health equity, 153–54
HealthierHere, 146–47
Health Information Exchange (HIE), 14, 422–28
HealthLink Diversion Tool, 226–28, 228*f*
Healthy Harlem of HCZ, 31–32
Healthy Steps, 213
Hendricks, Audrey, 422
Hickey, Daniel, 319–21
Higher Expectations for Racine County, 405
Ho, Daniel E., 413
Homelessness, 113–14, 271, 274–75, 425–26
Household surveys, 271–72, 273, 275
Housing and health improvements, 429–36; evidence building for, 430–35, 432–33*f*; reflections on, 435–36
Housing as a Platform evaluation tool, 434–35
Human Synergistics (HS), 232–34
Hurricane Katrina (2005), 209
Hybrid learning models, 81–84
Hypothesize after the Results are Known (HARKing), 297

ICEP (Interagency Council on Evaluation Policy), 387
IES (Institute of Education Sciences), 307

IGP (Impact Genome Project), 359–61
Impact economy, 13, 374–80; active impact management, 377–78; data collaboration, 378–79; impact management and capacity building, 376–77; stakeholder feedback, 376
Impact evaluations, 14, 391–98; evidence infrastructure and, 253; failure rates and, 394; finding progress in, 394–95; future of, 397–98; ineffective programs, harm of, 393; Iron Law and, 392–93; priorities vs. programs, 395–97; reimagining evidence and, 350
Impact evidence standards, 10–11, 317–26, 320f, 324f
Impact Genome Project (IGP), 359–61
Impact investing, 366–67
Impact management, 376–78
Impact Management Project (IMP), 367, 378
ImpactMatters, 378
Impact washing, 370
inBloom, 116
Incarceration. *See* Law enforcement
Inclusion. *See* Equity in data and evidence
Income measures. *See* Comprehensive Income Dataset Project
Independent Sector, 189–91
Indigenous peoples: data sovereignty of, 6, 137–43; education of, 139–40; guaranteed income and, 172; participatory research and, 133; research grants for, 168; tailoring programs to, 222
In Equality, 196–97
Infante, Pia, 293
Inflation Reduction Act (2022), 14, 399
Infrastructure for evidence, 252–55, 278–86
Infrastructure Investment and Jobs Act (2021), 14, 399
Innovation, 255. *See also* Scalable innovations and impact
In Other Words: A Plea for Plain Speaking in Foundations (Proscio), 205–06

Institute for Justice, 196
Institute of Education Sciences (IES), 307
Institute of Healthcare Improvement, 430
Institutional missions, 189–90
Institutional Review Boards (IRBs), 127, 130
Intangibles, 375
Interagency Council on Evaluation Policy (ICEP), 387
Intergovernmental Personnel Act (IPA, 1970), 419
Internal evidence teams and experts, 254–55, 263, 268
International Business Council, 378
International Declaration on the Rights of Indigenous Peoples, 139
Intrapreneurship: Managing Ideas Within Your Organization (Desousa), 290
"Invest in What Works State Standard of Excellence" (Results for America), 402
IRBs (Institutional Review Boards), 127, 130
The Iron Law of Evaluation (Rossi), 392

Jackson, Angela, 164
Jargon, 205–06, 330–31
Jean-Louis, Betina, 1, 4, 5, 28
Joint powers agreements (JPAs), 195–98
Jolin, Michele, 14, 399
Jones, Archie, 3, 4, 21
J-PAL, 148
Justice: behavioral health and criminal justice, 225–30, 228f; civil court data, 133–34; data justice and data sharing risks, 8–9, 194–200; evidence-based policy and practice (EBP) and, 5, 89–93; juvenile justice system, 190; social justice, 144, 155, 162
Just philanthropy design principles, 6, 103–10; actionable evidence and, 105–07; assessment and accountability, 107–09; for teachers, 104–05
Juvenile justice system, 190

King, Heather, 13, 356
King County Metro Transit, 147–48
Kingsley, Chris, 6, 111
Komar, Brian, 13, 374
Krause, Heather, 5, 87, 94, 107

Lab @ DC, 146, 148
Lakota nation, 137–43
The Language Warrior's Manifesto (Treuer), 139
La quen náay Kat Saas Medicine Crow, Liz, 6, 140
Lashua-Shriftman, Erin, 180
Law enforcement: body-worn cameras, studies on, 146, 300–301; Civil Legal System Modernization project, 132–34; Criminal Justice Lab of New York University, 9, 225–30, 228*f*; criminology RCTs, 307–09; employment services for convicts, 311–16; juvenile justice system and, 190; Pace Center for Girls and arrest reductions, 236–37; predictive policing and, 196; public safety vs. health, 9, 225–30, 228*f*; public school data sharing with, 194–200; The Bail Project and, 4, 67–73
Leadership: community organizers and leaders, 164–65, 201–02; culture of evidence and, 256; evidence building and integration of, 249–50
Leadership Conference for Civil and Human Rights, 116
Leadership to advance equity, 7, 161–69; funder support of leaders of color, 166–68; next generation evidence and, 164–65; next generation leadership and, 162–66
Learning agendas: community learning and, 204; compliance frameworks vs., 265, 267–68; culture of evidence and, 255–56; Evidence Act and, 146, 383–88, 386*f*; evidence fears and, 268; Summer Literacy Initiative of The Duke Endowment and, 219, 224; The Bail Project and, 69–70

Learning impact evidence, 10–11, 317–26; current state of, 318–19, 320*f*; in formative research with media, 320–22; impact evidence standards and, 322–25, 324*f*
Learning organizations, 255
LEO (Wilson Sheehan Lab for Economic Opportunities at Notre Dame), 148
L'Esperance, Gabrielle, 203
Levine, Michael H., 317
Levy, Matt, 327
Lewis, Michael, 420
Lindegaard, Stefan, 290
Listen for Good, 376
Living Goods, 378
Local and state governments: academic partnerships with, 419; building evidence and advancing equity, 6–7, 144–52; data analytics capacity, strategies to strengthen, 11–12, 278–86; data ethics and access, 130–31; data infrastructure improvements and, 14, 399–407
Logic models, 37–38. *See also* Theory of change
Lowenkamp, Christopher, 225

Mabry, Rhett, 9, 209
Machine learning, 337–40
Magnolia Mother's Trust, 174–75
Management integration and evidence building, 249–50
Manzi, James, 12, 256, 306
Margolis, Greg, 2, 29
Markovits, Zachary, 14, 399
Martin, Julie, 41
Martin, Ryan, 14, 391
Marx, Mary, 231
Mastercard Center for Inclusive Growth, 410
Maternal, Infant, and Early Childhood Home Visiting program (MIECHV), 57, 395
Mathematica Policy Research, 32, 36, 404
Maynard, Rebecca A., 4, 35, 46, 47

MBK (My Brother's Keeper) initiative, 351–52
McAfee, Michael, 5, 89
McBride, Christine, 422
McGhee, Raymond, Jr., 7, 153
McKinney-Vento Homeless Assistance Act (1987), 113
McMahon, Andy, 429
MDRC, 81
Means, Andrew, 374
Means, Tatewin, 6, 137
Medicaid, 429–36
Memorandum on Restoring Trust in Government through Scientific Integrity and Evidence-Based Policymaking (2021), 381, 388, 411
Menlo Report, 127
Mental illness assessment. *See* Criminal Justice Lab of New York University
Meta-analyses, 298, 358–61
Meyer, Bruce D., 11, 270
Michaux, Fabienne, 369
MIECHV (Maternal, Infant, and Early Childhood Home Visiting program), 57, 395
Miklasz, Kevin, 317
Milgram, Anne, 226
Milner, Yeshimabeit, 196, 198
Milway, Katie Smith, 231
Misleading evidence, 257–59, 296–97
Mission Measurement, 360
Mohammad, Amina, 292
Monitor 360, 104
Monitor Institute, 401–02
Morrisey, Jordan, 6, 119
Mozambique, health data collection in, 119–20
My Brother's Keeper (MBK) initiative, 351–52
Myrick Neal, 12, 287

National Academies of Science, Engineering, and Medicine, 131
National Affordable Housing Trust (NAHT), 434
National Center for Education Statistics, 167
National Commission to Transform Public Health Data Systems, 133
National Congress of American Indians, 140
National Education Goals, 317
National Implementation Research Network, 12, 212
National Indian Child Welfare Association, 168
National Institute of Standards and Technology (NIST), 128
National Institutes of Health (NIH), 128, 130, 307, 354, 360
National Research Cloud, 419
National Science Foundation, 131, 134
National Secure Data Service, 116, 419
Native Americans. *See* Indigenous peoples
Naturally occurring experiments, 336
Negative research findings, 246, 248–49, 251, 267, 298
Nelson, Dallas M., 6, 137
Nelson, Dusty Lee, 6, 137
Net Promoter Analysis, 82
Networked Improvement Community, 41
Newman, Robert, 6, 15, 119
New York University. *See* Criminal Justice Lab of New York University
New Zealand, digital government strategy of, 130
Next generation evidence: cash distributions and, 173–75; evidence-based policy and programs, 13, 356–63; evidence building and, 261–63; impact investing as use case for, 366–67; leadership for, 164–66; local and state government contributions to, 403–05; Nurse-Family Partnership RCT and, 64; principles of, 1–14; Project Evident and, 26–27; Rural Church Summer Literacy Initiative, 224
NFP. *See* Nurse-Family Partnership
Nickelodeon, 318, 325
NIH (National Institutes of Health), 128, 130, 307, 354, 360

NIST (National Institute of Standards and Technology), 128
No Child Left Behind Act (NCLB, 2001), 318
No Data About Us Without Us Fellowship, 199–200
No-evidence trap, 259–60
Noggin, 10–11, 317–26, 320f, 324f
Non-financial performance, 13, 374–80
Noonan, Kathleen, 422
Null results, 246, 248–49, 251, 267, 298
Nurse-Family Partnership (NFP), 56–66; application of results, 63; challenges and responses to, 60–61; data collection, 59; evaluation practices of, 23; findings, 61–62; innovation development at, 57, 58f; innovative programming of, 213; reflections, 63–65; research and practice integration, 60, 211–12; site selection, 58; study objectives, 58
Nyandoro, Aisha, 165

Obama, Barack: data integration and, 280; evidence-based policy, 2, 22, 29–30; My Brother's Keeper initiative, 351–52
Objective quantitative research, 99
Office of Education Technology, U.S., 319
Office of Management and Budget (OMB), 280, 381–84, 386–88
O'Hara, Amy, 6, 115, 126
Olazabal, Veronica, 13, 364
Olds, David, 4, 56, 57
Organizational culture, 233–35, 255–56, 260
Organizational health data, 204–05
Organization for Economic Cooperation and Development (OECD), 364, 368

Pace Center for Girls, 9, 231–39; challenges and responses, 235–36; evaluation practices, 233–35; partnership and capacity growth, 232–33; reflections, 237–39; results, 236–37
Pandemic. *See* COVID-19 pandemic
PAPs. *See* Pre-analysis plans
Parent advisory councils, 222
ParentCorps, 7, 180–86; challenges and responses, 184–85; evidence-building approach, 181–83; reflections, 185–86
Participatory action research (PAR), 233
Participatory evaluation (PE), 233
Patel, Nisha G., 7, 170
Pathways for Advancing Careers and Education (PACE), 47
Patton, Michael Quinn, 267
Pay for performance funding, 397
Pay for Success, 278–79, 283
Peer learning partnerships, 368
Pellegrino, James, 319–21
Per Scholas, 4–5, 80–86; impact, 84–85; participant and provider centered evaluation, 82–83; reflections, 85–86; results, 83–84
Pew Charitable Trusts, 132–34
Pfefferkorn, Marika, 8–9, 187, 194
P-hacking, 296–97
PHA (Public Housing Agency), 432
Philanthropy: community learning and, 203–04; community trust and, 9, 209–17; data ethics and access, 131–32; design principles of, 6, 103–10; evaluation, role in, 154–55, 201–08; evidence, donor and political intent shaping, 22; evidence-based policy and, 30, 35; organizational health data and, 204–05; power of community voice and, 189–90; racial disparities in leadership of, 161–69; role in evaluation, 9, 201–08; strategic, 204. *See also* Funders
Pine Ridge Reservation, South Dakota, 137–43
Platinum Seal of Guidestar, 378
Police. *See* Law enforcement
Policy. *See* Evidence-based policy and programs

PolicyLink, 89–93
Poverty: child tax credit payments and, 176; Comprehensive Income Dataset Project to measure, 11, 270–77; COVID-19 pandemic and, 236, 364; evidence-based policies and, 89–93; First Place for Youth program and employment outcomes, 327–34; health equity and, 154; homelessness and, 113–14, 271, 274–75, 425–26; housing and health improvements, 429–36, 432–33*f*; The Bail Project and, 67–73; welfare recipients and work requirements, 37–38. *See also* Career training; Cash distributions; Nurse-Family Partnership; Welfare recipients
Power imbalances: data subjects, controllers, and users, 128; decision making and, 109, 120; evidence building and, 29. *See also* Ethics
Practitioner-centered evidence building, 3–5, 19–86; actionable evidence, 35–45; COVID and participant feedback, 80–86; evaluation practices and, 21–27; full arc of evidence and, 28–34; Nurse-Family Partnership, 56–66, 58*f*; overview, 19–20; public school CTE outcomes, 74–79; The Bail Project, 67–73; Year Up Professional Training Corps, 46–55, 51*f*
Practitioner's Prayer, 111
Pre-analysis plans (PAPs), 12, 295–305; defined, 295–96; FAQs on, 301; HARKing and, 297–98; *p*-hacking and, 296–97; Project Evident and, 301; publication bias and, 298–301; purpose of, 295–96
Precision analytics (PA) process, 336–41
Predictive analytics, 195–98
Preschool programs: learning impact research for, 317–26; RCTs for, 307
Presidential Memorandum on Restoring Trust in Government Through Scientific Integrity and Evidence-Based Policymaking (2021), 381, 388, 411
Pretrial Fairness Act (Illinois, 2021), 72
Principal Project, 106
Privacy. *See* Confidentiality and privacy issues
Privilege, 92, 100. *See also* Racial disparities and inequalities
Project Evident: Actionable Evidence Framework, 5; anti-racist evaluation and, 92–93; Baltimore City Schools, consulting for, 74–76; *Data Equity and Evidence Guide*, 8; equity in evidence and, 5; founding of, 25–26; guaranteed income programs and, 173–75; guiding principle of, 3; ParentCorps early childhood intervention and, 183; Pay for Success and, 283; pre-analysis plans and, 301; strategic evidence plan of, 25; Summer Literacy Initiative, strategic evidence plan for, 214, 220–21; Talent Accelerator of, 29. *See also* Next generation evidence
Project management best practices, 298–99
Promise Neighborhoods, 29
Proscio, Tony, 205–06
Psychiatric care program for youth. *See* Gemma Services
Publication bias, 298–301
Public Broadcasting System, 318, 325
Public engagement, 130–31, 147, 405
Public good, data used for, 127–30, 198–200
Public health. *See* Academic-public health collaboration for evidence building; COVID-19 pandemic
Public Housing Agency (PHA), 432
Public/Private Ventures, 81
Public schools and students: Career and Technical Education programming and, 4, 74–79; COVID and participant feedback, 80–86; CTE outcomes, 4; data justice and data sharing risks, 194–200; Every Student Succeeds Act, 318–22,

395–96; expulsion and suspension rates, 99–100, 197; homeless students and, 113; No Child Left Behind Act, 318; Pace Center for Girls and, 9, 231–39; ParentCorps early childhood intervention and, 180–86; preschool programs, 307, 317–26; racial disparities and inequalities in, 99–100, 197, 404; researcher collaborations with, 32; State Longitudinal Data Systems and, 280; Summer Literacy Initiative of The Duke Endowment and, 9, 218–24; teachers, just philanthropy design for, 104–05; Year Up Professional Training Corps and, 46–55

Public trust: community voice and, 192; confidentiality and privacy issues, 198–99; COVID-19 pandemic response and, 416; data economy and, 115–16; ethics and access to data, 126–29; evidence-based policy and, 209–17; of law enforcement, 227; philanthropy and, 9, 207; of scientists, 102, 297

Quantitative research, 98–101

Racial disparities and inequalities: child poverty and, 171; COVID-19 pandemic and, 352, 414–15; data infrastructure improvements and, 404; data justice and data sharing risks, 194–200; democratizing evidence and, 410; diversity, equity, and inclusion efforts to address, 375; equitable data practices and, 115–17, 132–33; evidence as justice and, 89–93; Executive Order on, 7, 388, 411; faculty appointments in colleges and universities, 167; just philanthropy design principles and, 103–10; legacy of, systemic change to address, 352–53; local governments advancing equity and, 144–52; in organization revenues, 354–55; in philanthropy leadership positions, 161–69; power of community voice and, 189–93; in school expulsion rates, 99–100, 197; systemic racism and, 90–91, 153–54, 161–69, 171–72, 352–53; The Bail Project and, 67–73; unemployment rates and, 171–72

Racial equity analyses, 116

Racine County, Wisconsin, 405

Racism: anti-racism discussions, 191; systemic, 90–91, 153–54, 161–69, 171–72, 352–53

Raising the Bar: Building System-and Provider-Level Evidence to Drive Equitable Education and Employment Outcomes for Youth in Extended Foster Care (Van Buren, Schroeder, & York), 332, 342

RAND Corporation, 68

Random assignment evaluations, 81

Randomized controlled trials (RCTs): on adoption placement strategies, 210; alternatives to, 123, 256–57; on Center for Employment Opportunities effectiveness, 311–16; disadvantages of, 221–22, 231, 256, 435; effectiveness of, 12, 306–10; evidence-based policy and, 22, 211, 357–58; evidence building and, 256; on guaranteed income programs, 173; investments in, 3; of Pace Center for Girls, 231–32; on ParentCorps early childhood intervention, 181–82; preparation for, 25; unintended consequences of replication, 212–13. *See also* Nurse-Family Partnership

Recidivism rates, 311–16

Red Cross, 209

Regulation, Evaluation, and Governance Lab at Stanford Law School (RegLab), 148, 413–20

Reimagined evidence, 12–14, 345–436; academic-public health collaboration and, 413–21; data infrastructure improvements and, 399–407; democratizing evidence, 408–12; Evidence Act and, 381–90, 386f; evidence-based policy and, 356–63;

Reimagined evidence (cont.)
    healthcare and public health data integration, 422–28; housing and health improvements, 429–36, 432–33*f*; impact economy and, 374–80; overview, 345–47; in post-COVID world, 364–73; small-sample studies for dependent variables, 391–98; system changes for, 349–55
Reisman, Jane, 13, 364
Remote learning models, 81–84
Request for Information (RFI), 385
Research and development-like use of data and evidence, 10–12, 241–344; actionable evidence for practitioners, 335–44; COVID-19 pandemic and nonprofit sector, 287–94; evidence building, unfinished business of, 243–64; evidence fears, 265–69; federal strategies for data analytics, 278–86; learning impact evidence, 317–26, 320*f*, 324*f*; overview, 241–42; poverty measures, 270–77; pre-analysis plans and, 295–305; RCT effectiveness and, 306–10; re-entry services and, 311–16; strategy, data, and culture alignment for impact, 327–34
Research bias. *See* Biased data
Research integrity, 296
Results for America, 401–02, 404
RFI (Request for Information), 385
Riley, W. T., 358
Rivera, D. E., 358
Roadtrip Nation (PBS series), 81
Robert Wood Johnson Foundation (RWJF), 133, 153–60
Robichau, Robbie Waters, 203
Rockefeller Foundation, 410
Rodriguez-Lonebear, Desi, 141–42
Rossi, Peter H., 392
Rudman, Sarah, 416
Rural Church Summer Literacy Initiative, 9, 218–24
RWJF (Robert Wood Johnson Foundation), 133, 153–60

Safeguarding, 376
SAHF (Stewards of Affordable Housing for the Future), 434
Salesforce (organization), 367, 375
Santelises, Sonja Brookins, 74
Saul, Jason, 13, 356
Scalable innovations and impact: challenges in, 212; funding, 290–91; HealthLink Diversion Tool, 229; nonprofit funding for, 23–24; on-demand evidence building and, 340, 342–43; ParentCorp and, 182–83; Summer Literacy Initiative of The Duke Endowment, 219–21
Schaefer, Luke, 270
Scholl, Brian, 10, 11, 15, 241, 243
Schönrock, Philipp, 291–92
Schools. *See* Public schools and students
Schroeder, Jane, 327
Scriven, Michael, 267
SDGs (Sustainable Development Goals), 291–92, 366
SDOH (social determinants of health), 430
Secondary data uses. *See* Data ethics and access
Security. *See* Confidentiality and privacy issues
SEED (Stockton Economic Empowerment Demonstration), 173
Selection bias, 338
Self-determination for communities, 202–03
Senge, Peter, 203
SEPs (strategic evidence plans), 25, 146, 220, 383–85
Sesame Workshop, 317–18, 325
Sex education, 38–39
Sherman, Lawrence, 308–09
Shivji, Azim, 47
Show Me the Evidence: Obama's Fight for Rigor and Results in Social Policy (Hasking & Margolis), 2, 29
SIF (Social Innovation Fund), 53, 268, 349–51
SLFR (State and Local Fiscal Recovery) funds, 280

Small-sample studies for dependent variables, 14, 391–98
Smith, Michael D., 13, 345, 349
SNAP (Supplemental Nutrition Assistance Program), 275
Snipp, Matthew, 138
Social audits, 123
Social determinants of health (SDOH), 430
Social impact investment strategies, 431
Social Impact Partnerships to Pay for Results Act (2018), 396
Social Innovation Fund (SIF), 53, 268, 349–51
Social justice, 144, 155, 162
Social license, 6, 128–32, 134
Social Value International (SVI), 369
Societal Experts Action Network, 130–31
Society for Research in Child Development, 168
Socioeconomic status. *See* Poverty
South Africa, local social audit in, 123
Souvanna, Phomdaen, 47
Spencer Stuart (search firm), 233
Spera, Christopher, 11, 265
Springboard to Opportunities, 174–75
Squires, Catherine, 198
Stack, Kathy, 11, 278
Stakeholder feedback and engagement: democratizing evidence and, 409–11; Evidence Act and, 384–87, 386*f*; evidence generation and, 368; impact economy and, 376; local and state government programs and, 405; research design and, 221–22; transparency and, 77–78
Stanford Impact Labs, 420
Stanford RegLab, 148, 413–20
Stanford Social Innovation Review, 206
State and Local Fiscal Recovery (SLFR) funds, 280
State governments. *See* Local and state governments
State Longitudinal Data Systems, 280
Statistical significance, 101, 215, 296, 298–300, 321

Steinberg, Robin, 67
Stewards of Affordable Housing for the Future (SAHF), 434
Stock, Elisabeth, 41
Stockton Economic Empowerment Demonstration (SEED), 173
Stop the Cradle to Prison Algorithm Coalition, 196–97
Strategic evidence plans (SEPs), 25, 146, 220, 383–85
Straus, Stephanie, 6, 126
Students. *See* Public schools and students
Study design, 39–41
Substance use disorder assessment. *See* Criminal Justice Lab of New York University
Suicidality assessment. *See* Criminal Justice Lab of New York University
Summer Literacy Initiative of The Duke Endowment, 9, 218–24; challenges and responses, 221–22; evidence building for, 214; reflections, 224; results, 222–23; scaling strategy for, 219–21
Supplemental Nutrition Assistance Program (SNAP), 275
Sustainable Development Goals (SDGs), 291–92, 366
SVI (Social Value International), 369
System changes for reimagined evidence, 13, 349–55
Systemic racism, 90–91, 153–54, 161–69, 171–72, 352–53

T. Rowe Price, 205, 207
Tableau Foundation, 12, 288–92
Tax data, 273, 275
TBP (The Bail Project), 4, 67–73
TCIA (Twin Cities Innovation Alliance), 194–96, 198, 199–200
TDE. *See* The Duke Endowment
Teacher2Teacher, 104–05
Technical career training, 80–86; evaluation of, 82–83; impact, 84–85; reflections, 85–86; results, 83–84
Te Mana Raraunga–Māori Data Sovereignty Network, 138

Temporary Assistance for Needy Families (TANF), 171
Tendo, Kiribakka, 6, 119
Te Reo Irirangi o Te Hiku o Te Ika (Te Hiku Media), 140
Theory of change, 70–73, 246–49, 350
Third Sector, 283
Thrive!, 132
THRIVE East of the River (THRIVE), 173–74
Thunberg, Greta, 364
Thunder Valley Community Development Corporation, 137, 139
Tiered evidence designs, 2, 397
Transparency: American Rescue Plan Data and Evidence Dashboard, 404; data extraction and, 122–24; data justice and data sharing risks, 194, 199; in decision making, 77–78; ethics and access to data, 126–29; Evidence Act and, 384–87, 386*f*; local governments' research sharing and, 149–50; no-evidence trap and, 259–60; pre-analysis plans for, 295, 301–02; of research goals, 223; standards for, 129–30
Trauma-based approaches, 222
Treuer, Anton, 139
Truchil, Aaron, 422
Truhe, Nicole, 429
Trust. *See* Public trust
Trust-Based Philanthropy Project, 207, 293
Tseng, Vivian, 14, 408
Tung, Gregory, 4, 56
Twin Cities Innovation Alliance (TCIA), 194–96, 198, 199–200
$2.00 a Day: Living on Almost Nothing in America (Edin & Schaefer), 270

UK Anonymisation Network (UKAN), 131–32
Unemployment rates, 171–72, 173
UnitedHealthcare Community & State, 429–36
United Kingdom: Administrative Data Research Network in, 127; data sharing code of, 130; UK Anonymisation Network, 131–32
United Methodist churches. *See* Summer Literacy Initiative of The Duke Endowment
United Nations: COVID-19 Multi-Partner Trust Fund, 292; Declaration on the Rights of Indigenous Peoples (UNDRIP), 139; Human Rights Council, 270; Sustainable Development Goals (SDGs), 291–92, 366
University of Arizona Native Nations Institute, 138
University of California, Los Angeles (UCLA), 114
University of Cambridge (UK), 307–08
University of Chicago, 331–32
University of Dayton, 404
University of Denver, 113
University of North Carolina, 113
University of Pennsylvania: Actionable Intelligence for Social Policy, 130, 132; data economy research and, 113–14; Year Up program and, 4, 39, 46, 47
Upswell Summit, 190–91
UpTogether, 174
Urban Institute, 29, 90, 132, 173–74, 283, 378
US Digital Service, 115

Value-free evidence, 99
Value judgments of scientists, 299–300
Van Buren, Erika, 327
Villanueva, Edgar, 353
Vuong, Bi, 74

Walker, Darren, 204
Warfield, Garrett, 46, 47
Watford, Tara, 67, 68
Webb, Robb, 224
Welfare recipients: career training for, 170–71; data integration among programs for, 280; foster care, 209–10, 327–34, 342–43; homelessness and, 275; work or school requirements for, 37–38

Wentt, Taiheem, 81
What Works Cities, 401, 403
What Works Clearinghouse, 360
White-dominant narratives and culture, 90, 155, 157, 167. *See also* Colonialism
Whitt, Ahmed, 311
Williams, Koku Awoonor, 166
William T. Grant Foundation, 168
Wilson, Kate, 291
Wilson Sheehan Lab for Economic Opportunities at Notre Dame (LEO), 148

Wong, Carina, 6, 87, 103
Workforce Innovation and Opportunity Act (2014), 396
World Economic Forum, 378

Year Up Professional Training Corps, 4, 46–55; evidence building and, 39–41; practitioner-centered approach to evaluation, 47–50; reflections, 52–54; results, 50–52, 51*f*
Yokum, David, 12, 295
York, Peter, 41, 335